The Life and Legend of James Watt

Science and Culture in the Nineteenth Century

Bernard Lightman, Editor

the life and legend of James Watt

Collaboration,
Natural Philosophy, and the
Improvement of the Steam Engine

David Philip Miller

University of Pittsburgh Press

Published by the University of Pittsburgh Press, Pittsburgh, Pa., 15260
Copyright © 2019, University of Pittsburgh Press

Printed on acid-free paper

ISBN 13: 978-0-8229-6611-1
ISBN 10: 0-8229-6611-5

Cataloging-in-Publication data is available from the Library of Congress

Cover art: *James Watt* by Carl Fredrik von Breda (1792). National Portrait Gallery, London.
Cover design: Joel W. Coggins

For my grandson, Lawrence Albert Maxwell Miller, who likes machines

Contents

Acknowledgments

*G*etting to the point where I could attempt this biography has been a long process. In more than twenty years of work on Watt, I have accumulated many debts. My children and their families have happily accommodated the Great Steamer's presence, indeed encouraged it. My wife, Margaret, has given excellent editorial advice, caring because unsparing, and has been a sounding board for my ideas. I am very grateful to her. Lawrence, to whom this book is dedicated, has been a source of great joy to me in his early years and an object lesson in the miracle of human learning. Colleagues at UNSW aided and abetted my research and covered for my absences, and the university has provided invaluable facilities to continue my work into "retirement."

Friends around the world have assisted in many ways. I thank in particular Ben Marsden, who made numerous important and very useful suggestions for improvement of the manuscript. The book bears the mark of his collegial generosity and also that of Trevor Levere, who gave me the benefit of an expert reading. Readers for the press have my thanks for taking on a substantial task and giving much useful advice. Others who have inspired, assisted, and encouraged me in the course of my Watt studies include Larry Stewart, Peter Jones, Christine MacLeod, Patricia Fara, Malcolm Dick, Jim Andrew, Mike Gibbs, Jack Morrell, Arnold Thackray, Simon Schaffer, Steven Shapin, John Schuster, Rod Home, the late David Oldroyd, Richard Yeo, Robert Bud, Jeffrey Sturchio, Tom Carroll, and Crosbie Smith. I also thank the many archivists, curators, and librarians who have found, made available, and interpreted the materials from which this work is constructed. Bernard Lightman and Abby Collier believed in the project and have done their utmost to expedite it. I am in their debt. Des, Dave, Gary, Mick, Bob, Rob, Nader, and my other early-morning walking and swimming mates have been a source of healthy Aussie skepticism and predictable humor about the project: "What?!" The fact that the book would not have been completed without them is sweet revenge! I may well leave Watt alone for a while now. May he rest in peace!

SYDNEY
JULY 2018

Introduction

*T*he lone horseman approached Dumbarton on the north of the River Clyde in driving rain. It was a late September day in 1773. Exhausted and enveloped in a sense of impending loss, he could go no farther that day. The next morning, back on the road, he met a friend who had traveled by chaise from Glasgow bearing the news. They met silently, but from the old friend's face and his black coat, there could be no doubt of what words soon confirmed. James Watt, then in his thirty-seventh year, had lost his wife, Peggy, shortly after she had delivered a stillborn child. That child was their fifth, their third son, but only one of those sons— James junior, then four years old—and a six-year-old daughter, Margaret, had survived to share their father's grief when he arrived the next day at the family home in Glasgow.

The loss of his wife coincided with a crisis in Watt's business affairs. Surveying for civil engineering work of the sort he had been engaged in when called back to Glasgow was proving hard and unremunerative. His earlier attempts to build a business as an instrument maker and general merchant in Glasgow had proved unsatisfactory. His experimental steam engine, a pet project that had been neglected because of the imperative to support his family, was threatened by the financial troubles of Watt's partner in the venture. Eight months later, Watt traveled south to a new partnership with the Birmingham "hardware man" Matthew Boulton, and to a new life. Left initially with their grandfather, Watt's children subsequently joined him, as did their new stepmother, Ann McGrigor, with whom Watt had two further children. Watt never lived in Scotland again. The hero of Scottish invention made his home, his name, and his fortune in the English Midlands.

The move to Birmingham was the pivot around which Watt's life story turned. While his youth and early manhood had prepared his mind and hands in many respects for what was ahead, it was the partnership with Boulton and the chance to participate in the "Industrial Enlightenment" of the West Midlands that provided the means to bring his inventions to market and build the legend he was to become in the nineteenth century.

What, then, was the nature of his achievement? Watt is sometimes loosely referred to as the "inventor" of the steam engine. He was, rather, its improver in two crucial ways: first, he greatly improved the efficiency of the preexisting Newcomen engines with his invention of the separate condenser; second, he developed the means to convert these newly efficient engines from providing only reciprocating motion (useful primarily for pumping water from mines) to providing rotative motion (applicable to driving textile machinery and many other forms of industrial equipment). Numerous other subsidiary inventions contributed to the successful implementation of Boulton & Watt engines, as did the development of new "business models" for their development and sale. The increases in fuel efficiency, in theory, freed steam engines from locations where fuel was plentiful and therefore cheap. They could, on this score, now be located anywhere. This provided the basis for the rise of giant industrial cities powered by steam. However, while Boulton & Watt got steam rolling industrially, the steam engine only gradually substituted for waterpower. There was no overnight steam revolution in Watt's lifetime, and some suggest that Watt became a conservative force whose defense of his broadly cast patents held back the development by others of high-pressure engines and other steam innovations for some years.

The visible results of Watt's steam invention are well understood, but the springs of that invention less so. His work has often been characterized as empirical, the achievement of a master craftsman. But it has also been seen as a product, at least in part, of Watt's scientific understanding of steam and hence his involvement with the natural philosophy and philosophers of his age. I will explore these alternative interpretations and argue that some of Watt's steam innovations did indeed have a natural philosophical basis. Moreover, consideration of some of Watt's other inventions relevant to pottery manufacture, his development of a copying machine, and his use of various airs for medical purposes allows us to see a consistent set of philosophical (notably chemical) interpretations lying behind his inventive activity.

Watt was a remarkable man, there is no doubt, though usually a mild, guarded, and even dour personality, at least until his retirement from business in 1800. A versatile, inventive facility characterized his engineering and his natural philosophy—for Watt was a thinking engineer, and an experimentalist. Some admirers in the nineteenth century called him a "profound" engineer. Others claimed for him significant chemical discovery. We can admire his ingenuity, his style, and his fecundity as an inventor and experimenter. But invoking his "genius" is not very helpful in understanding him or the character and springs of his achievement. It mystifies more than it elucidates. Watt's successes, and his failures, were the products of a collective enterprise. Family, friends, business partners, and employees all played a part in generating the phenomenon that was Watt. A key aim in this biography is to do justice, in a way that prior biographies have not, to this collective dimension of Watt's achievements.

Yet the heroic Watt was, and is, a real and important phenomenon, a cultural product. Beginning before his death in 1819, through the nineteenth and much of the twentieth centuries, Watt was celebrated as a crucial contributor to the process of industrialization through which first Britain and then other countries passed. There were many versions of Watt, constructed to serve a great variety of purposes as his legend was seized upon to support views about the nature of engineering genius and its role in human affairs. Though multifarious and divergent in many ways, these accounts agreed in seeing Watt's contribution as a positive one. In the last few years, however, as concern has mounted about human environmental impacts and especially anthropogenic climate change, Watt's legend has taken on a darker side. He is now mentioned as a progenitor of "fossil capitalism" and consequently as an author of the Anthropocene, the era in which, for the first time, human activity registers itself, irrevocably and possibly disastrously, in the global geological record.[1] Of course, Watt could have no inkling of this, and it would be nonsense to in any sense blame him, but such shifting frameworks do shape the significance we see in past human lives. Even as it delves into and draws upon a vast documentary record, this biography of the man and the legend is inevitably, and only partly consciously, a contemporary interpretation. If we live on in the minds and memories of others, then those minds and memories help to determine who we were as well as who we are. Our questions then are: Who was Watt; Who did Watt become; Who is Watt?

The structure of this book is not simply chronological, and the complexity of the material varies between chapters. We begin straightforwardly in narrative mode in chapters 1 and 2 with Watt's early life in Greenock and his formative influences there, in Glasgow, and in London, where he was briefly apprenticed to a scientific instrument maker. We will see that Watt's family, including his grandfather, father, and uncle, provided exemplars of technically focused and mercantile careers. Their industry was in the service of the philosophy of "improvement" that underwrote Scottish development in the eighteenth century. The family's Presbyterianism inculcated an ethic of discipline and accountability, which remained with Watt even though he appears not to have been particularly observant in his religion. After his time in London, Watt's newfound skills were deployed in his nascent mercantile business, and also at Glasgow College to assist the professors there who required their instruments and demonstration apparatus to be kept in good order. In this environment, Watt's sponge-like qualities enabled him to acquire significant theoretical knowledge, especially of chemistry, and he befriended lifelong allies in the business of natural philosophy. His initial interest in steam engines was also sparked there. Upon his marriage Watt moved his growing business into the city. His ideas about steam engine improvement held out a prospect of a major breakthrough, but the necessity of earning a living for a growing family led him to pursue civil engineering projects that provided a more reliable and immediate return. Though he had family background and some skills in this area, his engineering practice involved him in a rapid learning experience. Apart from pursuing surveying and canal building work, Watt also became involved in a variety of ventures that applied natural philosophical insights and experimentation to possible business opportunities. I argue that Watt is perhaps best seen as emerging through all this as a Scottish "improver" in the family tradition. But the call of the steam engine project was insistent, even though neither the finance nor the facilities available to him in Scotland were proving adequate. He turned to the friends he had made in Birmingham who had assisted in drawing up his patent on the separate condenser in 1769. That brings us to the tragic loss of his wife and his move to Birmingham in 1774, a major turning point in his life.

In following chapters, in the service of a thematic approach, the reader is asked to jump around in time. Chapter 3 on the Boulton & Watt steam engine business and especially chapter 4 on Watt's natural philosophical inquiries involve significant technical discussion to support their central claims, and some acquaintance with eighteenth-century technology (industrial arts) and chemistry is assumed.[2] These chapters also both reach back to the 1760s and forward to about 1800, a year that marked Watt's retirement from business and also, in many ways, from natural philosophical inquiry.

Chapter 3 examines the formation of the Boulton & Watt partnership and traces the bases on which its steam engine business was established, first through Watt's patents and the production of the early pumping engines primarily for Cornish mines. We see that although their fortune was substantially based on these Cornish engines, they quickly became problematic, and by the early 1780s Boulton and Watt were contemplating additional markets. Boulton encouraged Watt in another bout of invention directed at the production of smaller rotatory engines designed to drive machinery and so capable of penetrating manufacturing enterprise of all sorts. This quickly became their main source of new business, which, despite efforts to develop overseas markets, remained almost entirely in Britain. Staying with the technicalities of steam engines, the final section of this chapter tackles the question of high-pressure steam engines so vital to the steam locomotion and fast, powerful engines of the nineteenth century, and whether Boulton & Watt was, as is sometimes claimed, an obstacle to their development.

In chapter 4 I reveal the extent of Watt's natural philosophical and experimental work over a number of projects. I argue that he developed a coherent "chemistry of heat" through which he investigated and understood a range of phenomena including steam production and the operation of engines, the nature of water and airs, and the therapeutic action of airs of different kinds in the human body in so-called pneumatic medicine. This work has not been well understood before, partly because most of it was contained in Watt's private notebooks and correspondence and unpublished. Our scientific culture frowns on such privacy, being dominated by the idea of open communication. But Watt's world was different, even though he was to a degree a participant in Enlightenment culture. I explore his attitude toward publication and his relationship with the institutions of organized natural philosophy. In many respects he, and Boulton, treated these as another forum through which to pursue business objectives.

In the service of understanding the extent to which Watt's achievements in the steam engine business and in natural philosophy were collective ones, long periods of his life are also dealt with in chapter 5, in which I examine the role that his family, partners, friends, collaborators, and employees played in his work. Watt's complex and important relationship with his son James Watt Jr., who was born in 1769, is not spread thinly through early chapters but rather dealt with in this chapter through the lens of the son's crucial role as the leading member, in many respects, of "Team Watt." An account of familial relations that gestated and evolved over long periods is also telescoped into this perspective on the social structures that supported Watt, often despite severe difficulties and conflict. The Boulton & Watt engine business became profoundly collaborative, and a number of its employees made significant inventive contributions. Watt found this environment difficult. Partly to humor his partner, Boulton insisted that, whatever the collaborative reality, Watt have the final say in such matters and also be attributed with the lion's share of credit for all inventive activity in the firm. We will see that this was done partly because from the beginning, Boulton & Watt adopted a competitive strategy that marketed its products as if they stemmed from the extensive and ingenious natural philosophical inquiries that Watt undertook. Watt took on the character of a natural philosopher—this remained the eighteenth-century term for investigators in the natural sciences, especially experimentalists, though historically it had a much broader meaning. Our term "scientist" was a nineteenth-century "invention."[3] This presentation of Watt was very important commercially to the firm, but its individualistic construction leads us astray historically.

The next three substantive chapters examine matters of money and fame. Watt's interest in material success is not usually stressed or even mentioned in most accounts of his life. His publicity, then and subsequently, tended to paint him as a philosopher and great public benefactor because of the transformative effect of his inventions upon industry. But Watt was focused upon personal material success on a grand scale. So, in chapter 6 I examine the fruits of his success, attempting first to assess just how much money he made during his working lifetime, how he invested it, and how he spent it. We will see that Watt made a substantial fortune, much larger than indicated merely by his last will and testament. He invested much of this in purchasing landed estates in the Welsh borders and had a range of other substantial investments. The fruits of Watt's success were not confined, however, to financial comfort and security. Rather uncharacteristically, during his retirement from 1800 on, Watt enjoyed a life of leisure and travel, spending time in resorts on the south coast of England, in London, and with family and friends in Scotland. He devoted considerable time and effort to the improvement and management of his gardens at Heathfield Hall (also known as Heathfield House), the elegant mansion he built for himself not far from Soho on the outskirts of Birmingham, and also of his Welsh estates. His technical interests did not disappear. He spent many hours in his famous garret workshop at Heathfield House, and he gave advice and assistance on engineering projects to a variety of individuals and public bodies. Despite the blow of losing in the space of ten years all but one of his four children who survived beyond infancy, Watt also appeared in his retirement to be happier than in his early life. He became the avuncular sage, known for his vast stores of knowledge and great powers of entertaining conversation.

Fame becomes our focus in chapter 7, which deals with Watt as a living legend during the last two decades of his life. I argue that his legendary status was not simply granted to him by others but was, significantly, seized by Watt himself and by Watt Jr. through sometimes brazen but often carefully camouflaged acts of self-aggrandizement. Watt himself habitually expressed diffidence about matters of fame, but this was, I think, a cultivated persona. Watt Jr. was more openly his father's reputational promoter. He was also an attack dog when, as happened from time to time, rivals, disgruntled employees, or even "misguided" friends had the temerity to question the nature or the extent of Watt's achievements.

Watt's posthumous fame is the subject of chapter 8. The "filial project" through which Watt Jr. continued to promote and defend his father's reputation is an important part of this story, and it involved not only literary works and statues but also the preservation of heirlooms and the management of the business that still bore Watt's name. But there were many independent public commemorations of Watt, from the colossal Chantrey statue of him placed in Westminster Abbey to the great variety of memorials in Birmingham, Glasgow, Greenock, Edinburgh, Manchester, Leeds, and elsewhere. I suggest that he was a "man for all causes" in that many groups in British society—the industrial class, mechanics, inventors, patent law reformers, some groups of scientists, civic leaders—not only commemorated Watt but also used his memory, or particular versions of his character and what he represented, to advance their own objectives. In the twentieth century, the age of electricity and atomic energy, we will see that Watt's symbolic value was attenuated, a process itself symbolized by the difficulties in placing elsewhere the Chantrey statue that was ejected from the abbey in 1960. His name remained a byword for inventive genius in a generic sense, deployed, for example, in advertising, often with that famous steaming kettle. In the twentieth century general awareness and knowledge of his work probably declined.

The documentary base for a biography of Watt is enormous, indeed overwhelming, and I have been overwhelmed. The most comprehensive documentary treatment of Watt remains the three-volume work by Richard Hills, upon which I have relied a great deal in what follows.[4] Hills's unparalleled knowledge of the Watt archive makes his work the place to go to understand details of the multifarious inventions and projects in which Watt engaged. Hills's account of the development of the engines remains unsurpassed. But the very comprehensiveness and complexity of that biography involves the danger of losing the man in the multiplicity of his actions.

Other recent books devoted to Watt cannot be accused of succumbing to the search for comprehensiveness. Astute distillation has been the order of the day. Ben Marsden's short, very readable work presents an imaginative and reliable synthesis, brings in the mythmaking dimension, and offers a number of novel insights.[5] Written for a series devoted to popular exposition, the book sits at the other end of the spectrum from Hills's contribution as an object lesson in compressed synthesis informed by recent scholarship. Ben Russell's account of Watt is another excellent work.[6] Inspired by the task of reassessing and re-exhibiting Watt's famous garret workshop, held at the Science Museum since its removal in 1925 before the demolition of Heath-field House two years later, Russell emphasizes Watt as a "maker," as a craftsman, providing a modern take on Dickinson's development of that theme, which was also inspired by Watt's workshop.[7] Russell places Watt in the context of a transformation in the world of artifice and the consumer revolution of the eighteenth century so richly depicted by Maxine Berg and Celina Fox, among others.[8] Russell provides an entry to a context of economy, art, and consumption, of which there is too little in the present study, but he can only gesture to the complex interweaving of the engi-neering, philosophical, and social dimensions of Watt's work with which I am mostly concerned.

Building on these earlier important Watt biographies, and on the work of other Watt scholars, I have pursued what I see as a middle way between the comprehensive-ness of Hills and the concision of Marsden and Russell. I have developed major themes lacking or underdeveloped in prior treatments, and I have done this at a length that enables extensive exemplification and allows for exploration of the interconnections between Watt's technical, philosophical, business, social, and family lives.

In doing this I am extending and consolidating, for biographical purposes, the work on Watt that I have undertaken over the course of twenty years or more and published in two monographs and numerous papers. I began with an interest in why it was important not only to Watt's family and collaborators but also to many others in the mid-nineteenth century that Watt be characterized as a "philosopher." This led me to an analysis of the "water controversy" of the nineteenth century and the competing symbolisms of Watt that were mobilized through it in the service of con-trasting ideologies of the relationship between science and technology.[9] This work is reflected here in my treatment of Watt's legend and in the general caution that I observe about taking past stories about Watt at face value. I then pursued a more specific contention that much historical mischief was done by nineteenth-century characterizations of Watt that hid his chemically based natural philosophy behind a thoroughly misleading depiction of him as a "physicist" of heat and even a kind of "proto-thermodynamicist," neither of which he could have been without revers-ing time's arrow.[10] These arguments are further developed here also, with the larger purpose of freeing Watt's biography from the weight of legend formation and myth-making, which can only be done by integrating the story of those processes into the biographical picture.[11] That liberation allows us to bring out Watt's human qualities, to see the collaborative nature of his achievement, and to make a realistic assessment of his contribution.

There is much important Watt scholarship outside the bounds of biography, and it will be examined and deployed as we proceed. Three scholars in particular beyond those already mentioned have stimulated my thinking about how best to characterize Watt by their own efforts to do so. Margaret Jacob has depicted Watt as "entrepreneur" within the framework of the Enlightenment.[12] Peter Jones and Ursula Klein have in their different ways attempted to capture the personification in Watt, and others, of both a natural philosopher and an exponent of the industrial arts, Jones in the term "savant-fabricant" and Klein through the notion of a "hybrid expert."[13] I explain in the conclusion why I think it best to see Watt as an "improver" in the Scottish tradition.

In the course of elucidating my themes, a picture of Watt emerges. The man we will meet in these pages was above all, I think, an obsessive analyst and improver of everything with which he became involved. This was true as much of such things as his tourist activities and decisions about land purchases in retirement as it was of steam engines. He was also a model of pragmatism in his natural philosophy, his inventive activities, his religion, and even, we might judge, in his family relations. In his politics he feared in equal measure the irrationality of the mob and what he regarded as the mindlessly unproductive hereditary aristocracy. Partly because of this rationalist, pragmatic persona, Watt was, for much of his life, an emotionally cold, rather dour individual, obsessed by his work and financial worries—though on some occasions, especially in retirement, a more congenial character emerged. But as the activities of Team Watt illustrate, he was also capable of attracting many who loyally assisted him and felt privileged to know him for his human qualities as well as for his technical prowess. The man remains something of an enigma, but the elements of that enigma will, I hope, come into focus in the following pages as the great improver of the steam engine takes shape.

The Life and Legend of James Watt

One

The Making of a Scottish Improver

⚜ ⚜ ⚜

*T*he port of Greenock, where James Watt took his first salt-laden breath on
January 19, 1736, was then a small coastal town but very much on the rise.
Located on the southern bank of the mouth of the River Clyde, twenty-five miles
west of Glasgow, Greenock was in fact a recent settlement, with a population of
about three thousand in 1741, rising sixfold by the end of the century. It was one of
those towns created by landowners granting charters by which a group of local worthies could form a burgh council and raise taxes to develop the town. The local land-
owner in this case was Sir John Schaw. The granting of charters in this way indicated
that landowners were seeking to turn their estates to business and profit rather than,
as in the past, to military power and authority. The improving impulse that was to be
so important in Watt's life came from the top of his local community.[1]

The development of the town, the building of its docks and harbors, churches,
schools, and civic buildings, occurred in the decades around Watt's birth. His grandfather and father played an active role in this development. Greenock grew on account of trade, partly in agricultural products and linen (mainly with England) but
increasingly, and rapidly from the 1740s, through the import of tobacco from the
American colonies. By 1769 Greenock, Port Glasgow, and Glasgow handled the bulk
of the Scottish tobacco trade, which in turn accounted for 52 percent of the total
British trade in that commodity. The American colonies received in return a vari-

ety of basic needs channeled into Greenock by local industry and launched from its docks.[2] Between 1730 and 1750 Glasgow's colonial traders invested heavily in other manufacturing enterprises, including leather production and boot and shoe making. Textiles were another key area, especially the finishing processes such as bleaching fields and textile print-works, which required significant injections of capital. This expansion as well as trade in iron products made possible key initiatives in iron production, notably the Carron Iron Works, established in 1759 near Falkirk in Stirlingshire. Both the investors in those works and the skilled workmen often imported from England to establish and operate them represented links between Scottish economic development and the industry of the English Midlands that were to be important in Watt's life trajectory.

Twenty years before Watt's birth (in 1715) and again when he was nine years old (in 1745), the political settlement on which the prosperity of Greenock and Scotland was being built was challenged. The Jacobite rebellions of those years challenged the political and economic union between Scotland and England that had been sealed in the Act of Union of 1707. The Watt family, like other enterprising inhabitants of Greenock, were firm in their political, economic, and religious support of the Union. For them the Jacobites threatened a return to a less prosperous and less promising past. The future, as they saw it, lay with expanding trade and industry, enterprise and thrift. The politics of the Union, and the philosophy of "improvement" and Presbyterian accountability, ran deep in Watt's background.[3] They were thoroughly ingrained in the lives his family led and the futures they imagined for themselves. Watt's paternal grandfather, Thomas Watt (1642–1734) was an import to Crawfordsdyke near Greenock in its early days, coming from Aberdeenshire and carrying skills as a teacher of mathematics, or "professor" of the subject as he was sometimes styled. His skills were valuable in a town with an economy based on shipping, trade, and construction.[4] Whether teaching basic mensuration or high astronomical and navigational skills, Thomas Watt found a strong demand for his services. He had a house in Crawfordsdyke, then a separate burgh but subsequently part of Greenock. The fact that he owned another house besides his family residence evidences his prosperity. He had been an elder of West Parish of Greenock since 1685 and was for several years Baron Bailie (or chief magistrate) of Crawfordsdyke, denoting authority in the community as well as prosperity. Thomas Watt married Margaret Shearer in 1679, and they had six children, only two of whom reached maturity. One of these was Thomas's elder son, John (1694–1737), who was also a teacher of mathematics and had been educated as a mathematician and surveyor. He produced a survey and map

of the Clyde that was later published by the other surviving child, his younger brother James (1699–1782), the father of our protagonist.

James Watt Sr. was trained as a "wright," or house carpenter, having been apprenticed to a John McAlpine. He married Agnes Muirhead in 1729, and between
1730 and 1734 two sons and a daughter were born, but all died in infancy; James,
also a sickly child, arrived two years later. Tragedy in the loss of children in infancy and childhood haunted generations of Watts, as was common in those days even
though mortality rates were improving. Amid all this, Watt Sr. became a prosperous
builder. He was engaged by Sir John Schaw to extend his Greenock mansion house.
Since Watt Sr. would have required financial reserves and access to credit to undertake that venture, we can conclude that he was a man of some substance. Watt Sr. also
went into business as a ships' chandler, which involved him in fitting out ships and
carving their figure heads, as well as attending to the working of their instruments,
compasses, and quadrants. This suggests that the family decision in 1756 to have
Watt train with an instrument maker was perhaps not unrelated to the requirements
and possibilities of his father's business. Like his father before him, Watt Sr. was
an important figure in the community. In fact, he was one of a select few who were
entrusted with the funds raised by the community for public improvements. Thus in
1741, under the charter of the town issued by Sir John Schaw, Watt Sr. was one of
nine residents selected to manage the funds raised by a tax on the malt ground in the
mills west of Greenock. Ten years later, Sir John and the inhabitants of Greenock
applied to the British Parliament for an act to levy a duty on every pint of ale and
beer brewed, brought into, or sold in Greenock. The funds raised were to be applied
to harbor repairs and to the construction of a new church, poor house, school house,
marketplaces, and a public clock. Watt Sr. was one of the trustees of this fund. He also
held a number of other public offices in later years. On a tombstone erected long after
his death by his son, Watt Sr. was described as a "benevolent and ingenious man, and
a zealous promoter of the improvements of the Town."[5]

The Watts were Presbyterians, adherents of the Church of Scotland. Watt Sr. was
buried not in the grounds of the church that he helped to construct but in those of the
Old West Kirk that dated back to 1591 and of which his father, Thomas, had been
an elder.[6] We know little about the family's religious observances, though it seems
likely that their enthusiasm for the Kirk declined over the generations. Watt was to
attend Kirk when in Glasgow; his payment for a pew there seems to be evidence of
that. He also on one occasion told a business partner that he would be happy to experiment with him but would prefer to avoid sacrament days, "as we cannot decently

try any experiment upon them."[7] After moving to Birmingham, Watt made public his Presbyterian religion seemingly only on convenient occasions, attaching himself, by all appearances, to the Church of England.[8] But the old family pew, for which Watt continued to pay rent after his father's death, was abandoned only in the early nineteenth century along with other property in Greenock.[9] In a cultural sense, the family's Presbyterianism died hard, and it left an important and lasting mark on Watt's life and work through the moral precepts it supplied.

The harsh Calvinism of seventeenth-century Scotland softened in the early eighteenth century, rendering it reconcilable for many with thrusting commerce and, later, moderate Enlightenment views.[10] Presbyterianism supported commercial activity in various ways in Scotland during the period of Watt's youth. Not the least of the supports was the impulse that it provided to order and accountability. The inordinate influence of Scottish accounting texts in the eighteenth century has been noted. So too have the high levels of financial literacy in the population, especially among those of the "middling sort."[11] There was a related strong commercial aspect to school education. Accountability practices were central to the Scottish Presbyterian churches in their forms of governance but also, crucially, formative of individual behaviors that the church encouraged, notably confessional diaries. Also part of such individual behaviors was the careful, routine, and accurate keeping of business and personal accounts, the latter in particular as a form of moral discipline inculcated by habitual personal practices.[12]

This accounting habit was impressed upon Watt as a child and young man, and he in turn impressed it upon his children. We will see that such accountability issues lay at the heart of some of Watt's struggles with his family and with his business partner, Matthew Boulton. More positively, in a way parallel to findings concerning some of Watt's nineteenth-century coreligionists, we will see that Watt's concerns with engine efficiency and the improvements that they impelled were also underwritten in part by this cultural Presbyterianism.[13] Suspicion of trifling entertainments as a waste of valuable time, and the stoic invocation of Providence at times of loss, also came naturally to this son of Greenock.

Watt Sr.'s carefully accounted-for prosperity, then, was built significantly by servicing the ships that used the port of Greenock in plying the English and Atlantic trades. He had an interest in virtually everything that contributed to that trade, whether it was the infrastructure to support it or the commodities involved in it. From the time Watt first left home for Glasgow at the age of sixteen, through his time in London and then back in Glasgow, his correspondence with his father includes a

FIGURE 1.1: Marcus Stone, *Watt Discovers the Condensation of Steam* (1863). Engraved by James Scott (1869). Reproduced from A. L. Baldry, *The Life and Work of Marcus Stone, R.A.* (London: The Art Journal, 1896). Author's collection.

miscellany of topics showing that both men were alert to, and concerned with, any commercial or industrial development affecting the prosperity of their community and offering an opportunity for themselves: tobacco, coal, iron, pottery, instruments, and all manner of merchant goods included. Father and son were enterprising and opportunistic to the core.[14] Ventures in shipping were financially risky, and there are signs that Watt Sr. suffered a severe financial reversal at some stage, which Williamson describes as telling heavily "upon a fortune, till then in all respects adequate to the maintenance of an easy respectability."[15]

Watt's younger brother John (b. 1739), known as "Jockey," was also directly involved in business with his father. Marcus Stone's well-known nineteenth-century painting of the Watt family at a table shows Jockey at the window, or counter, of his father's business, oriented to the world of ships and the sea (see figure 1.1). Despite family trepidation Jockey went to sea as part of a trading venture to the West

Indies—this at a time, the early 1760s, when there were moves to diversify a trade heavily reliant upon the increasingly fractious American colonies. Jockey drowned at sea on 30 October 1762. News from distant parts traveled slowly and uncertainly in those days. The correspondence between Watt and his father in which Jockey's death is first feared, and its circumstances gradually revealed, shows a deep tenderness between the two men in their common loss. His father wrote: "I pray God may comfort you under the loss of so deir a brother who if he had been spaired would have been a credit & support to both you & me, but since Providence has Deprived us of that comfort we must Indivor to support it & confort on[e] another."[16] Watt, who had already lost his mother in January 1753, was to exhibit similar stoicism in the face of many other losses in his life.

What manner of child had the young James Watt been? Stone's painting famously depicts him playing with a steam kettle under the watchful, admiring, and slightly perplexed gaze of his parents. This is nineteenth-century mythmaking rendered graphic, but the little evidence we have does paint the young Watt as a studious, often sickly, rather gentle child, teased by his more rough-and-tumble schoolmates at Mr. M'Adams' commercial school.[17] He spent much time at home with his mother, Agnes, who reputedly created a "genteel" and "orderly" domestic environment.[18] At about the age of fourteen, Watt attended Greenock Grammar School under its first headmaster, Robert Arrol, a classicist and author of some reputation who published translations of Cornelius Nepos and Eutropius, the Roman historians, and also some of the colloquies of Erasmus. Whilst at that school Watt was instructed in mathematics by John Marr, who in the early 1750s was retained by the lord of the manor, Sir John Schaw, in some capacity and also appears to have had a salary from the town.[19] Marr presided briefly over a very well-equipped schoolroom containing, among a great variety of instruments, a brass telescope and quadrant, and a "pair of Mr Neals Globs" as well as a good collection of books on geometry, arithmetic, trigonometry, navigation, pilotage, and annuities. At this stage of his life, his early teens, Watt was thoroughly exposed, both at home and at school, to mathematics and its practical applications. An interest in electricity, a popular preoccupation in the 1740s and 1750s, may have been encouraged in Marr's classroom, which also contained a "frame for Electricity."[20]

Until the year 1753 Watt had spent most of his time in Greenock, apart from brief trips to Glasgow to see his mother's brother, Uncle John Muirhead, whose country estate Watt also visited. The Muirhead family was an old and distinguished one: it had contributed a number of men who held high office over the centuries and provided the bodyguard of the king at Flodden Field. More immediately, Agnes Muirhead's father was Robert Muirhead, a merchant in Hamilton and Glasgow of whom we know little. Her brother John was later recalled as being a builder and timber merchant in Glasgow.[21] The signs are that the Watt and Muirhead families were close after Watt Sr. married Agnes in 1728. Watt Sr. engaged in joint trading ventures with John Muirhead, those we know of involving salt and herrings. The two men pursued their joint activities without need of formal partnership arrangements—such was the trust between them. After John Muirhead died in 1769, the family relationship remained strong, with Watt maintaining close contact with John's son and heir, Robert, throughout his life. Although John Muirhead's fortunes took a turn for the worse late in his life, his son Robert inherited from him the estate of Croy Leckie in the parish of Killearn, Stirlingshire. That estate was located on the River Blain (or Blane) near its junction with the River Eldrick and about four miles east of Loch Lomond. It had been owned by John and Agnes's father, Robert, who had been responsible for substantial planting improvements, especially on the fifty acres around the mansion house.[22] His mother's family both historically and during his young years were clearly a cut above the Watts, economically at least. We know from the testimony of John Muirhead's daughter Marion (later Mrs. James Campbell) that the young Watt had substantial exposure to their way of life, spending some summers at her father's house near Loch Lomond, presumably on the Croy Leckie estate.[23]

Many of the stories told of Watt in his younger days rely upon Marion Campbell's recollections. She describes him as a "sickly, delicate child" taught reading by his mother and mathematics by his father. She tells a number of stories that address the capabilities of the young Watt. First, there is the tale of the visitor to the Watt household when the boy was six who reproached his parents for not sending Watt to public school, only to find that the child's seemingly idle chalking on the hearth was in fact a precocious exercise in trigonometry. The visitor withdrew his criticism and concluded that this was "no common child." Then, there was his practical manual prowess: "His father gave him a set of small carpenters tools & one of James's favourite amusements was to take his little toys to pieces, reconstruct them, & invent new playthings." His powers of "imagination & composition" were illustrated by a story of him staying for a while with some friends of his mother's when he was about thirteen. The friend was relieved when his mother recovered him because the whole family was deprived of sleep, having been kept awake until late at night by Watt's humorous, dramatic, and compelling storytelling. Then, of course, there is the famous kettle anecdote in which Watt, aged fifteen, is having tea with his Aunt Muirhead, who scolds him for his idle playing with the kettle, lifting its lid, capturing steam in a cup, letting it play upon a spoon, and watching it condense.[24] The account continues: "It appears that when thus blamed for idleness, his active mind was employed in investigating the properties of steam. . . . Once in conversation he informed me that before he was [fifteen] he had read twice with great attention Gravesand's Elements of natural Philosophy, adding that it was the first book upon that subject put into his hands, & that he still thought it one of the best." The book must have been the recent popular English translation of Willem 's Gravesande's *Elements of Natural Philosophy* that was widely admired for its presentation of the foundations of Newton's mechanics through experimental demonstrations.[25] We are told that when a little older and in Glasgow with his Uncle Muirhead, Watt read and studied a great deal in chemistry and anatomy, showing particular interest in "the medical art."[26] A strange, gory tale has him carrying off the head of a dead child for dissection. In a more bucolic vein, the time Watt spent in the summer months at Croy Leckie on the shores of Loch Lomond saw every excursion as an occasion for research into the botany and mineralogy of the area. He also studied the poor and their traditions, songs, and superstitions, and read indiscriminately all kinds of literature. These interests were to reemerge in his later years.

When it comes to Watt's character and demeanour, Mrs. Campbell documents numerous traits. From a very young age he displayed "manly spirit, a retentive memory, and strict adherence to truth" and was "docile, grateful & affectionate" in relation to his indulgent but judicious parents. He suffered from violent headaches that left him "for days, even weeks, languid, depressed, & fanciful," and at such times there was a "roughness & asperity" in his manners. He was given to long periods of high activity and of apparent indolence, and he "was subject to occasional fits of abstraction." Although the young Watt was "modest & unpretending yet like other great men, he was conscious of his own high talents, and superior attainments, and proudly looked forward to their raising him to future fame & honour."

This, then, is the young Watt that his cousin Marion remembered. Or did she? Like other retrospective accounts, it is likely much embroidered with a mix of distant memory and family folklore. To that extent it is part of the extensive mythmaking that went on around Watt.[27] Yet, as we will see, it is also the case that some of Marion's account rings true, given what we learn about the man from better-documented periods of his life. Watt himself indirectly vouchsafed Marion's memories when he noted on her death that he was "in great measure brought up" with her.[28]

Our other major sources for the very young Watt's life and character are the memorials assembled by George Williamson, the president of the Watt Club of Greenock, in the nineteenth century. He lamented the paucity of information but relied upon gathering the oral testimony of Watt's old school friends and of others who knew him in his young days. There are many points of coincidence between George Williamson's stories of the young Watt and those offered by Marion Campbell. Indeed, it may well be that he relied on the latter to some extent. But he did differ in important respects. He was very skeptical about the story of six-year-old Watt as trigonometer, regarding it as apocryphal, and regretted the fact that it had already received wide circulation thanks to the efforts of an "Academician." This must be a reference to François Arago's eloge of Watt, first published in English in 1839, which had relied upon Marion Campbell's recollections as supplied to him by Watt Jr. Williamson's skepticism about that anecdote is based on the testimony of Watt's schoolfellows, who found in him in his early years an apparent mental dullness: "The truth in regard to young Watt's first years in the public school is, that, owing doubtless to infirm health, to the suffering and depression which affected his whole powers, he was unfitted for a considerable time for displaying even a very ordinary and moderate aptitude for the common routine of school lessons; and that during those years he was regarded by his schoolmates as slow and inapt."[29] Williamson inclines to the view that Watt's genius did not begin to reveal itself until his thirteenth year, when, as one of his schoolmates testified, he was put in a mathematical class and began to excel in that field. This was where John Marr played his role. Watt began, according to Williamson, not only to love geometry and mathematics but also to study the skies, perhaps taking advantage of instruments that his father traded in. One further outdoor pastime of the young Watt that Williamson's sources revealed was angling, the suggestion being that the house in which Watt grew up on the high street in Greenock then had direct access to the river, so that Watt might have dangled a line almost out of the back window!

Williamson also confirms Watt's early love of handiwork and mechanism, which was naturally encouraged by the environment that Watt Sr.'s business provided. The evidence here comes from the testimony of John Rodger (1754–1827), a master shipwright and blockmaker of Greenock, who had been apprenticed to Watt Sr., as also many years earlier had been his father, James Rodger. John Rodger recalled that when he was a boy, he was sent to clear out the attic of Watt Sr.'s house, which contained miniatures and ingenious models of numerous devices. Watt Sr. told him that they had been made by his son, who was then in business in Glasgow.[30]

The sudden death of Watt's mother in January 1753 at the age of fifty-two must have been a serious blow to the seventeen-year-old. It seems to have been the occasion for Watt to achieve more independence. Early biographies often stated that in 1753 Watt was sent to Glasgow to be apprenticed to a mathematical instrument maker. But there is little evidence for this. Richard Hills has argued very cleverly, partly upon the evidence of the kind of clothes that young Watt took with him to the city, that the intention was rather that he train as a merchant with his Uncle John Muirhead so that he could act as a representative of his father's business. This is convincing, not least since letters exchanged with his father during this period are preoccupied with innumerable instances of sourcing supplies and settling accounts.[31] (In contrast, subsequently when Watt really was training in London, he reported regularly to his father on his accumulating knowledge and skills.) Also, Watt clearly had enjoyed a good and close relationship with his Uncle Muirhead during his younger years, so that learning from him would be a comfortable arrangement, another example of close family trust.

The move to Glasgow also brought Watt into his first close contact with a number of prominent members of the university there, usually referred to in those days as Glasgow College. This medieval institution entered a remarkable new period of dynamism during Watt's early life, becoming an important part of the expression of the Scottish Enlightenment.[32] There was a family connection here, too, in the form of George Muirhead (1715–1773), another relative of Watt's mother, who was appointed to the chair of Oriental Languages at Glasgow in 1753 and that of Humanity (Latin) in 1754. Though in many respects George Muirhead, reputedly the best Greek scholar in the country at that time, might seem a long way distant from his young relative from Greenock, his activities too were part of the improving mentality.[33] The seventeen-year-old Watt was welcomed by George Muirhead and introduced to a number of key figures in the college's renaissance, including the famous mathematician Robert Simpson; Robert Dick, the younger (who succeeded his father as professor of natural philosophy in 1751); and William Cullen (1710–1790), the prominent medical teacher then in the early years of professing chemistry at Glasgow before his move to Edinburgh University in the mid-1750s. Another important acquaintance for Watt at this time was Adam Smith, who had beaten George Muirhead to the Glasgow chair of logic in 1751. These men were part of a cultural vanguard of the Glasgow Enlightenment. Under the watchful eye of major patrons such as the Duke of Argyll, they were as much concerned with "improvement" as were the likes of Watt and his father, but in a different way. George Muirhead, for example, was an editor of classical works issued by the local Foulis Press. These editions, renowned for their accuracy and the quality of their production, helped to make Glasgow well known in the classical literary world. The more astute among Scotland's economic and political elite saw great value in such cultural celebrity as well as in more practical forms of improvement. Edinburgh was becoming known as "the Athens of the North" at this time, but Glasgow had its place and its ambitions too.[34]

The Duke of Argyll's control over academic appointments in Scotland saw him favor those philosophers and chemists (Cullen among them) who could turn their academic knowledge and expertise to practical improvement. Even bodies such as the Philosophical Society of Edinburgh were significantly shaped by this impulse, a third of its members being connected into the networks of Argyll and Lord Milton. The latter also worked through the Board of Trustees for Fisheries and Manufactures, established in 1727 out of patriotic concern to promote the linen, woollen, and fishing industries, to encourage technical expertise and development at all levels. Many of the men with whom Watt associated, and from whom he learned, as he built his various early careers had been aided by the Duke's patronage. The political management of Scotland was pursued in significant part through this encouragement of economic improvement.[35]

During these early years in Glasgow, Watt's associations with professors and students, as well as merchants, encouraged him to cultivate learning of all sorts. In a letter to John Craig in 1805, Watt recalled an "irregular club" that met in the early 1750s at Mrs. Scheid's (who had a tavern in the Trongate), whose members included John Millar (from 1761, professor of law at Glasgow College), William Morehead of Herbertshire, John Allan of the Row, and Craig's father. Watt recalled, "Our conversations, besides the usual subjects with young men, turned principally upon literary subjects, religion, morality, belles-lettres &c.; and to those conversations my mind owed its first bias to such subjects, in which they were all my superiors, I never having attended a college, and being then a mechanic."[36]

The John Craig to whom Watt wrote this letter provided another angle on this grouping in his edition of one of John Millar's key works, there recalling that Millar often visited the "house of Mrs Craig whose eldest son had a taste for literary conversation and philosophical experiment," where he met a number of young men including Watt.[37] The Craig family in question here were probably the timber merchants whose yards were located at one time at Waterport, the southern gateway of the city, a little west of the bridge and in line with Clyde Street. The John Craig (d.1765) we will come across financing Watt's mercantile ventures from 1759 was probably a member of this family.

Watt is also mentioned as a member of the "Literary Club" of Glasgow, which was founded in 1752 and included most of the professors of the college, and which some of the students attended. Watt seems to have partaken of some of the "clubability" that was characteristic of the period both among young men of his own age and among more senior figures in the college and city. He was perhaps representative of the nexus between Glasgow's merchant community and the college, which was a strong one. One assessment of that connection has found that of 166 tobacco merchants identified in Glasgow between 1740 and 1790, no fewer than 36 had matriculated at Glasgow College, and many others attended college classes without matriculating.[38]

Despite the commonalities of improving purpose between his life in Greenock and in Glasgow, Watt's new Glasgow connections and the society he enjoyed at his Uncle John Muirhead's house began to draw him ever so slightly out of the orbit of his father's business in Greenock. Watt's abilities were no doubt recognized by his new acquaintances, and the idea that the young man might be trained to take advantage of his mathematical and craft skills must have occurred to them. It appears that it was Professor Dick who suggested that Watt go to London for training as a mathematical instrument maker and bring back skills that, while they would certainly be useful to his father's business, would also be of potential value to the college, and to Scottish development more generally. Dick provided Watt with an introduction to the instrument-making community of the English metropolis in the form of a letter of introduction to the celebrated telescope maker James Short. Armed with this Watt set out for London in June 1755.

London

Watt was not the only son of Greenock who left Scotland to find ways to develop his talents. For all the aspirations of the economic and cultural improvers, there were still limits on what even Glasgow could offer bright and enterprising young men. Watt did not make his journey to London in June 1755 alone. John Marr, the son of his former mathematics teacher of the same name, accompanied him. Marr the younger was going to London to take an examination for a post as instructor in the navy. Leaving on 7 June, the two of them traveled on horseback via Coldstream, Berwick, Newcastle-upon-Tyne, and then along the Great North Road, arriving in London on 18 June.[39]

Marr at the time of this journey was in his early thirties and was probably assigned by their fathers to watch over the younger man.[40] He was instrumental in gaining Watt temporary employment with John Neale, a watch and globe maker, almost certainly the fabricator of the globes that Watt had used in his Greenock classroom and by that channel a business acquaintance of the Marrs. Soon Watt was working informally for the instrument maker John Morgan (d.1758), and after a month or so terms were suggested on which Watt might receive instruction from Morgan. The latter was to be paid twenty guineas for nine months of instruction. With involvement from Watt Sr. and a surety offered by James Short, a payment was made to Morgan, and Watt's training began.[41]

Although Watt later became well known in scientific circles, his London situation in itself gave him few possibilities for contact with the scientific community of the metropolis. Instrument making was a large industry in London in the mid-eighteenth century but also highly stratified. A small number of leading makers supplied natural philosophical instruments to prominent fellows of the Royal Society, and to collectors among the aristocracy, gentry, and rich merchant class. They also supplied specialized metrological instruments to institutions such as the Royal Observatory. The general metrological instrument market was supplied by a host of other makers, and it would have been among this group that Watt's instructor John Morgan sat.[42] As a short-term, informal trainee rather than a long-term formal apprentice to such a maker, Watt was very marginal to the metropolitan instrument-making community. His work for Morgan reflected this. And yet Watt would have gained some awareness of the more elite business of instrument making for, and within, natural philosophy. When he returned to London in 1767 on canal business, Watt clearly already knew Jesse Ramsden well enough to use his business in Haymarket as a postal address and may well have stayed with him at that time.[43] They were of a similar age, and it is possible that they had become acquainted during Watt's training. James Short, himself a Scot, clearly did what he could to assist his young compatriot,[44] and Watt would have become aware, if he was not already, of these higher possibilities. Short was celebrated for his reflecting telescopes and, like most English makers, for the excellence of workmanship rather than theoretical sophistication. As well as supplying the apex of the market, Short made large numbers of high-quality instruments for the burgeoning consumer market among wealthy dabblers in astronomy. Short, like much of the industry at this time, was moving from a craft to an industrial mode of production. In these circumstances of the industry, it would be likely that a youngster eager to learn would find himself engaged too much in routine and repetitive work rather than ranging over a variety of skills. The latter was what Watt and his father wanted. So, in a sense, the position with the rather marginal Morgan was fortunate in the variety that it offered. Watt reported to his father that Morgan was "very ready to show me anything I want to know."[45] So it was on Morgan's premises on Finch Lane, a small north-south lane running between Threadneedle Street at the northern end and Cornhill to the south, with the Royal Exchange a short distance to the west, that Watt knuckled down to learn the instrument trade.

What skills did Watt acquire during his time with Morgan? Quite quickly, he was learning how to make brass rules and quadrants, and, crucially, how to "divide" them, that is to mark scale divisions accurately. Subsequently, he worked on making azimuth compasses, a brass sector, and a theodolite. By his own assessment, the quality of Watt's work was as good as that of most journeyman makers, though he lacked speed, an important consideration when making articles for sale at a competitive price. While with Morgan, he worked hard and long hours, both he and his employer hoping to extract the most from the situation. He had time, however, to act as his father's agent with London suppliers, compare prices, shop around for supplies, and gain experience in shipping items to Scotland. In this way his business skills were undoubtedly also developed by the London experience.

Watt's health, which was to be such a preoccupation through much of his life, was actually quite good in London for most of the time he was there, contrary to stories that circulated later and were taken up influentially by Samuel Smiles. Smiles has Watt essentially fleeing London wracked by colds and headaches, seeking recovery back in Scotland. This, as Richard Hills has argued, is pure fantasy.[46] So is the idea that Watt was miserably isolated in the metropolis. His correspondence with his father shows him having an active time and crossing paths with others of their acquaintance who had business in London. This was a situation bolstered by Watt's attention to his father's business affairs. He sought out good prices for goods that his father wanted shipped up to Greenock, buying a great variety of things from drinking glasses and punch bowls to compasses and stoves to quadrants and telescopes. He also on occasion cut out the middleman in finding these supplies. He spent a good deal of time on these activities. Watt was conscious that his father was extending himself financially to support his training with Morgan, and he must have been keen to do what he could to help his father's business affairs.

Watt was careful with his expenditure, finding that he needed to spend eight shillings a week to feed himself. He sought to economize where he could, and he was able sometimes to undertake a little paid work on his own account in the morning before he started on the tasks that Morgan had laid out for him. He kept careful accounts regarding expenditure for his father, although his personal accounts were not kept consistently, either through lack of time or perhaps as a small act of rebellion. But Watt must have felt that he was doing his best to live very frugally.

The extent to which Watt explored the metropolis is difficult to judge. The errands for his father took him to a range of businesses around the city, and he had to visit the docks in order to load goods for shipping to Scotland. There were times in the spring of 1756 when he was very wary about going out because of the concern that naval press gangs were active in the city. Between 1755 and 1757 over seventy thousand men were recruited to the navy, and over half of these were victims of impressment, at least initially. While a landsman (someone without sailing experience) in times of peace would usually be let go, in wartime that was less likely. Even in wartime a legitimate apprentice or tradesman, or a "gentleman," would have a good chance of successfully fighting impressment.[47] Watt was particularly vulnerable because he was not officially an apprentice. Apart from making it less likely that he would escape being pressed into naval service, his situation, once discovered, would likely get him into trouble with the trades.[48] So, Watt had reasons for keeping his head down.

The years 1755 and 1756 were eventful ones for Britain, and Watt was curious about the centers of power, those distant fulcrums about which Scottish improvement turned. Military tensions had been building for some time, and war against France was formally declared on 17 May 1756. Watt witnessed the public declaration. There is little evidence beyond this, and his witnessing of King George II's return to London in September 1755 from one of his habitual visits to Hannover, that Watt involved himself much in the wider life of the metropolis.

Throughout his time in London, Watt clearly had in mind putting the skills he was acquiring to use in his father's business. He also had an eye on his new friends at Glasgow College; before he left London he was contemplating communication with Robert Dick to see if the professor had any assignments for him in London before he left.[49] Dick had already been very useful to Watt in gaining James Short's support in London, and Watt would have been acutely aware that continued sponsorship from college people would be valuable when he returned to Glasgow. We know from the nature of the supplies that Watt purchased to take back to Glasgow that he contemplated producing Hadley's quadrants, barometers, spirit levels, and the like, all to a quality finish.[50]

Glasgow and the Business of Natural Philosophy

It is unclear exactly when Watt journeyed from London back to Glasgow. His engagement with Morgan according to the agreed term would have ended in June 1756. Thereafter he spent time on business for his father, to whom he wrote his last letter from London on 19 August 1756. By late September we can place him in Glasgow. Watt returned without his former traveling companion, John Marr. Marr had passed his examination to become an instructor in the navy and had gone to sea shortly thereafter. Watt reported that his friend had been in Lisbon at the time of the great earthquake on 1 November 1755. Marr went on to a career as an engineer in the army, helping to fortify Quebec during the American Revolution. They remained friends, and they became relations when Marr married Agnes Millar, known as "Nancy," a sister of Watt's first wife, Peggy. The two engineers were on different paths. Watt's, the less adventurous, with no particular sign of the celebrity to come, led back to Glasgow.

Watt appears to have returned to Glasgow rather than going straight to Greenock to see his father and brother Jockey. He had resumed work for his father's business, especially efforts to secure money from creditors. But he also had an opportunity to deploy his new knowledge of instruments, thanks to Professor Dick, who likely persuaded the college to hire Watt for a fortuitous assignment in October–November 1756. This related to a collection of instruments, including an astronomical clock, reflecting telescopes, and transit instruments, that had been bequeathed to the college by Alexander Macfarlane F.R.S., one if its graduates. Macfarlane was a merchant, long based in Jamaica and engaged in the sugar trade. He had decided that the college should have his observatory instruments after his death.[51] The instruments arrived in a rather disheveled condition, and Watt's assignment was to clean them and to ensure that they were properly reassembled and set up. He was paid the sum of five pounds for the job, which he must have finished by early December because at that time he finally returned to Greenock. The instruments were deployed in the Macfarlane Observatory erected on the college green by 1760 (see figure 1.2).

It was while working on these instruments that Watt first met John Robison (1739–1805), who was to be a lifelong friend.[52] Robison was instrumental in introducing Watt to the steam engine, and he also became an important, if not always reliable, source of historical information about key periods of Watt's life. Robison came from a Glasgow mercantile family and was educated at Glasgow Grammar School. He entered Glasgow College in 1750 and graduated with a Master of Arts degree in

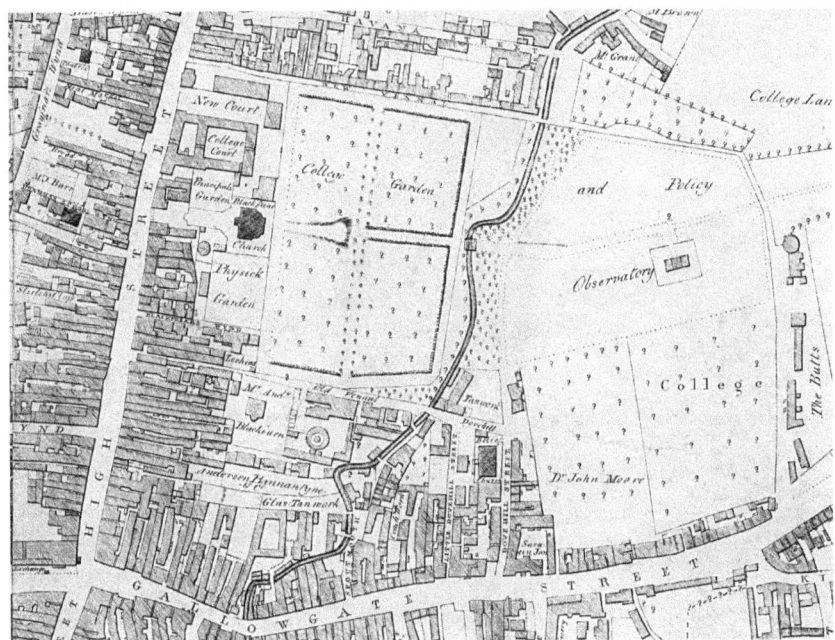

FIGURE 1.2: Map of Old Glasgow College (circa 1778). Watt's workshop was in College Court, and his shop on High Street was near the gate to the Principal's Garden. Extracted from John McArthur's *Plan of the City of Glasgow: Gorbals and Caltoun* (1778). Image courtesy of the University of Glasgow Library.

1756. Not long after that he published an improved design of the Newcomen steam engine in *Universal Magazine*. His subsequent career was to include a brief career in the navy as tutor to an admiral's son and as a member of an expedition to Jamaica to test John Harrison's famous chronometers; the chair of chemistry at Glasgow in 1766; a period in Russia on the coattails of his naval patron the admiral; and then election to the chair of natural philosophy at Edinburgh University in 1774, where he remained for the rest of his career. Robison recalled of his first meeting with Watt that he "saw a workman, and expected no more; but was surprised to find a philosopher, as young as myself, and always ready to instruct me."[53]

During the six months after he finished with the Macfarlane instruments, Watt appears to have been laying the foundations of an instrument-making business, undertaking smaller repair projects, selling the occasional Hadley's quadrant, and expanding his own collection of tools. His father's business and contacts provided an

FIGURE 1.3: North side, inner court, Old Glasgow College, High Street Glasgow before its demolition in 1870. Watt's workshop was on the first floor above the right-hand turret. From Thomas Annan, *Photographs of Glasgow College* (Glasgow: T. Annan, 1866). Reproduced courtesy of University of Glasgow Library, Special Collections.

outlet for his productions and a source of commissions. The local Greenock book-seller, a William M'Dowal, also took one of Watt's quadrants at this time. It is not clear whether Watt ever considered that he would be able to build a successful busi-ness along these sorts of lines. But in mid-1757 he seized the chance to relocate his workshop and tools to the college, where he was appointed mathematical instrument-maker.

The college was in the habit of hosting other ventures like the Foulis Press and the Foulis Academy of Art and Design, which was granted quarters above the library from 1752. It also hosted a type foundry run by Alexander Wilson, which supplied the Foulis Press and other printers with high-quality type. The foundry was housed in a building abutting upon the "College Open."[54] It is possible that granting Watt a role and space in the college was part of an effort to balance these earlier concessions to the literary and visual arts by providing assistance to the mechanical arts and natural philosophy. Whatever the case, Watt was given generous accommodation. He was provided with a workshop on the first floor of "the north-west side of the inner quadrangle, immediately under the gallery of the natural philosophy class, with which it communicates."[55] The building pictured in figure 1.3, which was located in "College Court," as the inner quadrangle was also called (see figure 1.2), was where his work for the college and its professors would have been done. But, like the Foulis brothers and Wilson, Watt was enabled to pursue his private trade, and for this purpose he had a shop that "formed the ground floor of the house situated next to the Principal's Gate." The house had an entrance directly from the high street.[56] So, the move to the college was not to mean that he abandoned his instrument making and general mercantile activities on his own and his father's account. Rather, the college appointment provided a source of income and accommodations around which his private business could be gradually fostered. It was later suggested that Watt's desire to establish a business in Glasgow had been stymied by the opposition of the guilds and that the college had provided a haven for him in that sense.[57] Whether or not that is true, the college certainly provided a haven for him in terms of building a viable business. Watt's workshop was to be based there until he relocated it in the early 1760s during his partnership with John Craig, which saw Watt gradually withdrawing from the shelter of the college.

Aside from its official purpose, according to Robison, Watt's workshop at the college became an important gathering point for young men interested in natural philosophy:

All the young lads of our little place that were any way remarkable for scientific predilection were acquaintances of Mr. Watt; and his parlour was a *rendezvous* for all of this description. Whenever any puzzle came in the way of any of us, we went to Mr. Watt. He needed only to be prompted; everything became to him the beginning of a new and serious study; and we knew that he would not quit it till he had either discovered its significance, or had made something of it. No matter in what line,—languages, antiquity, natural history,—nay, poetry, criticism, and works of taste; as to anything in the line of engineering, whether civil or military, he was at home, and a ready instructor.[58]

This image of Watt, though put forward many years later in a context where hyperbole was at a premium, tends in the same direction as stories of him as an even younger man and also accounts of him in his old age. There is likely much truth in it.

The workshop, however, was not intended to be a "drop-in" center for the young men of the college. It was to serve a serious purpose. As Professor Dick may have foreseen when he first suggested that the young Watt seek training in London, there was certainly an increasing need for the services of a versatile instrument maker and mechanic at Glasgow College. At that institution and at Edinburgh University in particular, the mid-eighteenth century saw the efflorescence of a distinctive teaching regime in the sciences that attracted students from far afield, especially from dissenting communities in Britain but also from the American colonies and from Europe. Centered upon medical education but not confined to it, the teaching of the natural sciences assumed greater prominence.[59] Old teaching methods using a tutorial system were superseded by large-scale lectures. Charismatic professors attracted large numbers of eager students to lectures in chemistry and natural philosophy, and they informed and enlivened their teaching through sometimes spectacular and often impressively deft experimental demonstrations. Those demonstrations required equipment of various sorts, which had to be kept in good repair and set up as occasion demanded. The professors also conducted research very often; this was more of a private matter. However, advanced or particularly favored students might be inducted into the mysteries. Watt's function, then, was to keep the college's instruments, particularly those used in teaching, in good working order. Given the extent to which the reputation of professors as teachers—and hence their salaries, which depended heavily on student fees—and the success of the institution relied upon proficient demonstrations in lectures, Watt's function was an important one.

Watt's arrival at the college coincided with some important shifts in personnel. Professor Dick died in May 1757, which would have been a personal blow for Watt since the professor of natural philosophy had been an important guide and support. Dick was replaced by John Anderson (1726–1796), who also proved supportive in the sense that he gave Watt considerable employment in cleaning and repairing the full gamut of apparatus that was used in his lectures to demonstrate principles across the whole range of natural philosophy. Anderson, who was at one time known to his students as "Jolly Jack Phosphorus" because of his predilection for using explosives in demonstrations, also employed Watt as an assistant and demonstrator in his lectures. Hills considers that the education Watt received from Anderson as a result of these activities was invaluable and has not been fully appreciated in its relevance to his later work on steam engines.[60] Watt's marvellous powers as an autodidact were given considerable play by the opportunities that Anderson gave him. The impulse that later led Anderson to institute lectures on the principles of natural philosophy for working men and to bequeath his estate for the education of workers may well have guided his treatment of Watt, who was also given the run of Anderson's personal library.[61]

Another major field of teaching at the college was chemistry. This found much of its rationale in its medical connections and had been given a boost by the appointment of William Cullen at Glasgow. Cullen was part of a movement that sought to expand chemical teaching beyond the field of pharmaceutical chemistry, the original rationale for its inclusion in the medical curriculum. Eventually Cullen taught chemistry as a "liberal art" in itself, creating a curriculum that combined "philosophical chemistry"—by which was meant a form of pure inquiry into the nature and combinations of matter and heat—with the applications of chemistry in agricultural and industrial improvement.[62] With Cullen, academic chemistry and practical improvement achieved close proximity in a way that greatly pleased the impresarios of Scottish academic development, especially the Duke of Argyll. Cullen left Glasgow College for a chair at Edinburgh in 1755.

Cullen's successor at Glasgow was Joseph Black (1728–1799), who had himself studied with Cullen at the college before moving to Edinburgh to take his MD. Black was appointed to the chair of medicine at the college in April 1757 and probably took up the position at about the time that Watt transferred his workshop there. The two men were to become close associates, business partners, and lifelong friends. For historians and others, the nature and significance of their relationship has been of great importance and sometimes controversial, particularly the question of how much Watt relied on Black's scientific ideas about heat in conceiving his improvements to the steam engine.

Black was born in France, but his family were long based in Ulster, which had of course very close ties with Western Scotland and Glasgow in particular.[63] He studied at Glasgow College from 1744 and was inspired to an interest in chemistry by Cullen, whose lectures he attended and who employed him as a laboratory assistant for his lectures. Black was to be a chemist in much the same "improving" vein as Cullen, combining an experimental philosophical chemistry (in Black's case concerned with causticity, the production of airs, and the chemical nature of heat) with close involvement in the applications of chemistry to agriculture and the industrial arts. Professionally, Black had decided to enter medicine and he transferred to Edinburgh University to pursue his MD. It was for the dissertation required for that degree that Black undertook his studies of *magnesia alba* (magnesium carbonate).[64] The decision to work on this material was informed by the search for a remedy for urinary stones but also related to the industrial context, in which alkalis were already of considerable importance in the Scottish economy. But the famous outcome of the work was Black's discovery that magnesia alba, when treated with acid, released an air, which Black called "fixed air" (and which we call "carbon dioxide"). For reasons to be explained later, this discovery placed Black as a founder of the study of airs—or "pneumatic chemistry"—a field that was of great importance to changing chemical ideas in the later eighteenth century, and one in which Watt would also claim a role. Black was to be reticent about publishing his work in later years, but this early discovery was published in the journal of the Edinburgh Philosophical Society.[65]

When the two men first coincided at Glasgow College, Watt was twenty-one and Black twenty-eight years old. There is little direct record of their early acquaintance. Understandably, their daily proximity meant that their correspondence only began when Black moved to the Edinburgh chair in 1765, where he was to spend the rest of his career. There are some historical testimonies to their early relations (by Black and by John Robison), but these date from the mid-1790s and were produced as evidence in patent trials in which Watt's character as a philosopher was an important issue, and so are not necessarily reliable. Given Black's work on heat during his years as professor at Glasgow, and given the supposed importance of an understanding of heat to Watt's steam engine innovations, the philosophical relations of Black and Watt have taken center stage. But all that was in the future when they met. How might they have regarded each other at the time?

There was, of course, a significant difference in status: Black was a physician and a professor; Watt was a bright, resourceful, but only partly trained mechanic and instrument maker. However, in some respects their backgrounds and preoccupations were similar. Black's father, though undoubtedly wealthier than Watt Sr., was in the ship-victualling business and a merchant. Also, in the hiatus between his Edinburgh studies and Glasgow professorship, Black ran a shop of some sort. Black shared fully in the "improving mentality" of Watt and his family. He did a great deal of work as a consultant, both formal and informal, on the chemistry of agriculture (for example, the role of lime and marl in soil fertility), mining, and manufacturing. Black also invested in various industrial projects. Watt was to be a partner with Black in some of these, and a recipient of his financial support in a number of improving ventures. Black clearly thought very highly of the slightly younger Watt, both as a "projector" and as a philosopher.

During his time as the Glasgow chair, Black continued his research but became increasingly reluctant to publish it, at least in printed form. He did report on his research in the lectures that he gave, which numerous prospective medical men attended, and no doubt regarded this as a form of publication. His Glasgow research pertained to heat and mixture. On the modern map of knowledge, heat is seen as the preserve of the discipline of "physics," but in the eighteenth century it was part of chemistry, not least because many philosophers regarded heat as a substance that combined chemically with other materials. Black's experiments on heat, which were also inspired by the research agenda handed down from Cullen, almost certainly began at or around the time that he met Watt. Black claimed that he discussed his ideas about what came to be called "latent heat" in his first series of lectures at Glasgow in 1757–1758. Watt later maintained that he pursued his heat experiments of the early 1760s independently of Black, only learning of Black's notion of latent heat after his own rediscovery of it. This seems very unlikely given the close relations of the two men, Watt's location at the college, and his involvement in servicing the lecture programs of the professors. There were also good reasons for Watt, and his friend Black, to subsequently maintain a fiction of Watt's independent philosophical work on heat. But we have no direct evidence to contradict their claims, except John Robison's remark that Watt did attend Black's lectures. However, this testimony, too, was long after the event and was contradicted by Watt himself. It must also be admitted that Robison was not the most reliable of witnesses.[66]

So, whether Watt and Black were close in this later-to-be-consequential respect remains uncertain, but it is known that they shared company and also practical ventures. Indeed, the hardware shop that Watt established in the Trongate, Glasgow, in 1763 with John Craig as his main partner was partly financed by Black and by Alexander Wilson, the type founder and professor of practical astronomy. From the late 1750s Black had lent Watt various sums of money that helped launch and maintain the young man's activities. Black was even to help fund the steam engine project that Watt entered into with his first major collaborator and backer in that field, Dr. John Roebuck. Watt's debts to Black were not finally cleared until 1784, to Watt's evident embarrassment. But in the early days Watt lacked investment capital and had to rely on others, among whom Black was prominent.

Watt undoubtedly saw natural philosophy and chemistry in particular as important to business, indeed as a crucial aid to it. Only later, and then to a limited extent, did Watt regard himself as part of circles of the learned in scientific and other academies. Unlike the professors of the college, he had no formal student audience, despite the attractions of his workshop for some of the collegians. Thus it was the business of natural philosophy that primarily concerned Watt—the market that natural philosophical inquiry and teaching provided for the products of his skills, and also the use of philosophical acumen and experimental inquiry in the prosecution of business ventures. A key example of the latter was a scheme involving Black, Watt, and Roebuck for the production of alkali from salt. This scheme was also a clear bridge between the trading world from which Watt had emerged and the business of natural philosophy in which he was increasingly engaged.

Salt was an important commodity in mid-eighteenth-century Scotland in both the traditional economy and in the emerging industrial one. The Watts, father and son, traded in salt for use in preserving herring, a traditional use, and also traded in herring. Salt had long been produced in Scotland by evaporation of seawater, but when made in this way the salt contained other impurities (such as magnesium salts) that rendered it unsuitable for use in fish preservation. In consequence, much of the salt used for curing fish was imported. This was where the potential of chemistry came in. William Cullen had developed technically successful but economically unviable ways of purifying Scottish sea salt. John Roebuck also pursued purification techniques, apparently with more practical success; in the mid-1760s he was supplying Watt Sr. with salt for use in preserving herring that were coming into Greenock from the Western Isles.[67]

Another crucial commodity for a number of industries was alkali—used in making glass and soap, as a fertilizer, and in bleaching. The traditional sources of alkali were natural ones: wood or kelp were burned to produce potash. Much alkali was imported in the form of barilla (soda ash, that is sodium carbonate in modern terminology) that was produced by burning halophyte plants. These plants were common in the Mediterranean countries, and Spain was the major source of very pure soda ash. If a chemical way could be found to produce sodium carbonate from salt, then there was a real possibility of significant economic advantage in supplying the user industries with pure alkali. Indeed, the freeing of alkali production from reliance on biological sources was one of the keys, if not *the* key, to the foundation of the heavy chemical industry, or what has been called the "palaeotechnic transition."[68] Although this was not fully realized until the adoption of the Leblanc process in Britain beginning in the 1820s, John Roebuck and our Scottish improvers, including the Watts and their circle, were one of the key groups trying to make improvements in that area.

Watt's own account of the beginnings of this episode credited Joseph Black with the initial chemical insights suggesting how alkali could be produced by treating salt with lime. This was in 1765. Thus Watt informed his Birmingham friend William Small in 1769: "As to the Alcali affair you know Dr. Black first invented the theory which he communicated to me I tried experiments & found it succeed. After I had given it up, I went on with Experiments till I brought it to a probability of succeeding in practice."[69] Watt recalled that Roebuck had been brought into the venture shortly thereafter. Eric Robinson suggested that there was a division of labor between the three men, Black being the laboratory chemist, Roebuck the entrepreneur and capitalist, and Watt the industrial chemist and practical man-of-affairs.[70] Although this is broadly correct, it was not quite so simple. For one thing, the intention was that both Black and Watt have a financial interest in the venture, not just a technical role. Watt also exhibited tactical business insight. There was no immediate outcome, and the affair carried on through the next decade and beyond. There were initial plans to take out a patent on the process, perhaps in Black's name, especially when they learned that another of the Birmingham group, James Keir, was working in a similar direction, as was George Fordyce in London. But then secrecy seems to have been preferred. The scheme effectively ground to a halt as Watt became preoccupied with other things, notably steam, and with earning a living. The project was apparently left in Black's hands, but his efforts were desultory before they ceased.[71] Throughout, though, Watt was a key figure and active experimentalist. We need not explore the story further here, since the important point to be made is how quickly and how expertly Watt was operating, by the time of his thirtieth year, in the business of natural philosophy.

Another venture in which Watt was involved beginning at this time was the Delftfield Pottery in Glasgow.[72] This episode illustrates further how he brought technical expertise and insight to bear on business. It also shows us how Watt relied upon others, in this case members of his extended family, for necessary investment capital.

The Delftfield Pottery had been operating for twenty years or so when Watt became involved in it. Like much investment in manufacturing in the West of Scotland in the mid-eighteenth century, it came from tobacco merchants, in this case Laurence Dinwiddie of Glasgow and his brother Robert, who had traded in London before becoming governor of the colony of Virginia, no doubt a useful posting for a tobacco trader. Together with two other minor partners, they founded the pottery in 1748. Although they had to import technical and managerial expertise, the original partners did have a clear business plan and rationale. This was to produce tin-glazed delftware, inferior of course to porcelain but a cut above crude earthenware pottery. Such produce appealed to the middling classes of Glasgow and was intended also for export to the colonies. The pottery experienced numerous early difficulties but did manage to produce a saleable commodity before a major change in ownership took place in the mid-1760s. At around the same time, Watt came into the picture. What inspired his involvement is unclear. Richard Hills has suggested, very plausibly, that the fact that improvements in pottery were being made as a result of chemical analysis of new clays and glazes, and that innovative kilns and firing techniques were being used, may have encouraged Watt to think that his growing chemical expertise could help make a breakthrough for what had been up to that time a rather lackluster company.

Much of Watt's expertise had been acquired from Black, and Black himself had a history of consulting on ceramics for others. He took a close interest also in the design and development of kilns. Watt had manufactured kilns for Black, and the two men discussed experiments with ceramics in those kilns. At the time he joined the pottery and shortly thereafter, Watt experimented on glazes and on cobalt production from local supplies. He also experimented on new kiln designs, including kilns fired by coke and coals instead of wood, which promised more economical operation. Importantly, it appears that in all his experimental work, questions of economics, competition, and business practice were ever-present. Watt's command of the chemical and other technical details of pottery manufacture was impressive, but he was no mere technical advisor. He also had a strong business and managerial sense, and kept up with the latest developments in the industry and among competitors. His talents enabled these broader concerns; his financial interest in the pottery impelled them.

When we consider the financial side of Watt's involvement in the pottery, we are reminded of the extent to which he relied upon family and other members of his close circle. Though it is not certain, it appears from Watt's "Journal" of June 1772, where he recorded details of the capitalization of the company and his share of it as £474.4.2, that he was receiving a good return on his investment, but he notes that he owed £120 of that capital to "Mr Moreheid" and £110 to Nancy Millar.[73] This was almost certainly the capital loaned to Watt for his initial investment in the company, on which he paid interest. There was other family investment in Delftfield, by his cousin Robert Muirhead the younger and later by Gilbert Hamilton, who became Watt's brother-in-law by his marriage to the sister of Watt's second wife, Ann McGrigor. When Watt left for Birmingham in 1774, Hamilton became his agent in Scotland and subsequently a partner in the pottery with a close hand in its management.

Watt's involvement with the Delftfield Pottery, which continued long beyond the period concerning us here, shows us again how Watt turned his knowledge and his experimental capability to business use for himself and others. He relied in these early years upon financial support from his family, and indeed upon other forms of support; Watt Sr. took a keen paternal interest in the fortunes of Delftfield. Watt's family, however, were not engaged in charity here. They were investing in the capabilities of a young man already noted for his knowledge, skill, and drive, a son of Greenock who was expected to do well in the world through the business of natural philosophy.

We must not, however, as David Bryden insists, allow our knowledge of what was to come to warp our understanding of Watt's Glasgow years or lead us to treat them as merely background to the main story.[74] Despite the variety of projects with which he became involved during those years, Watt was, until at least the late 1760s, primarily engaged as an instrument maker and merchant. The fact that there was limited demand for mathematical instruments encouraged Watt to diversify his products. He and workmen he employed made musical instruments as well as mathematical ones.[75] There are intimations of Watt's mechanical ingenuity and talent for improvement in some of these activities, especially his improvement of a drawing aid—a "perspective machine." He had obtained such a machine, but it was heavily and clumsily constructed. Watt adapted the double parallel ruler in the construction of his machine and made other changes that rendered it both lighter and easier and more convenient to use.[76] These items of his own devising and manufacture were sold through his own outlets but also supplied to retailers in other cities. Watt purchased supplies for his manufacturing activities but also bought goods such as cutlery, other ironmongery, and toiletries directly for sale in his shops. Again, as Bryden says, Watt was not a closeted instrument maker but rather "a business man with vision, taking on employees and trainees, buying in a broad range of merchandise to realise his ambition of significant commercial success."[77] On occasion the pursuit of profits may have involved Watt in dubious practices. Michael Wright has discovered among the tools from this early period left in the garret workshop that Watt established in his Birmingham residence, Heathfield (which is now held by the Science Museum), a stamp bearing the letters "*T LOT.*" This he recognized as an imitation of the stamp placed on his instruments by the Parisian flute maker Thomas Lot. It appears that Watt may have been engaged in passing off his own instruments as of more prestigious pedigree, though the evidence is circumstantial.[78]

It is likely that Watt's business ambition both attracted and necessitated the substantial financial help he received from the merchant John Craig, with whom he entered a partnership in 1759 and whose financial support enabled Watt to employ more apprentices and journeymen as well as open his shop on the Trongate in 1763 and diversify the range of goods that he made and sold.[79] Watt drew himself a salary of £35 per annum, no doubt intent upon trying to build up his capital in the business. Craig invested heavily. By the time he died in December 1765, he had over £750 owed from the business. Paying that sum to Craig's estate proved a major headache for the young merchant and shopkeeper who had lost his business partner. The debt was to be paid and taken over by John Roebuck as part of his steam engine partnership with Watt. As we have seen, Watt also owed money to Joseph Black, who had invested in the shop and other projects. He had also accumulated debts to others in connection with his varied enterprises.

During the decade after his return to Glasgow from London, Watt developed many strings to his bow. His position as instrument maker at Glasgow College gave him a good start, which acquired other dimensions as he aided the professors in other ways and built up his hardware and general mercantile activities. His contacts and friendships at the college as well as the work itself expanded Watt's knowledge, skills, and capabilities. Watt was a young man who could absorb, recall, and turn to effect all of his experiences. He continued to assist, and to a degree rely upon, his father and his father's style of business, always alert to trading opportunities. But gradually he began to take a part in other ventures that gave play to his ever-expanding natural philosophical and technical repertoire. He was ambitious for success and no doubt hoped that one or more of his projects might begin to pay in a significant way and make his fortune.

The making of that fortune, or at least a reliable sufficiency, became more urgent as Watt contemplated and then acquired his own family. He married a cousin, Margaret Millar, always known as "Peggy," on 16 July 1764. Peggy was the daughter of Daniel Millar, a wright of Hamilton, and his wife Margaret Muirhead (1700–1774), herself a daughter of Robert and Agnes Muirhead, and so sister of Watt's mother. Peggy and James had known each other from childhood. It seems likely that Peggy would also have been a regular visitor at their grandfather Robert's Croy Leckie estate at Killearn near Loch Lomond during some of Watt's summer sojourns there. Given the evident closeness of the Watt and Muirhead families, James and Peggy must also have met at many family gatherings. A preserved item of correspondence between them from 1761 or 1762 (in which Peggy begs Watt to "take cair of yourself for know my happiness depends upon your health") perhaps betrays at least the beginning of a romantic relationship.[80] It was not unusual for cousins to marry, and they seem to have been well suited. For all his ambition, perhaps even because of it, Watt could be easily discouraged and was not of a very sanguine temperament. By all accounts Peggy made up for this and provided the aid and encouragement her husband sometimes needed.

Marion Campbell's recollections, as relayed by her daughter Jane, conveyed a little

about James and Peggy's "early & constant attachment": "My mother even considered it [their relationship] as having added to Mr Watt's enjoyment of life & having
had the most beneficial influence on his character. Even his powerful mind sunk occasionally into misanthropic gloom from the pressure of long continued nervous headaches & repeated disappointments in his hopes of success in life. Miss Millar from her
sweetness of temper, & lively, cheerful disposition, had power to win him from every
wayward fancy; to rouse & animate him to active exertion. She drew out all his gentle virtues, his native benevolence, & warm affections."[81] Watt and Peggy reportedly
lived initially in a substantial house belonging to the Delftfield Pottery Company on
Delftfield Lane, which was at the western edge of Glasgow, beyond the main dock at
Broomielaw. This would have placed them between central Glasgow, where much of
Watt's business was located, and his family in Greenock.

Their first child was born in July 1765, and they named him John after Watt's
recently deceased brother, Jockey. But young Jockey survived only five or six months.
A daughter, Margaret, arrived in 1767. Known like her mother as "Peggy," she lived
to adulthood, married, and had children of her own but died at the age of twenty-
nine. The only one of their children who was to survive Watt, James Watt Jr., was
born in 1769, while another daughter, Agnes, born on Christmas Eve 1770, lived for
just over a year. As we have anticipated, Peggy's final pregnancy was to end in late
September 1773 with a stillborn child and her own death.

After a couple of years on Delftfield Lane, the young couple moved to central
Glasgow, living in a number of properties there. This was a period when Watt was
conducting experiments on steam, which we will examine in detail later, and it is
interesting to note that a topic of discussion between Watt and Peggy when accommodation was being considered was whether the property had a "garret," preferably
with a fireplace.[82] This perhaps suggests that long before Watt established his famous
garret workshop at Heathfield House in Birmingham, he needed a domestic workshop with a fire available. When Watt moved out of the college in 1763, he maintained a number of premises in central Glasgow for his shop and workshop, where
experiments were conducted.[83] But it looks as if these sites were not enough, and that
Watt's incessant experimentation habitually intruded on the domestic space. By May
1767 Peggy, who was pregnant, faced "flitting" again while Watt was in London on
canal business: "As to my Garrit," he said, "I beg you would speak to Mr Macintosh
to let the things stay in it for a month or so till I come down as they cannot be removed
by any body but myself rather offer him a years rent of it for that time."[84] Watt was

obviously very concerned that his workshop or laboratory not be disturbed by un-comprehending others.

Besides the shop, the pottery, and the alkali venture, Watt had been drawn into an investigation that was eventually to make his fortune, the working and improve-ment of the Newcomen steam engine. But once again financing and prosecuting this proved difficult, and in an effort to bolster his income to support his wife, two young children, and the prospect of more, he turned to yet another line of business that ran in the family—surveying and civil engineering work.

Two

Improving Ventures

※ ※ ※

Civil Engineering and Steam

*D*uring *the decade after his return to Glasgow, Watt had created a varie-* gated pattern of work. Like his father, he traded as a general merchant in goods produced by others. Much more than his father, he also sold products of his own making, including Hadley's quadrants and musical instruments of various sorts. Watt also traded on his mechanical skills by repairing and maintaining instruments and experimental and demonstration apparatus for the professors of the college. He deployed his experimental and organizational skills as a technical advisor to manu-facturing enterprises. Here, he sold his capacity to develop or improve products and processes while also often taking a financial interest himself in the industry in ques-tion. As we have seen, Watt borrowed money from relatives and friends to fund his investments. It thus seems likely that his patchwork of employment was generating a sufficiency and not much more. His marriage and growing family responsibilities led Watt to consider a change of scale in his work in order to exploit potentially more lucrative possibilities.

Watt had become interested in steam and steam engines, as is well known. We will examine shortly, and in detail, the origins and nature of that interest. The point to notice initially is that Watt's interest in engines is based on a vision of an alto-gether grander way of making money—the production of a significant invention of generic importance (that is not connected to a particular industry or particular product) and so with a potentially wide range of application. Such an invention could be controlled either by secrecy or by taking out a patent, and could then be either manufactured or licensed to others to manufacture. Properly managed, such an innovation, coming at a time of rapid industrial growth, held out the prospect of making a great deal of money. However, partly because of the scale of such a project, the improvement of the steam engine would not be an immediate source of income. On the contrary, it would be a drain on resources, both Watt's own and those of the people he persuaded to back the project. So, some more immediate source of substantial income was required. Work as a civil engineering consultant looked promising. There was a tradition of family involvement in such work. Road, canal, and other infrastructure projects were gathering pace as Scotland expanded its trade and industrial development. Here, perhaps, was a way that a smart young man with skills in surveying and mechanics, as well as good organizational sense, could extract the more substantial living that his family circumstances, and his other ambitions, required.

In the late 1760s and early 1770s, then, Watt juggled work on the steam engine on the one hand and various more immediately rewarding employments in civil engineering on the other. There is little doubt that steam really captivated him. But he also periodically despaired of success in that area. Even as he prepared to apply for the key steam patent of 1769, he was taking on other work, including the Monklands Canal and also a study of improving the navigation of the Clyde at Port Glasgow. Writing to William Small in September 1769, Watt said of the latter work that he would not have "meddled with" it "had I been certain of being able to bring the engine to bear, but I cannot on an uncertainty refuse every piece of business that offers."[1]

The role of civil engineer had only very recently originated when Watt took it upon himself. While a number of individuals had taken up from mid-century the surveying, design, and construction of roads, canals, river navigations, lighthouses, and the like, the definition and promotion of the "profession" of civil engineer is usually credited to John Smeaton (1724–1792), who was responsible for establishing the Society of Civil Engineers in 1771. Smeaton's key conception was that the civil engineer should model himself upon the physician; the role should be a consulting one.[2] Ordinary engineers and workmen would take on the construction work, but the civil engineer would consult to the promoters of a scheme on its route and its design, and only occasionally might supervise construction. This was the role into which Watt decided to insinuate himself. But the character of a civil engineer was very labile in these early days, and there was no formal means of qualification for it. Those seeking the mantle had to persuade their potential employers that they were fit to take on this kind of role.

Watt had a number of characteristics that might persuade him, and others, that his attempted "self-construction" as a civil engineer, Smeaton-style, made sense. First, as we have noticed before, there was his family background, which gave him both confidence through familiarity with a range of relevant activities and also specific skills. The fact that Watt's father (and his father-in-law) were "wrights"—his grandfather a teacher of mathematics and his uncle John a surveyor—gave him a good start. These were among the constellation of existing occupations from which many civil engineers came. Watt's training and experience as an instrument maker was also useful. In fact, Smeaton himself came from that background. This ensured familiarity with mensuration generally and with surveying instruments in particular. But important as these background associations and acquirements were, when it came to surveying in the field, the practical selection of routes and specifying viable and economical methods of construction, there was no substitute for experience. As in so many other areas of his life, Watt was able in this phase of it to learn by harvesting the experience of others.

Hills identifies a number of individuals who collaborated with Watt and from whom he acquired practical experience.[3] Alexander Wilson, the professor of practical astronomy at the college, involved Watt in a survey of the Clyde that he (Wilson) undertook for the Glasgow magistrates in 1757. James Barrie, who became the City's first paid surveyor in 1773, had been involved with the surveying activities of Watt's Uncle John long before. When Watt was employed in late 1769 by the magistrates to check various levels and soundings of the Clyde, Barrie assisted by also checking Watt's measures. It is likely that Barrie used his influence on other occasions to get work for Watt. Two years earlier Watt had undertaken joint work with Robert Mackell, a millwright and engine builder who was also engaging in more civil engineering work.[4] They surveyed a new route for the Forth and Clyde Canal in early 1767, one commissioned by Glasgow's tobacco merchants, who wanted a canal routed close to or through the city. Mackell and Watt's report was printed, and Watt was asked to go to London with his collaborator to help see the passage through Parliament of a bill for the canal. Though this came to nothing, the experience, and the journey, were valuable to Watt. He also received advice from John Smeaton himself, who was a rival to the extent that he had designed a competing route for the Forth and Clyde Canal. Smeaton helped Watt by appraising some of the young man's work on the Monkland Canal project, and the two men were to become long-term friends.

So, once again, in a new field Watt was able to learn from collaborators and pro-moters. There must have been aspects of his personality that prompted people to as-sist him. Perhaps an admixture of clear ability and acuity with a degree of diffidence prompted others to back his talents and encourage him to exercise them.

We need to put Watt's civil engineering engagements (mainly relating to canals) into a broader context. Between 1770 and 1840 over twenty canals were built in Scotland. More than a dozen others were surveyed, even financed with construction begun, before being abandoned. Four canals—the Caledonian, the Crinan, the Ed-inburgh and Glasgow Union, and the Forth and Clyde—remain today, while there are still remnant stretches of the Monkland Canal.[5] In the course of his eight-year involvement with such projects, Watt first mapped out routes for the Forth and Clyde Canal (1766), the Monkland (1769), and the Crinan (1771); then, in a second burst of activity in 1773–1774, he searched out routes for the Caledonian, the Campbel-town to Machrihanish, and the Crieff to Tay canals. As was typical for a consulting engineer, Watt also provided costings and work plans for these projects. His situation with the Monkland Canal was more complex since, unusually, he was both consulting and resident engineer and so was involved directly in the actual construction of part of the canal. This further broadened his experience but reinforced a resolve that con-sulting roles (whether in civil or in steam engineering) were his preference. Watt did not enjoy the hurly-burly of dealings with contractors and workmen. The tensions it induced in him affected his health.

Canal projects were complex, protracted, and often frustrating affairs. Virtually all were driven by a widely dispersed set of private interests in improvement of the transport of raw materials, particularly coal and goods. The Caledonian was an ex-ception in that it was driven primarily by the broadly political objective of integrat-ing the highland and lowland economies. Canals offered opportunities for integrating markets by joining the Forth and the Clyde, the cities of Edinburgh and Glasgow (in the case of the Union Canal), the Western Isles and the Clyde (the Crinan Canal) and coal deposits and industry (the Monkland Canal). The Scottish canals were not an epiphenomenon of the canal mania that had begun slightly earlier in England. There were some English investors in Scottish projects, but most of the money was raised from Scottish sources. Intense local interest was important to ultimate success, but it also meant struggle and delay. Contending engineers drew up competing routes on behalf of divergent interests. For example, Glasgow City interests sponsored the route mapped by Watt for the Forth and Clyde. It contended against an earlier route mapped by John Smeaton. The city interests were not impressed that Smeaton's plan bypassed the city to have the canal issue farther down the Clyde. The Carron Iron Company also sponsored an alternative plan for part of the Forth and Clyde Canal that would have run it directly through the site of its operations. In the end the company accepted £1,500 compensation for agreeing to withdraw its alternative proposal.[6] When it came to seeking passage of a parliamentary bill for the Glasgow plans, the struggle crystallized as one between powerful Edinburgh interests and the Glasgow merchants.

So what was involved in Watt's participation in these projects? His work with Mackell surveying two alternative routes for the Forth and Clyde canal took place in February and March 1767. Both routes shared long sections with Smeaton's survey, but both brought the canal to the city, the main objective of the Glasgow people who hired them. One, coming into the city from a northerly excursion through Dunbartonshire, would terminate west of the Broomielaw; the other, beginning east of the Broomielaw proceeded east through the coal mining areas of Camlachie and Shettleston. By opting for smaller barges and a canal only twenty-four feet wide and four feet deep, Mackell and Watt kept costs down, estimating them at £44,510 6s. 8d. and £55,563 11s. 8d. respectively, less than two-thirds of Smeaton's costing. Watt had earlier written a report critical of Smeaton's scheme, contending that if it was to involve seagoing vessels traveling directly from each firth through the length of the canal (as Smeaton envisaged) then it would need to be nine feet deep rather than the planned depth of five feet. Watt calculated that at this necessary depth, Smeaton's scheme would actually cost almost £200,000. For their part, the proponents of Smeaton's scheme described Mackell and Watt's much smaller scale waterways as "ditches" or "puddles" fit for little.[7]

Watt seems to have entered into the spirit of these contests quite readily. Perhaps the application of ingenuity to technical jousting was more to his taste than the toe-to-toe arguments involved in managing construction, which he dreaded. As we will see later, the stereotypical picture of Watt as averse to business is quite mistaken; his attitude depended very much on the *kind* of business in question. Certainly Watt was happy, with the significant inducement of a £100 fee, to go to London to support the Glasgow merchants' canal bill that was to be introduced to Parliament by Lord Frederick Campbell. The journey down to London, which began on 10 March 1767, became a canal study tour as Watt took in the Calder Navigation at Brighouse and then the famous Bridgewater Canal. The latter was a level waterway, which avoided the need for expensive locks and which also shipped materials and goods in containers that could be easily unloaded and placed on other modes of transport at the canal's termini. These experiences informed his later canal designs in Scotland, especially the Monkland Canal.

Watt arrived in London on 17 March and was still there in early May, having found parliamentary dealings conducted by "wrong headed people on all sides."[8] The kind of rational business contest that Watt enjoyed, and at which he excelled, was not what he found in London, but rather opportunistic shenanigans and the exer-cise of naked power. (We will see that he became more used to this during his, and Boulton's, parliamentary lobbying for the steam engine cause.) There was a growing expectation that the larger project capable of taking seagoing vessels would win the day as the Edinburgh interests behind it garnered widespread support in other parts of Scotland. The Glasgow-sponsored bill was due to be presented to Parliament on 14 May, but it was withdrawn. Eventually, the Edinburgh-sponsored version of the Forth and Clyde Canal was approved in early March 1768.[9]

Interestingly, during at least part of his stay in London, Watt told Peggy to direct his mail care of "Mr Ramsden, Instrument Maker, Haymarket."[10] Watt was still very much in the instrument business and used the trip to London on canal affairs to main-tain those contacts. Ramsden was, of course, one of the major figures in the London instrument making trade. Watt also used the trip to purchase various items including tea kitchens and walking canes for his hardware business in Glasgow.

In 1769 Watt began involvement in what was to be his most consuming canal proj-ect, the Monkland Canal.[11] The Monklands was a coal-bearing region some ten miles east-southeast of the city of Glasgow. Its coal supplied both domestic hearths and industrial furnaces in a buoyant local economy, but the price of coal was increasing, partly because of growing demand but also, many thought, because of exploitation by the colliery owners. As always, the price of coal depended very much on the costs of transporting it to where it was needed. The prospect of greatly reducing those trans-port costs led the Glasgow trade and merchant interests and the magistrates to decide that a canal should be built. They commissioned Watt to survey a route. By November 1769 Watt had completed the surveys and suggested two possible routes. The most ex-pensive, which he costed at over £20,000, would have brought the canal to the River Clyde near Glasgow Green. One reason for the expense of this route was the fact that it involved constructing twenty-five locks, always a costly aspect of canal projects. The other route had no locks and was about half the cost, but it would not reach the center of the city. This proposal involved the construction of a wagon way from the canal terminus into the city. Watt's inspiration here was almost certainly the Bridge-water Canal in Manchester, which he had studied on his recent trip to London. A public meeting was held and a committee appointed to consider the Monklands proposals, which favored the cheaper, lock-free option. A call for subscriptions went out on that basis. The Monkland Canal Act named the subscribers, who included numerous owners of land along the route of the canal and three of the four major tobacco merchants of the city, together with a large number of other Glasgow merchants. The council and various trade organizations also supported it. After passage of the act— Watt declined a trip to London in connection with that (perhaps he had had enough of Parliamentary affairs)—Watt agreed to supervise construction as "resident engineer." This was to be a major preoccupation between May 1770 and July 1773.

The work of digging the canal was contracted out, but problems with the supply of labor, and the performance of workers and contractors as well as the level of wages and fees being paid, all ultimately fell at Watt's feet. It was in these circumstances that Watt expressed his discomfort to his Birmingham friend William Small: "I have been cheated by undertakers [i.e., contractors] and clerks and am unlucky enough to know it. The work done is slovenly, our workmen are bad and I am not sufficiently strict. . . . I would rather face a loaded cannon that settle an account or make a bargain—in short I find myself out of my sphere when I have anything to do with mankind. It is enough for an engineer to force nature and to bear the vexation of her getting the better of him."[12] This letter, cited by virtually every Watt biographer since, has been widely taken as evidence of his distaste for business at "the sharp end." At the same time, Watt made repeated confessions of inadequacy in the engineering role that he was occupying. Thus he told Small that people had "conceived a much higher idea of my abilities than they merit," and implied that they did so because they wanted to employ a local rather than a stranger.[13] He also confided that one reason he took on the construction as well as the design of the Monkland Canal was so that any errors he had made in the latter would not be revealed to another practitioner. Watt was also concerned about the report writing involved. As late as 1772, when producing such a report he sent it to Joseph Black and to other friends, apparently so that they could check his written expression.[14] Watt, with some justice, regarded himself as still untutored in this respect, a reminder perhaps of his uneven education. His father's orthography was certainly very irregular, but the young Watt's less so. Nevertheless it was another cause of worry for him.

It is tempting to see in all this a feature of Watt's basic character, and this is a temptation to which most of his biographers have succumbed. But at least some of the diffidence that Watt exhibited is surely due to the fact that he was assuming a role of which he, and the wider world, was still uncertain. He was feeling his way and being honest about that. Consider the following declaration from Watt in March 1770 addressed to an unidentified person (but clearly someone associated with Watt's employment on the project):

> I have not yet taken upon me to determine in what Rank I stand as an Engineer & shall only observe that what knowledge I have in that way has not been acquired without great loss of time & money & that it is not everyday that a man is fit either for the Business of the Field or Closet. These things being laid before them I refer the price of my labour to the gentlemen concerned as I do not wish them to pay me in such a manner as may deter then from employing me afterwards & trust they will do it so that I shall remain willing to serve them.[15]

Watt can be taken here as saying a number of things about his capacities as an engineer: that they are as yet undetermined, that he has put much time and effort into learning, that he does not want to price himself out of the job, but neither does he want to be exploited through underpayment. What does he mean by the phrase "it is not everyday that a man is fit either for the Business of the Field or Closet"? I think that Watt here continues to adumbrate his qualifications for the work, so that the meaning is that someone like himself capable of working both in the field (as resident engineer) and in the closet (as consulting engineer) is an unusual find and a valuable asset. Thus we see Watt being assertive of his value to the project and perhaps not so diffident after all.

Whether confessing deficiencies or asserting strengths, Watt was engaged in all these exchanges with his canal employers and with his friend Small in the construction of an image of his character as an engineer. It is perhaps important to note that Watt's steam engine project was being pursued in parallel with his civil engineering, as we shall see shortly. During the years that he worked on the Monkland Canal, Watt was coming to the view that the steam engine was his main chance. Though he was, as we will see, in a steam partnership with John Roebuck in Scotland, he was also in contact with Birmingham, and Small in particular was helping with advice on the steam engine patent of 1769—how to draw it up and exploit it. So, when Watt built a picture for Small of himself as an engineer, his strengths and his weaknesses, it is likely that he had in view the future possible arrangements for his work on steam. Watt was perhaps sending out signals to those who might be involved in a steam venture about what his own role might be.

Returning now to the Monkland Canal, construction under Watt's supervision began in June 1770 and involved not only organizing the dig but also constructing bridges and tunnels. He was also constantly troubleshooting. He often devised new machines and methods for prosecuting the work. Supervising the workforce was undoubtedly the task that he found most challenging and galling. Nevertheless, by the time the money ran out in the first half of 1773, just over seven miles had been completed, leaving the canal two miles short of its intended terminus. Watt was still optimistic that it would, even then, reduce the price of coals for Glasgow. But only when new proprietors took it over in 1781 and works were completed did it pay its way, ultimately becoming very profitable.

While he was building the Monkland Canal, Watt took on a good deal of other civil engineering work, including surveys for the Crinan and the Caledonian Canals as well as for a number of smaller projects. He also undertook work related to roads, bridges, and harbors. The extent and variety of this work is surprising and was, as Richard Hills argues, quite consequential. This activity confirmed Watt's preference for the role of consulting engineer and his dislike of (and perhaps ineffectiveness in) construction work. He was spending a lot of time in the field and was away from home a good deal. He also often had difficulty in getting paid what he was owed for his services. On the positive side, this work expanded his acquaintance with the range of Scottish improvers, and with a number of industrial enterprises that the civil engineering works were intended to promote. There was thus much grist for the mill of Watt's autodidactic habits. Exposure to this panoply of industrial activity would be a great fund of intelligence once he entered the business of selling power to drive it.

Watt had eased out of instrument making to pursue his career as a merchant through the Glasgow shop. Then he had wound down the shop as his civil engineering career began to develop. By 1773–1774 he was unsure whether Scotland could sustain him in his preferred role as consulting engineer. As he contemplated the possibility of a move to England, or perhaps even overseas, work as a "surveying engineer" was still in his mind. But there had been for some time another string to Watt's bow, a project that captivated him even as it had to be repeatedly put on the back burner when opportunities to earn quick money appeared. This, of course, was his improved steam engine.

The Steam Engine

Watt was not, as he is sometimes described, the "inventor" of the steam engine. That title, if it is to be given to anyone, should be awarded to Thomas Newcomen

(1664–1729).[16] Watt is better described as the improver of the Newcomen engine. He was not the only improver—a number of individuals, including John Smeaton as well as numerous lesser-known or totally anonymous people had contributed improvements since the erection of what we believe to be the first Newcomen engine in 1712. By the time the device reached Watt's hands, its efficiency had already greatly increased. It is true, however, that Watt's improvements were to amount collectively to a step change in efficiency.

Newcomen's engine of 1712 was a crucial development in using fire to pump water. Yet, as Dickinson long ago emphasized, it was "little more than a combination of known parts."[17] The steam devices of Otto von Guericke, Thomas Savery, and Denis Papin had all used the cylinder and piston. The pumping apparatus was also well established, albeit on a smaller scale. Scaling up of a brewer's copper provided the boiler. These known elements were combined in a machine with one major innovation, the use of an injection of cold water to condense the steam in the cylinder. How it was that Newcomen came to combine these elements and to do so in proportions that produced a successfully operating engine remains mysterious. J. T. Desaguliers, to whom we owe much of our information about machines in eighteenth-century Britain, painted a picture of Newcomen (and his collaborator John Cawley) as enjoying a good deal of luck: "After a great many laborious Attempts, they did make the Engine work; but not being either Philosophers to understand the Reasons, or Mathematicians enough to calculate the Powers, and to proportion the Parts, very luckily by Accident found what they sought for."[18] Here Desaguliers creates a dichotomy that has informed the accounts (by actors and by historians) of steam engine development and improvement ever since—a dichotomy between reasoned, philosophically informed work on the one hand and empirical tinkering relying heavily on luck on the other. Desaguliers' own career was built on his advocacy of the practical utility of natural philosophy, and so we should expect him to interpret the history of the steam engine, as much else, in a way that reduced any other approach to good luck. We will see that similar dichotomies have shadowed interpretations of the origins and success of Watt's own improvements.

The attractiveness of steam engines and their diffusion through the economy—that is, the transformation of inventions into innovations—was not simply a matter of their technical "sweetness" or ingenuity, however that ingenuity was informed. It was an economic question with a number of dimensions. Economies depended first on purpose, and the purpose of the vast majority of steam engines through most of the eighteenth century was to pump water from mines. Pumping technologies were already available, of course, and were most commonly powered by horses. So, the key economic issue affecting the diffusion of the Newcomen engine was often the relative capital cost and operating cost of such an engine compared with the costs of traditional pumping methods (unless an engine was capable of a pumping job to a depth unreachable by traditional means, in which case the economics depended directly on the cost of pumping versus the economic value of the material that could then be accessed).

The capital cost of a Newcomen engine was initially quite high, partly because of the materials that were used and the quantity of them needed. In the first Newcomen engines, as we saw, the boiler was a scaled-up brewer's copper, and the quantity of copper required to make it, and the working of that copper, rendered it expensive. The early cylinders were made of brass, another expensive material, especially on the scale that the Newcomen Engine required. In the period between Newcomen's and Watt's first engines, improvements in iron production enabled the use of iron rather than copper and brass for the cylinder, boiler, and pipes. (It should be noted that those improvements in iron production were themselves affected by the cost of coal, which was in turn affected by the use of Newcomen engines to drain mines. Thus, there are many feedback loops in this regime of coal, iron, and steam).

The operating costs of a Newcomen engine depended crucially on the cost of fuel, usually coal. Early engines required vast quantities of coal, so their ideal location was at a coal mine, where large amounts of poor quality fuel, otherwise of little economic value, were available. Whereas engines spread readily to coal mine operations, they did so less readily to the tin mines of Cornwall, for example, where coal had to be transported in.

The calculation of capital and operating costs was, however, not an easy business, not least because they interacted in various ways. While the use of iron, rather than copper and brass, in the construction of engines did greatly reduce the capital cost, it had consequences for operating costs. In fact, Desaguliers opposed iron construction on these grounds:

> Some People make use of cast Iron Cylinders for their Fire-Engines; but I would advise nobody to have them . . . none of them can be cast less than an Inch thick, and therefore they can neither be heated nor cool'd so soon as others, which will make a stroke or two in a Minute Difference, whereby an

eighth or a tenth less Water will be raised. A Brass Cylinder of the largest Size has been cast under ⅓ of an Inch in Thickness; and at long run the Advantage of heating and cooling quick will recompense the Difference in the first Expense; especially when we consider the intrinsick Value of the Brass.[19]

This is a good argument on Desaguliers' part: within the confines of the problem as he sets it, he is correct that the operating efficiency of a brass/copper construction would in the long run outweigh the lower initial capital cost of iron construction. But few followed him in this reasoning. The lower capital cost of iron construction seems to have carried the day, perhaps because in the most common context of use of early steam engines (at the pit head), the real economic cost of fuel was so low that operating costs could be heavily discounted. There is a familiar process in technological change, known technically as "path-dependence," where the early diffusion of a technology can establish suboptimal design as unmoveable.[20] An initially irrational design becomes so tightly locked into a network of mutually dependent elements that it becomes almost unthinkable to abandon it. By Watt's time this had happened with iron construction of cylinders and boilers.

A couple of broader points need to be made here that profoundly affect how we treat accounts of Watt's steam engine improvements. First, the potential users of a technology can have a variety of reasons for adopting (buying) it. They may be attracted by its technical novelty, or "sweetness," especially if they are not overly concerned about cost. They may, on the other hand, make very complex assessments of capital and operating costs relative to benefits (increased efficiencies). Or they may prioritize immediate economic advantage over the long term. Being up to date can be valued in itself, as we know from the success of Apple's latest models of its devices in the modern marketplace. A second, related, point is that technical superiority can be used, and taken, as a shorthand way of expressing or symbolizing an economic superiority that springs from a more complex agglomeration of causes. "Ingenuity" is an easy way of describing and marketing efficiencies that in fact derive from more complex causes. In this way, stories of origins can be important real-life actors in the marketing of technologies. This, I believe, was the case with Watt's steam engine improvements: the efficiencies of Watt's engines are usually attributed to his invention of the separate condenser, when they were actually due, as we will see, to a wide range of improvements. Boulton & Watt sold its steam engines in part in terms of ingenuity, of which the separate condenser was a useful symbol. Perhaps use of that symbol avoided having to explain in full the sources of advantage that their engines brought, thus simplifying marketing and maintaining a proprietary secrecy concern-ing the full range of improvements responsible for an engine's efficiency.

So, let us return to our narrative, bearing in mind that it may have been strongly shaped by later marketing and reputational questions. When and how did James Watt become involved with steam engines?

The Scotland of Watt's youth already had a number of steam engines in operation. The first seems to have been at Stevenston Colliery at Saltcoats in Ayrshire, and by 1733 eight Newcomen engines were pumping in Scotland. Their numbers grew with the coal industry itself promoted by transport improvements that lowered the cost of coal to consumers both domestic and industrial. One estimate has fourteen or fifteen engines in Scotland by 1769, the year that Watt took out his initial patent.[21] We have no evidence, however, that the young Watt had any contact with these real-world engines before he authored his improvements. We have to rely on what Watt and his close friends of the time, John Robison and Joseph Black, had to say on the matter subsequently. According to these accounts, Watt's initial forays into steam were in miniature and the result of his friend John Robison's interests. We know for certain that Robison studied Newcomen steam engines at about the time that Watt relocated his workshop to Glasgow College, for in a letter dated November 1757 and published in the *Universal Magazine*[22], Robison suggested an improvement upon Newcomen's engine. This improvement involved dispensing with the lever, or beam, by inverting the cylinder directly above the machinery to be worked. So, this would be directly above the pit pumping apparatus in the case of an engine powering a pump, or direct-ly above pinions (gears) attached to an axle in the case that Robison envisaged of a wheeled carriage driven by steam.

Watt recalled in a number of sources this early involvement with Robison and steam. In his "Plain Story," written many years after the event, Watt recollected as follows: "My attention was first directed, in the year 1759, to the subject of steam engines, by the late Dr Robison, then a student in the University of Glasgow, and nearly my own age. He at that time threw out an idea of applying the power of the steam-engine to the moving of wheel-carriages, and to other purposes; but the scheme was not matured, and was soon abandoned on his going abroad."[23] In another ac-count, written in a notebook, ostensibly in 1765 but perhaps much later, Watt told a slightly different story: "About 6 or 8 years ago My ingenious friend Mr. John Robi-son having conceived that a fire engine might be made without a Lever by Inverting the Cylinder & placing it above the mouth of the pit proposed me to make a model of it which was set about but never Completed he going abroad & I having at that time little knowledge of the machine."[24] So here Watt was saying not only that Robison's interest in steam engines drew his own attention to them but also that he was asked by Robison to make a model of the new design, which he began but did not complete. This latter aspect was repeated in yet another account, given at the time of Robison's death in 1805. Writing of the period before Robison left the college in 1758, Watt stated:

> During this period he turned my attention to the steam engine, a machine of which I was then very ignorant, & suggested that it might be applied to giving motion to wheel carriages, & that for that purpose it would be most convenient to place the cylinder with its open end downwards, to avoid the necessity of using a working beam. . . . In consequence I began a model with two cylinders of tin plate to act alternately by means of rack motions upon two pinions attachéd to the axis of the wheels of the carriage but the model being slightly and inaccurately made did not answer our expectation, new difficulties presented themselves.[25]

There seems no particular reason to doubt the general thrust of these accounts. But the context in which they were drawn up does give some pause. With the possible exception of the account in Watt's notebook, these recollections are reconstructions of events developed within a justificatory context when Watt's originality in his inventions was at issue. Thus the Plain Story had its first outing during the patent trials of the 1790s, while the recollection at the time of Robison's death was sparked by Robison's published remarks on the debt that, he claimed, Watt owed to Black's ideas on latent heat. Such a direct debt Watt was keen to repudiate. In the 1790s, early 1800s, and beyond, Watt's own determination to claim the laurels due to him was supplemented by his son's ardent campaign on his behalf.[26]

The documents containing these accounts of Robison's involvement in Watt's initial foray into steam engineering also devote considerable effort to delineating Watt's experiments on steam as occurring independently of Black. As Watt puts it in his recollections of Robison: "I . . . went on with some detached experiments on steam until 1763, when I set about the matter more seriously, and discovered the principles upon which my improvements on the steam-engine are founded."[27]

Another element of Watt's own story of the origin of his improvements is, of course, the occasion in the winter of 1763–1764 when he was asked by Professor Anderson to repair a model Newcomen engine (pictured in figure 2.1) that he used in teaching his natural philosophy class. Watt managed to get the model working again

FIGURE 2.1: Model Newcomen engine reputedly repaired by James Watt at Glasgow College, now held in the Hunterian Museum, Glasgow. © Science & Society Picture Library.

but was puzzled by what he found to be its very large consumption of steam and condensing water. Richard Hills has produced a detailed account of this work but from sources that, as Hills points out, have a definitely retrospective and possibly reconstructive quality. Nevertheless, it does appear plausible that contemplating and experimenting with this model engine helped Watt to formulate what he considered to be the central problem accounting for its remarkable inefficiency. This concerned the repeated heating and cooling of the cylinder. Watt began to understand what had to happen if no steam was to be wasted: an engine would, to quote Hills, "use only one cylinder full of steam each stroke and the steam would be condensed to a perfect vacuum."[28] This concept of the "perfect engine" gave Watt a standard against which to gauge any configuration of machine.

How Watt arrived at the idea of the separate condenser as a central part of the solution to the conundrum of the perfect engine involves yet another story. This is also retrospective, and it is second hand. In 1859 Glasgow engineer Robert Hart published in the *Transactions of the Glasgow Archaeological Society* an account of a meeting that he and his brother John had had with James Watt back in 1813 or 1814. As Hart reported it, the young men were themselves building model engines and trying to make improvements when they met a tall old man who turned out to be Watt. In the midst of a wide-ranging conversation, they asked the great man where he had the idea for the separate condenser: "To the question if it was *in the College* that he experimented on the engine and invented the condenser? (as we had been told it was there by persons connected with the College) he said—'*No, it was not there. I believe the Faculty would very willingly connect the invention with the College, now that it has been of some use to the world.*'"[29] When pressed about where the idea of the separate condenser came to him, Watt told what became the famous story of his Sabbath walk on Glasgow Green:

> I had entered the Green by the gate at the foot of Charlotte Street—had passed the old washing-house. I was thinking upon the engine at the time, and had gone as far as the Herd's-house, when *the idea came into my mind, that as steam was an elastic body it would rush into a vacuum, and if a communication was made between the cylinder and an exhausted vessel, it would rush into it, and might be there condensed without cooling the cylinder.* I then saw that I must get quit of the condensed steam and injection water, if I used a jet as in Newcomen's engine. Two ways of doing this occurred to me. First the water might be run off by a descending pipe, if an offlet could be got at the

depth of 35 or 36 feet, and any air might be extracted by a small pump; the second was to make the pump large enough to extract both water and air. . . . *I had not walked farther than the Golf-house* . . . when the whole thing was arranged in my mind.[30]

If Watt really did say these things in 1813 or 1814, then perhaps they were true. But there are other possibilities, as we will see in a moment.

Considering these revelations about Watt's steam invention overall, a number of points need to be made. First, if the account in the "1765" notebook was written not in circa 1765 but in the 1790s or even early 1800s, then the stories concerning Robison's engine, the Glasgow College Newcomen model, and the Glasgow Green inspiration about the separate condenser, were given in the last decade of the eighteenth century or the first two decades of the nineteenth. This was a period when Watt and Watt Jr. were very busy constructing stories of invention for legal and reputational reasons. The patent trials of the 1790s were a very active period for trying out various stories on judges and juries, particularly stories that emphasized Watt's scientific credentials and his character as a *thinking* engineer when compared to his rivals.[31] Published accounts of Watt's achievements, including Robison's contributions, did not always comport with the emerging mythology. Correcting them was another occasion for Team Watt to retail their favored stories and images.[32] Thus, at the time that he encountered the Hart brothers, Watt was engaged on other fronts in retrospective accounts of his inventions and clearly actively "correcting" the record. Perhaps an old man, in a period imbued with romantic notions of genius and sudden inspiration (who was at the same time, as we will see, experimenting with machinery that copied busts of himself deep in thought for distribution to friends and, eventually, a wider adoring public), was presenting himself to the Hart brothers as conforming to the type of the inspired inventor? Perhaps, as his reported words about the college suggest, the story of Glasgow Green was yet another way for Watt to put distance between his invention and the personnel of the college, especially Joseph Black. The suggestion that his invention was directly indebted to Black's work on heat (as Robison claimed) greatly annoyed Watt. We are justified in wondering about the account given to the Hart brothers since there is no other evidence to support the story of Glasgow Green.

So there are good circumstantial reasons, and a solid prima facie case, for seeing the stories emerging in the 1790s and early 1800s as in some part convenient fictions, half-truths, or simplifications. This belief is further supported in two other ways.

First, the accounts of Watt's inventions that were used to market the early Boulton & Watt engines bear no trace of these later stories.[33] Second, recent and convincing historical reconstructions of how the efficiencies of Boulton & Watt engines were in fact achieved suggest that the separate condenser was only one contributing element among a number.

Jim Andrew argues convincingly that there were a number of important sources of the efficiency that Watt's engines were able to achieve.[34] One was certainly the separate condenser, but not for the reason usually given. Virtually all accounts of Watt's innovations suggest that the separate condenser achieved efficiency improvements because it meant that, unlike in the Newcomen engine, fuel was not wasted reheating the engine's cylinder after it had been cooled by injecting cold water into it to condense the steam. Andrew observes that given the bulk of the engine cylinder, the cooling and reheating involved when cold condensing water was injected directly into it were not very great. The "specific heat" of the material from which the cylinder is made is a measure of the amount of heat needed to change the temperature of a unit mass by a given amount, and given the specific heat of iron, cooling water would only change the temperature of the cylinder slightly. With his knowledge of specific heats, Watt would have known this. From a modern thermodynamic perspective, it is the temperature range within which the engine operates that is the crucial factor. Watt did not have any premonition or prevision of thermodynamics. His understanding of what was happening in the engine was, as we will see in a later chapter, a chemical one. But Boulton and Watt did recognize that keeping the cylinder as close as possible to the temperature of the steam all the time was crucial. Besides this, Andrew points to a wide range of innovations in Watt engines that also contributed substantially to their superior efficiency; Andrew's account shows the complex sources of the engine efficiencies that have usually been attributed solely to the separate condenser. Either Watt was himself deceived about the sources of his engine efficiencies (which seems not to have been the case) or he engaged in the simplification of a complex story into an easily told form. If Jim Andrew is correct about the varied sources of efficiency in the early Watt engines, then Watt's experience with real-world Newcomen engines in the late 1760s and early 1770s assumes greater importance in tracing the sources of those improved efficiencies. His investigations in miniature and his steam experiments of the early 1760s perhaps assume proportionately less importance.

Having examined these important issues about the nature of Watt's invention, we return now to our narrative of the place of the steam project in his life.

We know that Watt engaged sporadically from the late 1750s in experimentation

with model engines and also in experiments on heat and steam. These were conducted in the interstices of his mercantile and civil engineering activities. But Watt also
worked on full-scale engines in the mid- to late 1760s and early 1770s. He was involved in the erection of a number of Newcomen engines in Scotland on his own account and in partnership with others. He also experimented with full-scale engines,
in partnership with Dr. John Roebuck, in an effort to produce a working version of
his own novel engine design.

Richard Hills has shown that Watt's first attempted excursion into real-world engine business was at Leadhills in late 1765 and early 1766.[35] The Scots Mining Company was looking for a pumping engine so that it could work its coal seams at deeper
levels. It is unclear why Watt would have been seen as a possible supplier of a steam
engine at this stage in his career, but he was certainly involved in providing estimates,
and in suggesting a novel design of Newcomen engine for the job. But a rival bidder
won the contract. Watt later recalled erecting a very small engine at Carron wharf
to supply water to a turpentine distilling operation there.[36] This was almost certainly the turpentine still of Samuel Garbett & Co. opened in 1763, and it represented
another example of work sent Watt's way by members of the Carron Iron Company circle, of which Roebuck was of course a crucial part.[37] Thereafter, Watt was in
partnership with Robert Mackell (the same engineer with whom he conducted civil
engineering business) for supply of engines. The first was for the coal mine of Lord
Kennet in Clackmannan, which supplied coal to local distilleries. In this mine, as in
many others at the time, waterwheel-driven pumps and even human-powered ones
were proving inadequate to the task, and steam was being looked to as the solution.
The estimate that Watt and Mackell provided was twofold—one for a conventional
Newcomen engine with beam, and one in which an otherwise common Newcomen
machine used a wheel instead of the beam. This was a design that Watt contrived that
had been suggested also for the abortive Leadhills tender. Watt's design was costed
as cheaper than the conventional one, though it was noted that since the design had
not been constructed before, there would likely be unforeseen expenses. From the
small size of the engine house that was built, the mine owner seems to have plumped
for Watt's design. It is interesting to note that Watt was bringing his innovative flair
to bear successfully in these early full-scale engine projects.[38] Watt struggled to touch
anything without seeking to improve it

A second Mackell and Watt engine was built for the Carron Company. It exem-
plifies a very common use of steam in the early phases of industrial development and
even well into the nineteenth century—to pump water back up to a reservoir supply-

ing a water wheel. As demand for power multiplied at works such as Carron, water to supply waterwheels became insufficient. The use of steam pumps to augment the supply made a good deal of sense. It was a conservative use of steam engineering, one that added value to the capital already sunk into waterpower, and one that involved no change to the drive mechanisms of a factory of the sort that direct use of steam power as a motive force would have required. Some extant records reveal that this engine was a large one, with a seventy-two-inch cylinder, and that it was probably one of Watt's wheel engines. It may also have incorporated novel steam valve gear—a sign once again of Watt's constant impulse toward improvement.

The third and final Mackell and Watt engine project that we know of was for a Mr. Dick of Glasgow at Newton of Ayr. In this case they were to make use of a cylinder and other engine parts taken from an old engine and to supply the remaining necessary parts and then construct the engine. This exercise in engineering thrift was undertaken in December 1766 and January 1767.

Subsequently, Watt was involved on his own account in constructing engines for Peter Colville, a prominent colliery owner at Torryburn in Fife, for Robert Montgomerie (an engine just to the north of the harbor at Ayr), and for Lord Cochrane of Culcross. This latter project shows, as Hills demonstrates, that Watt tried to work ideally as a consulting engineer on engine projects too.[39] He designed the engine and its housing, contracted out the production of its parts (in the Culcross case to Carron Company), and contracted engine erectors to put it all together. This way of working was not always feasible, but where it was Watt preferred it, foreshadowing a style of work that he was to carry with him to Birmingham as an ideal.

A continual presence in Watt's engine erection and repair work, as well as in his experiments on his own engine, was John Roebuck. Recall that Roebuck was involved with Joseph Black and Watt in a scheme for the manufacture of alkali from salt. Perhaps some time before this, Watt and Roebuck came to know each other through a common interest in steam. When they met, Roebuck (b. 1718) was in his late forties and an experienced industrialist. His first venture in Scotland, the manufacture of sulphuric acid at Prestonpans, was in partnership with his Birmingham friend Samuel Garbett. In 1760, with Garbett and others he established the Carron Iron Works, a major landmark in the industrial history of Scotland.[40] Roebuck's strength lay in conceiving and initiating entrepreneurial ventures. He tired of day-to-day management of enterprises, always looking forward to the next project. After cofounding the Carron Company, he decided to venture into coal, partly to supply the Carron works, and so leased a colliery at Borrowstoneness. It was there that he employed Newcomen engines, one of which Watt used to carry out trials. As we have seen, Watt and Mackell supplied an engine for the Carron works.

Clearly impressed by Watt, Roebuck decided to support him by financing the construction of an engine of Watt's design. The idea was that the engine would be built at Roebuck's residence, Kinneil House, and to that end a large workshop on the grounds of Kinneil gave the necessary space. Kinneil had the additional advantage of privacy; espionage had to be guarded against. Roebuck recalled in a later letter to Boulton in 1768 that a small engine had been built at Kinneil a few years earlier that had convinced him to help Watt with his finances.[41] Not only did Roebuck fund the building of a large engine at Kinneil but he also took on Watt's debts when the latter's partner in the shop, John Craig, died. This was not a charitable act on Roebuck's part—he must have seen great potential for profit in Watt's engine design. Over the next few years, the impatient and restless Roebuck urged Watt to bring his project to a conclusion rather than to indulge in endless experimentation, as was his tendency.

Watt's progress was sporadic. A burst of activity in late 1765 and early 1766 with the first engine at Kinneil came to an end when the cylinder, which had been cast and bored at the Carron Iron Works, proved inaccurately constructed and unusable. Engine work seems to have ceased for a year or so as other projects took priority. In January 1768 it resumed. This time Watt worked not at Kinneil but in Glasgow, with progressively larger models. It appears that this was when he used premises near King Street and then a larger workshop (some reports mention Delftfield Pottery as the site of this). Watt was experimenting now with different versions of the separate condenser, with varied ways of packing the piston (to ensure that its fit in the cylinder was air- and steam-tight), and with valve gear that would render the engine self-acting. To this point his experimental engines had been worked manually. These experiments, together with Watt's prior experience with engines, in miniature and in the large, provided him with the knowledge and ideas from which he was to construct, in the second half of 1768, his application for a patent.

The Patent of 1769

The patent James Watt took out in 1769 is among the most famous documents in the history of technology. It is usually referred to as Watt's "patent on the separate condenser," but its full title, as figure 2.2 exhibits, was "New Invented Method of Lessening the Consumption of Steam and Fuel in Fire Engines." It concerned a number of improvements in the design of steam engines that had been suggested to Watt's mind by his experiments with steam and with engines both miniature and full scale. Rather than being a full template for the engines that Watt and Boulton were to produce from the 1770s onward, it was a kind of interim promissory note. Much of crucial significance in engine design and development was yet to come. The central feature of the patent, the provision of a separate condenser, was a very general idea of a kind that the law, strictly interpreted, might easily have regarded as non-patentable because of the lack of details. The patent, and his willingness to try to enforce it in all its generality, gave Watt considerable power in the technological affairs of the late eighteenth century. It is debated whether in the end and on balance the patent promoted steam innovation or whether, beyond a certain point, it proved a conservative force and thus delayed important technological changes, such as the development of high-pressure engines of the sort that made steam locomotives possible.[42] The fact is, of course, that the Watt patent did both, and as a result present-day partisans in arguments about the value or otherwise of patents can and do use the case to argue both for and against the limited monopoly protections that patents can offer.

When we take ourselves back to the 1760s, we are dealing with a patent system still very much in the making,[43] and we must ask why Watt decided to patent his early steam inventions. Inventors who sought reward for their inventions (and not all did) had a number of possibilities available to them. One was secrecy in combination with an attempted monopoly of use. Watt (and Boulton) did make use of this option on some occasions. If secrecy could be maintained, then this approach certainly offered control, but espionage had to be constantly guarded against. Another possibility was to seek direct reward. The Board of Longitude, established in 1714 to reward the solution of the problem of determination of longitude at sea, was an example of the government seeking to promote innovation through the direct incentive of reward. The Society for Encouragement of Arts, Manufactures, and Commerce, established in London in 1754 just prior to Watt's training there, was the chief example of a "premium society." The society sought to promote invention by rewarding it through premiums and promulgating the inventions for the benefit of the wider society. It was partly inspired by the prior Dublin Society for Promoting Husbandry; in Edinburgh, also in 1754, the Select Society of Edinburgh for Encouraging the Arts and Manufactures of Scotland was established, which also operated a premium list.[44]

Which of these approaches to realizing reward for invention might Watt have taken? On some occasions he did opt for secrecy, but that required that the inventor take on use (manufacture) also if the secret was to be guarded. Watt did not contemplate

Mr. W A T T's Specification of his Method of leffening the Confumption of Steam and Fuel in Fire Engines.

HIS prefent Majefty having granted to Mr. James Watt His Royal Letters Patent, dated the 5th Day of January, 1769, for his, the faid James Watt's, *new invented Method of leffening the Confumption of Steam and Fuel in Fire Engines*, the faid James Watt, in Compliance with a Provifo contained in the faid Letters Patent, caufed a Specification or Defcription of his faid Invention to be enrolled in the Court of Chancery, on the 29th Day of April, 1769, in the following Words, viz.

My Method of leffening the Confumption of Steam, and confequently Fuel, in Fire Engines, confifts of the following Principles:

Firft, That Veffel in which the Powers of Steam are to be employed to work the Engine which is called *The Cylinder* in common Fire Engines, and which I call *The Steam-Veffel*, muft, during the whole Time the Engine is at Work, be kept as hot as the Steam that enters it; firft, by inclofing it in a Cafe of Wood, or any other Materials that tranfmit Heat flowly; fecondly, by furrounding it with Steam, or other heated Bodies; and, thirdly, by fuffering neither Water, or any other Subftance colder than the Steam, to enter or touch it during that Time.

Secondly, In Engines that are to be worked wholly or partially by Condenfation of Steam, the Steam is to be condenfed in Veffels diftinct from the Steam Veffels or Cylinders, although occafionally communicating with them; thefe Veffels I call *Condenfers*; and, whilft the Engines are working, thefe Condenfers ought at leaft to be kept as cold as the Air in the Neighbourhood of the Engines, by Application of Water, or other cold Bodies.

Thirdly, Whatever Air or other elaftick Vapour is not condenfed by the Cold of the Condenfer, and may impede the Working of the Engine, is to be drawn out of the Steam Veffels or Condenfers by Means of Pumps, wrought by the Engines themfelves, or otherwife.

Fourthly, I intend in many Cafes to employ the expanfive Force of Steam to prefs on the Piftons, or whatever may be ufed inftead of them, in the fame Manner as the Preffure of the Atmofphere is now employed in common Fire Engines: In Cafes where cold Water cannot be

FIGURE 2.2: Extract from Watt's 1769 patent specification. Reproduced by permission of the Museum of Applied Arts and Sciences Collection, Sydney.

such a closely controlled approach to his steam engines. He needed to embody the invention in something other than hardware. Seeking reward by way of a premium was not really on Watt's horizon. The premium societies tended to have a strong agricultural focus and were more attractive to the inventive independent gentleman or the less ambitious mechanic than to those who were intent (like Watt) on making a fortune. In any case, at the time that the patent was taken out, Watt's improved engine was still more in the nature of an idea. He did not have a standard, settled material form for his engine. In these circumstances, Watt, Roebuck, and Watt's Birmingham friends must all have realized that a patent was the way to go. Given that innovation was coming from other quarters, there was some urgency in trying to protect the basic idea. A patent could create a defendable claim without necessarily requiring that the invention be a "finished" one. This was ideal for the circumstance in which Watt found himself.

English and Scottish patent law required at this time (1769) that once a patent was sealed, a specification of the patented subject matter be provided within a period of four months. This requirement created a tricky situation for the thoughtful would-be patentee. First, it had to be decided when a patentable invention had been achieved. Once this stage was reached, it was desirable to set the patenting process in train by initiating the application for the grant of a patent. But once a patent was "sealed," the clock began ticking, as it were, on the deadline for submission of the specification. Drawing up a specification was itself a delicate business, and too early an application (too early in the sense that it imposed an impossible deadline on the specification) was obviously to be avoided.

Watt appeared confident that his invention would be a paying proposition after he conducted his experiments on a model engine in May 1768. Those experiments showed that he had a viable form of self-acting valve gear (meaning that the cycles of the engine, unlike those of a standard Newcomen engine, would be substantially automated) and that the saving of fuel from the separate condenser arrangement would be substantial. He calculated that "every cubic foot of the contents of the cylinder will require only one cubic inch of water to be evaporated." It was in these terms that he presented the situation to his partner Roebuck, adding that he hoped "it will make you some return for the obligations I ever will remain under to you."[45]

It appears that at this point Roebuck agreed that there was a patentable invention. An early part of the application for a patent was the affidavit, which involved the applicant swearing an oath before a master in chancery that the invention was his and that it was, to the best of his knowledge, not already in use. Watt attended in person in London, leaving for the metropolis in early August 1768, and paid the fee for the patent oath on 9 August.[46] A patent agent, Thomas Handley, was engaged to undertake the complex process of application to various offices of government that was required. The patent was finally "sealed" on 5 January 1769, from which date the four month countdown to the deadline for submission of the specification began. The process was expensive—Hills calculates the cost, including not only the direct fees but also Handley's charges and Watt's traveling and other costs, at over £170. This was equivalent to a relative purchasing power in 2015 of £21,270.[47]

Watt left London on Saturday 27 August 1768 and traveled to Birmingham, where he stayed with William Small. It was during this visit that Watt met Matthew Boulton for the first time. There was doubtless much discussion of the patent and the engine project generally. Boulton raised at this stage the idea of his becoming involved in the Watt-Roebuck partnership, a promise (or a specter) that hung over Watt's relationship with Roebuck in the next few years. After Birmingham, Watt spent time in the potteries and visited Warrington and Liverpool before finally and belatedly returning via Kendal and Carlisle, reaching home on 11 October. Watt was not behaving like a man anxious to get to work on the specification. Almost as soon as he returned to Glasgow, he left for Torryburn, where he spent six weeks erecting the Newcomen engine for Peter Colville, the colliery owner.

Indecision and ambivalence appear to have haunted Watt about the future of the engine even as the climax of patenting approached. The various pressures he felt can be discerned from a letter written to Watt at about this time by George Jardine, who was Roebuck's secretary and also clearly knew the young engineer very well.[48] Jardine effectively gave Watt advice about how he might best deal with these pressures, and in so doing reveals what they were. First there was Watt's relationship with Roebuck and their different desires concerning the engine project at this stage. Watt had returned from Birmingham with the suggestion from Boulton and Small that bringing Boulton into the partnership would be a good idea, not only because it would relieve the financial situation but also because, as Watt saw it, Boulton could provide the manufacturing skills and facilities to work to great accuracy and so realize the engine properly. Roebuck saw no need for this. He wanted, according to Jardine, for Watt and himself to enjoy the harvest of Watt's ingenuity. He wanted Watt to build a full-scale engine at Bo'ness to ensure it could be specified properly and that they would gain the necessary experience to successfully build other engines under the protection the patent would provide. Watt was skeptical that an engine to the right standard could be built in Scotland. His extensive experience with ordering parts from

Carron Company had not boosted his confidence in that regard. In any case, Watt did not want the trouble of building engines. Watt's aims were clear in the words Jardine used to urge Watt to cede the point to Roebuck and have a go at the full-scale engine: "You have many motives, man, to have good spirits. You are surely very near to something that will be much to your advantage: the happiness, or at least the interest, of your family—your own ease and amusement—that life of ingenious indolence which you have so often figured out to yourself, are all within prospect;—not to mention the honour of a discovery of so much importance, a circumstance which few would think so moderately of as yourself, I would, therefore, have you resolve immediately to spend a great part of the summer here."[49]

The key phrase here is "that life of ingenious indolence." At bottom, this was the life that Watt sought for himself. We might call it, with some danger of anachronism, the life of the professional inventor as deliberately distinct from that of the innovator. Watt wanted to make a fortune by his wits and his wits alone.[50] This was the attraction of the patent as Watt saw it—the ability to license the product of his ingenuity, leaving to others the business of actually manufacturing and erecting engines. Yet this aim ought to have meant impatience on Watt's part with the myriad improvements necessary to produce an engine perfect enough to satisfy. This was the contradiction in Watt's position—he was impatient and wary of the business of producing the perfect engine even as he was drawn to it.

The production of the patent specification encapsulated this same problem in microcosm. Watt's life of "ingenious indolence" required that he be able to patent the *idea* of the separate condenser, leaving aside the details of the realization of an engine working on these principles. Specifications were an expected part of the process from mid-century, although the requirement of full disclosure of the invention via the specification was not stipulated until a decade or more after Watt's 1769 patent.[51] So Watt knew that he ought not be able to patent a mere idea. But the law stopped short at that stage of actually requiring enough detail in the specification to allow someone versed in the art to construct an engine working on that principle. This structural problem in the patent process and Watt's vision of his own future were both tied up in the delicate task of patent specification, which engaged him in late 1768 and early 1769.

Watt was at work on drafts of the specification in November 1768, as he reported to Roebuck.[52] But even after the patent was sealed on 5 January, he remained uncertain about how to handle the task. At the end of January, he wrote to Small explaining the dilemma as he saw it:

The plan I pursued in drawing it out, was in the first part to lay down my principles in as few words as possible, so as to be intelligible and clear from double meanings; so that they alone should ascertain my invention in so far as related to myself. This I have declared in the next sentence; but as I have been informed that some patents have been defeated because the specification was not clear enough to enable other people to execute the schemes, I have added descriptions of the machines with drawings. However, if something after the nature of the first part could be contrived to answer the whole purpose, it would be better.[53]

Watt clearly understood that the future security of the patent might well depend on it being defensible in court against the charge that it was not a full enough account of the invention. This advice may have come from Roebuck or Jardine, or possibly from the London patent agent Thomas Handley.[54] Watt had taken this point seriously, and the drawings mentioned, which would have made for a fuller (but thereby narrower in the sense of more detailed) specification, have survived.[55] But Watt makes clear to Small that it would be better if these details could be dispensed with. As we have seen, this was in part a necessity since no full-scale, working engine had yet been built. But it also squared better with what Watt considered to be his appropriate character and role as an inventor.

Support for Watt's view of the matter came in Small's reply, sent after consultation with Boulton:

Mr Boulton and I have considered your paper, and think you should neither give drawings nor descriptions of any particular machinery (if such omissions would be allowed at the office), but specify in the clearest manner you can that you have discovered some principles, and thought of new applications of others, by means of both which, joined together, you intend to construct steam-engines of much greater powers, and applicable to a much greater number of useful purposes, than any which hitherto have been constructed. . . .

As to your principles; we think they should be enunciated (to use a hard word) as generally as possible, to secure you as effectively against piracy as the nature of your invention will allow.[56]

Small proceeded to exemplify how the principles might be represented, and he expressed confidence that such a formulation would enable a skillful mechanic to make one of Watt's engines, though clearly not so good an engine as Watt might make himself. Such a situation was acceptable in law, Small opined: "You are certainly not obliged to teach every blockhead in the nation to construct masterly engines."[57]

In subsequent exchanges Watt expressed himself "much pleased" with Small's suggestions.[58] His next draft of the specification, however, was a modification of Small's.[59] The nature of the changes that Watt made indicates his sense of "truth-telling" in the patent game. In fact, the general reason he gave for the changes was that the specification "is delivered on oath, and declared to be a true description of the invention." He made other changes on the grounds that common knowledge need not be supplied but also that what was evident to casual inspection of the engine (such as the methods of keeping the cylinder warm) should be made explicit. This was a matter of telling the truth and being seen to tell the truth. Watt was obviously very concerned about the oath he would have to take respecting the specification. This, perhaps, reflected his Presbyterian upbringing, the taking of oaths being a widely discussed and contended issue in Presbyterianism in Scotland in the eighteenth century and beyond.[60] It would certainly have been an issue important to his family during his youth and clearly weighed on Watt's mind as he contemplated the patent oath. He could, however, be interpreted as being more concerned with being seen to tell the truth than actually doing so.

The whole truth was a different matter: Watt still played the patent game, exhibiting here a willingness to manipulate accounts of his invention for advantage in a manner that we will see on many subsequent occasions. Small was inclined to scold Watt in a joking fashion about his delays and the possible consequences. When he informed his friend about a "linen-draper called Moore" who had taken out a patent for steam driven carriages,[61] Watt replied, "If linen-draper Moore does not use my engine to drive his chaises, he can't drive them by steam. If he does, I will stop them." Watt was expressing confidence in the unique power of his engines and also in the patent process as giving him legal avenues to curtail others' attempts to pirate them. Watt was also aware, however, of the extent to which the patent process relied, in effect, on secrecy and obfuscation: "Nothing less than the experience I have had of steam and steam-engines will enable anybody to erect one of my engines so as to be perfect, unless they should see one of mine and copy it exactly."[62] Numerous aspects of Watt's engine improvements remained secret even after the specification.

Watt here reflected the points that his invention was so much more than the separate condenser and that to make the fuel savings possible, it relied upon numerous other coordinated changes and adjustments that were only appreciable through the experience Watt had had with steam and steam engines in experiment, in model forms, and "in the large." As discussed earlier, the analysis of the sources of efficiency in early Watt engines made by Jim Andrew confirms that this was indeed the case. There was a large, inevitably tacit aspect to building an engine.[63] Even when, in later years, Watt had established his engine designs and his engine erectors had been issued with far more detailed instructions than the patent specification provided, it was difficult to construct engines to realize the fuel savings that they were theoretically capable of. To do so involved numerous modifications and adjustments in a particular context. This fact was both the source of Watt's security over his inventions and the source of frustration, for many years, of his ambition for a life of ingenious indolence as a "philosophical" engineer.

After the discussion with Small and others, a final version of the specification was arrived at and Watt traveled to Berwick-on-Tweed, where on 25 April he witnessed it.[64] He declared that he had "after much labour and expense, invented a Method of lessening the Consumption of Steam and Fuel in Fire Engines." The principles set down in the body of the specification gave the patent very broad scope. This was to be an advantage in that the patent could be taken to cover all manner of engine designs, including Watt's later ones as well as competitors' efforts. But it also represented a risk if the courts were to decide in the future that the patent was void because it was a patent on principles.

The granting of the patent did not mean that Watt's engine experiments were finished. The issue of actually building engines and profiting from them remained. So, Watt continued to work on the design and operation of various aspects of his steam engine. Experiments with the large model engine in Glasgow proceeded until the end of May 1769, investigating the use of cylinders made of iron and of tin. He also tried different cylinder and condenser designs, different ways of extracting condensates, and various mechanisms for dealing with materials used to seal the piston.

After May 1769 Watt's steam engine work focused entirely on the construction of large-scale engines at Roebuck's Kinneil estate.[65] A safe place to conduct the work away from the prying eyes of the curious and potential engine pirates became more important as designs were perfected. In the summer of 1769 the walls of Watt's Kinneil workshop echoed with the construction of what became known as the "Kinneil engine." The experiments with that engine, which began in early September 1769, continued into the following year, but Watt's attention to them was interrupted by his canal work, as we have seen. In June 1770 Roebuck's long-mounting financial troubles came to a head, and his bankruptcy meant that there were no funds to continue. The Kinneil engine would eventually be dismantled in 1773 and shipped to Birmingham, where it was rebuilt after Watt's arrival there at the end of May 1774.

This was, of course, a major turning point for Watt. He had spent nearly twenty years of his adult life engaged in a varied pattern of employment as instrument maker, merchant, and civil engineer. Later he had seemed to envision steam enterprise as offering the best chance of a life of ingenious indolence, that is a chance to live by his wits, operating in the space between business on the one hand and natural philosophy and technique on the other. We need to understand how it was that Watt and his new friends in the English Midlands came to consider the move to Birmingham a good idea.

Three

Birmingham, Boulton, and Steam Enterprise

※ ※ ※

*W*att gained firsthand knowledge of Matthew Boulton's manufacturing enterprise before he met the man himself. In mid-1767, on his way back to Scotland from London after the Forth and Clyde Canal parliamentary business, Watt visited Birmingham and was shown around the manufactory by William Small. This happened almost by accident. On the way down to London, Watt had intended to meet in Birmingham with Roebuck's partner Samuel Garbett. But Garbett was away, and Watt had delayed his visit to the city for the return journey.[1] This time Boulton was absent, but William Small, who was meeting Watt for the first time, and Boulton's business partner John Fothergill, showed Watt around.

Soho Manufactory was a striking sight. Although barely six years had passed since Boulton had leased the thirteen-acre site at Soho in Handsworth, an ambitious building program had quickly produced a manufactory both architecturally grandiose in Palladian style (see figure 3.1) and technically up to date.[2] Watt recalled the visit many years later, remembering being impressed by the organization of the place and the ingenuity exhibited in the design and operation of the machinery, much of it Boulton's handiwork.[3] Although it was not long since Watt had gone into steam partnership with Roebuck, a recent attempt to have a steam cylinder made at Carron

FIGURE 3.1: *Soho Manufactory, Birmingham* (circa 1800). Engraved by Francis Eginton. © Science & Society Picture Library.

Iron Works had been a failure, and it must have crossed Watt's mind that the Soho Manufactory would be a fine place for a steam enterprise by comparison.

However, the manufactory did not then produce machinery but rather "toys," not children's playthings but small metal goods—belt and shoe buckles, buttons, watch chains, sword hilts, and plated wares of various sorts. Boulton had inherited this business from his father but resolved to expand it and take it "up market," that is to produce high-quality, fashionable items that would contrast with the cheap trade usually associated with Birmingham. Boulton expanded into production of items in silver, silver plate, and ormolu, and he campaigned successfully for the establishment of an assay office in Birmingham, which opened in 1773.[4] Boulton was certainly prominent among those who transformed Birmingham into one of the leading manufacturing towns in the world.[5] Peter Jones has argued that Boulton was a central figure, perhaps *the* central figure, in an industrial enlightenment in the West Midlands. He embodied the characteristics of the "savant-fabricant" who combined philosophical curiosity and surprisingly deep technical knowledge with the ability to get things done.[6] His flamboyance and energy encouraged others, and he was an essential driving force behind the meetings of the famous Lunar Society of Birmingham, a group that brought

FIGURE 3.2: *Matthew Boulton* by Carl Fredrik von Breda (1793). Engraved by Samuel W. Reynolds (1796). © National Portrait Gallery, London.

natural philosophy, engineering, and industrial development together, though not necessarily so directly at their meetings. The "Lunatics" included Boulton, Small, Erasmus Darwin, Joseph Priestley, and the potter Josiah Wedgwood besides Watt.[7]

When the Boulton and Watt partnership was finally established in 1775, it rescued Watt from the crisis caused by Roebuck's bankruptcy. What is not always understood is that in the longer perspective it was Watt's steam enterprise that rescued Boulton. To appreciate this we need to understand how Boulton the capitalist operated. The initial funding for Soho Manufactory had been provided mainly by settlements from his marriages, first to Mary Robinson, who died in 1759, and then to her sister, Ann. They were daughters of Luke Robinson, a wealthy mercer of Lichfield who had died in 1749, and both brought substantial sums to the marriage. Their brother, Luke Jr., who was not much impressed by Boulton's ardent pursuit of Ann so very soon after Mary's death, also died quite young. His demise meant that Boulton's second wife inherited most of the remaining family fortune. Boulton was on good terms with his two-time mother-in-law, Dorothy Robinson.[8] The money so gained as well as that subscribed by John Fothergill, his largely silent partner, enabled Boulton to build and equip Soho Manufactory. But the "toy" business, run in ever-grander fashion, was not profitable. Boulton was, despite all appearances, rather disorganized in the way he ran the works. In the future the silver and ormolu business was also to fail financially, if not artistically. Boulton spent too much time courting prominent patrons and succumbed to their common insistence upon custom-crafting articles that depended for economically viable production on standardized manufacture. The result of this, and Boulton's extensive borrowings to shore up his numerous businesses, was that in 1773–1774 his partner, Fothergill, despaired that financial failure was upon them. Boulton, ever the optimist, borrowed even more.[9] In time, as the steam business began to turn a profit, Boulton's financial situation was saved.

This rebalancing of our conception of the financial aspect of the Boulton and Watt partnership indicates that we might reconsider the standard picture of their relationship. Their partnership has often been depicted in terms of popular stereotypes, of the technical genius without business sense or capacity ideally teamed up with the thrusting capitalist and master of marketing. But convenient as this depiction has been to nineteenth-century mythmaking about inventors and twentieth-century conceptions of entrepreneurship, it is historically misleading.[10] While Watt was certainly a great technician with a penchant for philosophical experiment, he was also astute, and his real fear, on occasion, of business negotiation with others should not be elided into total dismissal of his business acumen. Similarly in the case of Boulton, his clear strengths in business and marketing should not blind us to his habitual recklessness in financial affairs, nor to his genuine technical capacity and interest.[11] Of course, Boulton's recklessness might be dignified instead as a willingness to take risks of the sort often necessary to realizing business success. Many years in the future, when Boulton and Watt were fighting the engine pirates who had challenged them in Cornwall, Boulton made his view of Watt and risk taking very clear. Boulton was then suggesting that they directly challenge claims about the performance of an engine built by Jonathan Hornblower Jr (1753–1815) at Tin Croft mine by building one of their own initially at their own expense, costing probably £1,000, and directly comparing the performance of the two engines. He had no fears of losing such a wager, and he vowed to make it on his own if, as he knew he would, Watt shied away from it. Boulton joked for those who, like his correspondent, carried figures on the performance of engines in their heads and could recognize a ridiculously low number when they saw one: "If a Wager was offerd to Mr Watt that one of our Engines was not capable of raising a Column of [water] to 10lbs pr Inch I verily believe he woud have so many doubts that he durst not accept the bet; & therefore I will risk it."[12] This perhaps illustrates in equal measure Boulton's wildness and Watt's timidity, or at least his Presbyterian caution.

When Small showed him around Soho Manufactory in 1767, Watt was "entranced" by its grandeur, but this was not the wide-mouthed gape of a business naïf. Watt had acquired a good deal of experience with the practical erection of steam engines. He had observed his partner Roebuck in action and had experienced business failure with Craig. He had become aware of the difficulties of obtaining well-made engine parts in Scotland. It seems reasonable to think that Watt saw that Soho, and Boulton's talent for building networks and cultivating connections, represented the sort of facilities that would be needed to make a successful engine business. Watt's evident reluctance to engage in large-scale engine work with Roebuck may have been owing not only to the need to earn a living from civil engineering but also to his underlying skepticism that Scotland in general, and Roebuck in particular, could supply what was needed. Watt was also impressed during his tour of Soho with Boulton's evident technical ingenuity.

Boulton and Watt first met face to face in the summer of 1768 when Watt was returning from his patent business in London. Watt recalled that Samuel Garbett or Dr. Small introduced them, and he "Had much conversation with Mr. B on [the manufactures at Soho] when he explained to me many things of which I had been before ignorant. On my part I explained to him my invention of the Steam Engine &

several other schemes of which my head was then full, in the success of which he ex-pressed a friendly interest."[13] Boulton's "friendly interest" was welcome, but so was the Birmingham hardware man's potential as another in that long line of friends and associates from whom Watt could learn. Watt did not have long in Birmingham, but when he returned to Glasgow, his correspondence with Small and Boulton concerned not only advice on drawing up the patent but also the suggestion that Boulton, and possibly Small, might enter somehow into the partnership with Watt and Roebuck.

For his part, Boulton, then aged forty, was undoubtedly interested in the impres-sive young Scot, who was some eight years younger. He had a very direct reason to be interested in Watt's schemes for rotary and reciprocating engines. He had acquired the Soho site in part for the waterpower to be had there by damming Hockley Brook and operating waterwheels from the resulting reservoir to drive some of the laminat-ing, grinding, and polishing machinery of the works. Boulton's waterwheel began operation in August 1761.[14] But there were problems with this source of power, not least the lack of water during droughts that Boulton experienced in each of the years from 1763 to 1765. He thought immediately that a fire engine might be used to pump water from Great Hockley Pool (where the water from the reservoir ended up after passing through the waterwheel) back up into the reservoir. A Newcomen- or Savery-type engine might be used for this purpose, but until canals delivered cheap coal, the inefficiency of these engines rendered the process uneconomic.[15] Nevertheless even an uneconomic process might be deployed in an emergency, and Boulton pursued the idea. He had a model of a Savery engine made and, looking for expert advice, Boulton submitted it to none other than Benjamin Franklin, who was then in London. Boul-ton's technical questions of Franklin elicited a disappointing response to the effect that "experiments will decide in such Cases."[16] Franklin returned the model engine to Soho and Boulton continued experiments on it. But Watt's visits to Soho in 1767 and 1768 promised a much more economic engine and even a steam wheel that might drive machinery directly. Boulton resolved to shelve his own engine project and wait for Watt's to come to fruition.

Boulton, however, was not a man to wait and do nothing, and his interest in Watt's engine was not confined to finding an economic engine for Soho Manufactory. He had told Watt during their second meeting that he wanted to be directly concerned in the entire business venture. We know this because Watt apologized, in a letter written shortly after returning to Glasgow, that he had been unable to make an immediate offer when Boulton was "so kind as to express a desire to be concerned in my fire-engine."[17] As he explained, this was because of his partnership with Roebuck, who held two-thirds of the invention in exchange for taking on Watt's earlier debts and covering the expenses of experimental work and "securing the invention," that is taking out the patent. Watt revealed that Roebuck's investment at the time he wrote was £1,200.

Watt had made clear to Roebuck his desire that Boulton be brought into the partnership. According to Watt, Roebuck agreed and would write to Boulton in "a few days" time: "If agreeable you may be a third part concerned, on paying the half of the cost [i.e., £600] and whatever you may think the risk he has run deserves."[18] Whether this was a misunderstanding with Roebuck on Watt's part, or whether his partner simply changed his mind we do not know,[19] but when Roebuck eventually wrote to Boulton, on 12 December, it was in very different terms. Anticipating a successful trial of the larger engine at Kinneil, Roebuck advised Boulton, "I shall with pleasure communicate a share of the property to yourself in the three Counties in your neighbourhood Warwickshire, Stafford[s] & Derby[r] and the terms you and I shall not differ about."[20] Whereas Watt had written of a third share in the whole partnership, Roebuck was suggesting essentially a license to make and sell Watt engines in a limited geographical area. Knowing Boulton's character, and having had discussions with Watt, Roebuck was surely making an offer here that he knew Boulton would refuse. And refuse it he did, if in the gentlest of terms. It was not until early February 1769 that Boulton responded, thanking Roebuck and Watt for thinking of him as a partner "in any degree."[21] He noted, however, how different Roebuck's offer was from that suggested by Watt. Boulton considered that he could not "meddle with" the proposed scheme "as I do not intend turning engineer." Two motives had led Boulton to offer assistance to Watt—love of him and "love of a money-getting, ingenious project." To succeed, that project in Boulton's view needed "money, very accurate workmanship, and extensive correspondence." By "correspondence" he meant the networks that he had cultivated through his other business ventures. He would employ the best workmen, keeping the project "out of the hands of the multitude of empirical engineers." The workmen would be located in a separate manufactory near his own operation and would be supplied with all the best machinery. This would result in engines of great accuracy, produced 20 percent cheaper than they could be otherwise, made in all sizes, "for all the world." All this had been Boulton's idea before he knew of Watt's "pre-engagement" to Roebuck. Now, however, given Boulton's reluctance to engage in a project that he could not supervise directly, the presumed impossibility of Roebuck relocating to Birmingham, and the fact that Boulton himself was "almost saturated with undertakings," the answer had to be "no." Boulton did hope, in concluding his letter, that there would be some way that he and Watt could be associated in Birming-ham in the future. He wondered whether Watt's "wheel-engine" invention might be separated from the reciprocating one. The wheel-engine (or steam-wheel) was a cir-cular chamber mounted on a horizontal shaft, fitted with a valve system, and driven by the pressure of injected steam on a weight of mercury. It had been included by Watt in the 1769 patent but was never developed as a commercial proposition.

This idea of developing the wheel-engine became an ongoing point of contact be-tween Watt and his Birmingham friends in the early 1770s, though it came to nothing in the end.[22] By September 1769 trials with the large engine at Kinneil had been disappointing, and Roebuck clove less jealously to the project, giving Watt hope, once again, that Roebuck, Boulton, and Small might come to some agreement. Watt tried again to persuade his partner that involving the Birmingham connection was the right course. He appears to have had some success because in November Roebuck himself wrote to offer them an agreement in which Boulton and Small would take one-third of the patent in return for a sum that they thought "just and reasonable" (but not less than £1,000) once the experiments on the Kinneil engine had been satisfactorily completed. The terms involved Small and Boulton making a final resolution twelve months hence. Small and Boulton agreed to this and Watt continued the experiments. It is not clear how close Watt was to having a viable engine, but trials ceased on 13 June 1770 when Roebuck declared bankruptcy and the prospect of Small and Boul-ton exercising their right to the third share collapsed at the same time.

Over the next two years, Watt and his engine were adrift. Roebuck could offer no further financial support to bring the engine to perfection, leaving Watt to find it himself. Although they continued some work on the steam wheel-engine, Small and Boulton were themselves stalling on the question of the reciprocating engine. We must remember that Watt had committed to canal and surveying business during this time, and so could give the engine only sporadic attention. Though Watt did not know it, or at least the extent of it, Boulton and Small were also in financial strife during this period, as Boulton's numerous projects and reckless borrowing hurtled toward a reckoning. Any partnership involving Roebuck was out of the question in such circumstances, even as Watt continued to urge it. Boulton and Small had, in any case, long wanted Watt for themselves. Further worsening of financial affairs, and with it disaster for Roebuck, was finally to break the deadlock.

On 25 June 1772, Douglas, Heron and Company, known as the Ayr Bank, failed, with liabilities of over £1,000,000.[23] This disaster came at the end of a long period of economic growth in Scotland that had underwritten, and was also a symptom of, the improvements and expansion of trade with which the Watt family fortunes had been tied up, in their own small way, throughout Watt's adult life. In the end speculative demand for credit and sharp financial practices led to a crisis of confidence and a run on the banks. The Ayr Bank had been established in 1769 to satisfy the high demand for finance that the chartered banks could not, or would not, provide. Its failure affected the whole economy: the linen manufacture, already in a decline, was severely hit. The Forth & Clyde Canal and the Monkland Canal were forced eventually to suspend operations. This was why Watt's oversight of construction of the Monkland Canal ended in July 1773. The Carron Company partners were nervous and in trouble. Francis Garbett failed to the tune of £193,000. By August 1775 only 112 of 226 direct partners in the Ayr Bank remained solvent. The creditors were paid in full and to enable this the partners paid up over £660,000. Many estates were sold as the landowners prominent among the shareholders in the Bank sought to liquidate their assets.

Before all this, Roebuck had managed to stabilise his financial affairs by means of a trust deed of 30 June 1770, but the financial crisis of 1772 proved his demise and in 1773 he was declared bankrupt.[24] As this was unfolding, Watt, who had seen where matters were headed, managed to extricate himself. The Watt-Roebuck partnership had not worked out quite as intended. Roebuck had originally paid £1,000 by way of clearing Watt's earlier debts to John Craig and also some of his debts to Joseph Black, but Roebuck had not always financed the experiments on the engine. In addition Watt had lent money to Roebuck, which placed him among the creditors. After a meeting of those creditors on 2 April 1773, Watt negotiated a deal with his partner. It remains unclear just what debts the £1,000 was taken to cover, but Roebuck gave up all claims to Watt's erstwhile shop and associated goods. Watt agreed to pay the remaining debt to Joseph Black and to give up any claim on Roebuck for all the money he (Watt) had laid out on engine experiments and also for the £180 he had loaned to Roebuck. The latter forgiven debt was accepted by Roebuck as payment for the Kinneil Engine. Watt was now free of him, except of course that his partner still held two-thirds of the engine patent.[25]

The speed and astuteness with which Watt moved on this occasion is another example of our supposedly "incompetent" businessman acting decisively when the cards were down. He saw a window of opportunity and went straight through it. He hastened to Kinneil House to recover the engine, dismantling it, packing it, and shipping it to Birmingham. This was done in the summer of 1773, before Peggy's death and long before there was any formal agreement with Boulton. Watt was rescuing the key fruits of his labors and depositing them in Birmingham, where, he believed, the best chance of completing the engine lay. However, it was also clear by now how the arrangement with Birmingham might be formalized.

Roebuck also owed Boulton money—officially £630, but actually twice that when a private debt was included. Watt, Small, and Boulton discussed how Roebuck might part with his two-thirds share of the engine patent to Boulton to discharge this debt. The situation was difficult, however. Roebuck and his family sought to retain a small interest in any future profits from the engine, but Boulton wanted none of this. It would, in any case, have complicated matters with other creditors. Watt was in limbo; sorry for Roebuck, serious about leaving for Birmingham, but unwilling to move or to conduct the necessary further experiments on the engine until the issue of property in it was settled. Watt was drowning in canal business, spending long periods away from home, and increasingly despondent when Peggy died in September 1773.

We began with that dread moment when Watt met his friend Gilbert Hamilton on the road and learned of Peggy's death. As Watt recorded in his journal, "by his black coat & his countenance I saw I had nothing to hope-We met without speaking . . . he informed me I had lost my dear friend on Friday morning. . . . I did what I could to force grief from my mind but feared to come home where I had lost my kind welcomer."[26] He lamented the loss of "the comfort of my life, a dear friend and a faithfull wife." She had borne him four live children, two boys and two girls, and she had died "of the 3 son," who was buried with her. Watt prayed that God would long preserve the son and daughter who were left to him. In a fragment of a letter to William Small, Watt described his unmarried friend as happy "to have no such connection" because he had no danger of the loss that he, Watt, had sustained. Watt struggled to think that matters might have been worse: "Yet this misfortune might have fallen upon me when I had less ability to bear it, and my poor children might have been left suppliants to the mercy of the wide world. I know that grief has its period; but I have much to suffer first; for if probity, charity, and duty to her family can entitle her to a better state, she enjoys it. I am left to mourn."[27] In his response, Small, who we must recall was previously a professor of moral philosophy in Virginia, while noting that he had "long cultivated the species of philosophy that is said to be the most consolatory" [Epicureanism or Stoicism?], encouraged Watt in "turning the attention to other objects, and never suffering it, even for an instant, to be engaged on the irretrievable one."[28]

Watt was to follow this advice, but not immediately. His journal reveals that on 1 October he went to Croy with his cousin Robert Muirhead, where he stayed for a week. This must have been a period when Watt mourned his young wife in a place where they had spent much time together as children. There is no record in the journal of the funeral. After the week at Croy, Watt began over two weeks of experiments on telescopes before, on 25 October, he resumed work with his field assistant, Mr. Morrison, pursuing canal surveying into November and beyond.[29]

As he sought to bury himself in work, however, Watt was despairing also about his prospects, as he wrote to Small in December: "I am heart-sick of this country; I am indolent to excess, and, what alarms me most, I grow the longer the stupider. My memory fails me so as often totally to forget occurrences of no very ancient dates. I see myself condemned to a life of business; nothing can be more disagreeable to me; I tremble when I hear the name of a man I have any transactions to settle with. What I am fittest for is a surveying engineer. Is there any business in that way?"[30] Even in this state Watt managed somehow to continue experiments in connection with trying to perfect the separate condenser. From mid-February 1774 he was working on thermometers and barometers in preparation for the course of experiments he began on 26 February "upon the heats with which [water] boils from vacuo to air." These experiments continued until 9 March.[31]

Perhaps taking refuge in work and experimentation had worked: by this time, Watt was more sanguine. He was optimistic that he was close to winding up his civil engineering business in Scotland and in May was finishing the last of his reports, packing his belongings, and settling his children, James and Margaret (Peggy), with their grandfather Watt in Greenock. After a last visit to Kinneil and Edinburgh, he journeyed south, setting out on 20 May 1774. By 31 May he was approaching Birmingham and what turned out to be a new life, if not quite the life of "ingenious indolence" that he craved. That remained a long way distant, much further off than a final acknowledgment that he had left Scotland for good.

Watt's Early Years in Birmingham

When he journeyed to Birmingham in May 1774, Watt had decided that Soho was the place to perfect his engine and make a commercial proposition of it, but there was as yet no partnership with Boulton. While it was fairly clear that scientific instrument making and a mercantile career were not for him, he had not completely severed his connections with them since he still held the stock of his former shop in storage. He had wound up his existing civil engineering projects before leaving Scotland (perhaps more accurately, they had collapsed beneath him as a result of the financial crisis), but this line of work, if restricted to consulting engineering, may have been a possibility in England, as his correspondence with Small revealed. Watt also retained an ongoing involvement with the Delftfield Pottery.

Yet, the attractions of Birmingham extended beyond the fact that Soho seemed a good place to prosecute his engine. Boulton, Small, Erasmus Darwin in nearby Lichfield, and others provided Watt with considerable encouragement. These individuals met frequently as members of the Lunar Society, one of the innumerable, informal, convivial gatherings that occurred in many parts of provincial England in the eighteenth century.[32] There were in fact many lunar societies; at or near the full moon was a popular meeting time when streets were otherwise dark, inhospitable, and unsafe at night. These societies were predicated on conversation, fine food and wine, and the exchange and cultivation of knowledge. They are often seen as a peculiarly English aspect of Enlightenment, and they attracted medical men and clergymen particularly—whose education and vocation predisposed them to an ongoing interest in learned and scientific affairs—but they could and did involve others who through trade, business, or merely inclination sought such company.[33] Some of these organizations were, or became, relatively formal, such as the Gentleman's Society of Spalding earlier in the century and the Manchester Literary & Philosophical Society (f. 1781),[34] which had formal elections to membership, an organizational structure, and published proceedings. Others remained informal to the point of evanescence and are barely traceable.

The Lunar Society of Birmingham was informal, but it left significant traces largely because its membership left papers and deeds behind that have been preserved, making a partial reconstruction of the society's history possible. We know from correspondence involving Erasmus Darwin, Matthew Boulton, William Small, and Thomas Day that "philosophical feasts" were sometimes held in Darwin's house in Lichfield from 1766. Thereafter, meetings were held in Birmingham, scheduled (in theory) regularly on the Sunday afternoon, and from 1780 on the Monday afternoon after the full moon. They came to involve, besides Watt, others including the former army captain and chemist James Keir, Richard Lovell Edgeworth, the Reverend Joseph Priestley (whose Sabbath duties prompted the changed day of meeting), Dr. William Withering, and to a very limited extent Josiah Wedgwood. In many respects the heyday of the society was in the 1770s and 1780s, but we know from various sources, including Watt's cash books, that meetings continued into the first decade of the nineteenth century and were attended by a second generation, including Matthew Robinson Boulton, James Watt Jr., and Gregory Watt.[35] The end had certainly come for the Lunar Society by 1813, when surviving members held a ballot to dispose of jointly purchased books from its library.[36]

The amalgam of interest in natural philosophy and practical affairs characteristic of the Lunar Society members was congenial to Watt. Since leaving the precincts of Glasgow College and losing Joseph Black to Edinburgh and his friend John Robison to the armed services, Watt had perhaps enjoyed less of such congenial company than he would have liked. While Greenock and Glasgow centered on maritime trade that squared well with his instrument making and mercantile activities, the scope of man-ufacturing was limited there, certainly when compared with Birmingham as a great metalworking town.

While the Lunar Society circle attracted Watt to Birmingham, the wishes of oth-ers also carried some weight in drawing and retaining him there. We have seen that Boulton (and Small) had clear business reasons for wanting to retain Watt's services and friendship. Watt's exchanges with them, Boulton in particular, had involved grand visionary schemes for providing engines to all the world. Even though Watt en-visaged himself as a consulting engineer on steam engines rather than as the manufac-turer of Boulton's vision, he must have been encouraged by Boulton's optimism and enthusiasm. In the end the Boulton & Watt partnership's way of working on steam was to conform to neither man's expectations. However, there was no partnership yet, and from Boulton's point of view there were good reasons why. First, there was as yet no working Watt engine. Further, even had there been one, the prospects for its exploitation to make money were overshadowed by the fact that five years of the patent's fourteen-year term had elapsed already. So the first order of business was to get the engines working.

Before Roebuck's property had been sorted out, the steam-wheel looked as if it might not be part of the Watt-Roebuck partnership, and the rotary motion possibil-ities of that device appealed to Boulton's power requirements at Soho. We have seen that, starting in 1770, Boulton and Small had taken steps toward constructing Watt's steam-wheel. As of February 1774 many of the components of the steam-wheel were ready, but Small told Watt that neither he nor Boulton had time to start trials of it or supervise the workmen. In any case they wanted Watt to be party to everything in its trials from scratch.[37] So this was among Watt's early business when he arrived, and by July 1774 there were favorable reports of its working. For a few years the steam-wheel remained a serious possibility: trials continued, and enquiries were received from potential customers. The steam-wheel was further described in the patent extension papers of 1775. But in the end, for rotary motion, the use of reciprocating engines to pump water for waterwheels, or the direct conversion of the motion of a reciprocating engine, proved the superior option. In any case, the dominant demand for steam engines at that time remained for pumping, and the road to business success was seen as lying, at least initially, with the reciprocating engine.

The Kinneil engine parts were used to erect a water-returning engine for the Soho Manufactory. Information is scarce because Watt was now dealing directly and daily with Boulton and Small rather than writing to them about progress. There is also a large gap in the extant correspondence with Joseph Black during these years, and a logbook of experiments on this first engine at Soho can no longer be found.[38] However, Watt's correspondence with Roebuck and with his father in late 1774 tells us that the engine was working by then; indeed, Watt told his father that "the fire engine I have invented is now going and answers much better than any other that has yet been made; and I expect that the invention will be very beneficial to me."[39] But there were still improvements to be made. The cylinder of the Kinneil engine, formed from tin, had eroded even before shipment to Birmingham, and there was surprise that it worked as well as it did before it failed just after Christmas. Boulton's friend, the ironmaster John Wilkinson (1728–1808), had devised new boring machines for cannon and also to produce hollow cylinders from cast iron. So a new cylinder was ordered from him that was bored to an unprecedented accuracy and finally delivered in April 1775.[40] Figuratively and literally the pieces were coming together, but there was still a long way to go.

As Watt, Boulton, and their workmen got their hands dirty modifying, trying out, and adjusting the emerging engine, moves began on the other key front, securing property in the invention for a period sufficient to achieve business success. In January 1775, accordingly, Watt consulted legal authorities in London. Alexander Wedderburn, the solicitor general at the time,[41] advised that it was possible to extend the term by surrendering the current patent and applying for a new one. Watt consulted others who suggested that obtaining an Act of Parliament extending the existing patent would be both a better process and cheaper! This appealed to Watt, and Boulton agreed.

Watt, as we know, had some experience of parliamentary workings through his earlier involvement with canal bills. He knew that alongside a formal petition for an engine act, they would need to lobby and persuade Parliament of the merits of their case. To this end Watt wrote an eight-page pamphlet titled "An Account of James Watt's Improvements upon the Steam or Fire Engine." This was composed with

William Small's assistance and adopted a strategy already familiar from their collaborative drafting of the original patent specification. They stressed the experimental
work on steam that Watt had done, portraying it as the basis of his engine improvements: "So very few experiments had been made even upon the most essential part of
the subject, that the real bulk of water, when converted into steam of a given heat,
remained unknown, until the author of these improvements determined it in the year
1764. . . . The boiling of water in an exhausted receiver at low heats was known; but
it was neither known what these heats were, nor what progression they observed under various pressures, before he made his experiments upon that subject."[42] Here was
no fly-by-night projector or empirical tinkerer, but rather a natural philosopher engaged over a long period in gaining a secure understanding of the properties of steam.
His deep understanding, it was claimed, underwrote the viability of his engines and
vouchsafed their economic value in terms of savings in fuel, making them far superior
to the common Newcomen engines.

The formal petition to Parliament focused more closely upon the hard work and
expense of making the experiments and trials that had enabled the development of
his superior engine.[43] Although successful engines had now been constructed, it had
taken a long time to get to that point. It would take more time to establish a manufactory for the engines and to market them. The likelihood was that the whole term
of the original patent would have elapsed before there would be any return. Beyond
that, the petition argued, the "great works & manufactures" of the kingdom would
benefit enormously from the cheap mechanical power Watt's engines would provide.
The parliamentarians were to be convinced that by granting the petition for an act
providing a further twenty-five years of protection to the patentees, they were also
ensuring public benefit.

The arguments honed, a long political battle was fought. The petition was read
before the House of Commons on 23 February and then referred to a committee consisting of Lord Guernsey, Sir Adam Ferguson, and other members.[44] Lord Guernsey
delivered the Committee's Report to the Commons on 3 March, in which we learn
that it had examined Watt's 1769 patent and had taken testimony from Roebuck,
Alexander Cumming, and Matthew Boulton.[45] This was hardly a random selection of
witnesses—the man who still held a two-thirds share of the original patent; the man
who would soon take over that share; and Cumming, an old friend of Watt's, who had
helped with the drafting of the specification![46] Unsurprisingly, their testimony confirmed the superior quality of Watt's engines; the extent of the ingenuity, labor, and
expense that had gone into them; the necessity of considerable further expenditure

to render them commercial; and the tremendous public benefit they held in store for the country. The committee ordered: "Leave be given to bring in a Bill for vesting in *James Watt*, Engineer, his Executors, Administrators and Assigns, the sole Use and Property of certain Steam Engines, commonly called Fire Engines, of his Invention, throughout His Majesty's Dominions, for a Time to be limited. And that the Lord Guernsey and Lord Frederick Campbell do prepare and bring in the same."[47]

Eric Robinson has shown that behind the scenes of this parliamentary process was a sophisticated lobbying exercise.[48] Boulton's fingerprints were all over the proceedings, and local Midlands interests were prominent. Lord Guernsey (Heneage Finch, 1751–1812, later 4th Earl of Aylsford) was a local Midlands figure whose mother, Lady Aylsford, had written to Boulton soliciting his support when the young man sought to represent Warwickshire in Parliament. More importantly, Guernsey had recently led the charge in another of Boulton's key parliamentary efforts—obtaining an assay office for Birmingham.[49] The fight for the Assay Office Bill, won in May 1773, involved a mobilization of Midlands political interests that was in many respects simply repeated for Watt's patent extension act. The work of the parliamentary committee on Watt's engine also bears the hallmarks of Boulton's orchestration, including the witnesses called before it. He recruited Guernsey again to lead the process. On the day Watt submitted the petition, Boulton wrote to Lord Dartmouth, president of the Board of Trade, seeking his support. The letter is a minor masterpiece, pleading Watt's ingenuity, his long hard work on the engines, his discovery of their true principles of operation, the danger of him being discouraged and abandoning them ("had I not assisted him"), and the great public loss involved if they were not brought to fruition: "I need not point out to your lordship's consideration the great utility of steam or fire engines in collieries, in lead, tin, and copper mines, and in other great works. . . . Mr Watt's intentions, if carried into execution, will very much extend the utility of fire engines by rendering them one-fourth of the expense usual, and by adapting them to a great variety of purposes and manufactures to which the present engines cannot be applied." Justice (to Watt) and public utility were the watchwords: "I have obtained the favour of Lord Guernsey to present the petition to the House of Commons, and . . . I doubt not but a communication of your favourable sentiments to Lord North, would greatly facilitate and ensure the success of it, as that communication would effectually convince his lordship of the justice and public utility of the measure, which seem to me to be the standards by which his lordship directs his public conduct."[50]

All did not go off without a hitch, however. The bill was given its first reading before the House of Commons on 7 March and the second on 13 March, after which it was again referred to a committee including Lord Guernsey, Edmund Burke, and others. Burke's intervention has been variously ascribed to his principled hatred of monopoly, his struggles with Wedderburn (who was advising Boulton and Watt) over American affairs, his representation of coal mining interests in Bristol, or his support for specific engine makers in competition with Watt.[51] Robinson makes a convincing case that the only explanation that holds up for Burke's interest in the matter is that he was representing the coal mining interests of his constituency. In doing so he was eager to latch onto rival claimants to Watt. In addition, on 20 March the House read a petition from an agent acting for one of those claimants, the engineer William Blakey, who claimed that Watt's 1769 patent had essentially stolen an engine design that he had patented in 1766.[52] Consideration of this petition was held over with the provision that its case could be argued when the House dealt with the committee report on Watt's Bill.

Watt took up the pen again to argue in defense; his role was very much the back-room boy generating the arguments. In one document he enumerated the basic differences between his own engine and that specified by Blakey, which was a type of Savery engine.[53] This seems to have done the job, since Blakey's claims were not heard again. In the ascendant once more, the Watt camp dominated a meeting of the committee on 11 and 12 April when it heard further from witnesses.[54] First there was Joseph Harrison, a smith at the Soho Manufactory and one of those usually anonymous characters who played a significant role in Boulton and Watt's engine work. He surfaced this time to attest to his personal experience of the performance of the Watt engines in comparison to common steam engines. Also heard from was a carefully briefed Robert Mylne, a well-known and eminent engineer who also testified to the distinctive workings of Watt's engine, its superiority to the common engine, and its economic advantages despite its initial capital cost.

Watt produced more pamphlets: one later titled "A Plea before Parliament" dealt again with the differences between his and Blakey's engines as well as claiming that even if they had been the same, Watt had priority of invention. Watt defended his original patent against Blakey's charge that it included no drawings by observing that none were required of him. In another document that was printed for circulation, Watt argued the great benefits that his engine would bring providing there was enough incentive to make the further substantial investments necessary.[55] The lobbying by Watt, Boulton, and their friends in the Birmingham interest proved successful. The act gained royal assent on 22 May. Watt now had protection of his engines for a further twenty-five years, certainly long enough to gain commercial advantage! That protection extended over England, Wales, Scotland, and the overseas "Plantations," the latter case preparing the way for some later engine sales to the sugar planters of the Caribbean.

Watching Boulton in action must have convinced Watt that his emerging alliance with his Birmingham friend was the right path to take. Though Boulton was the master lobbyist, Watt had also mobilized a significant number of Scottish parliamentarians, some of them as a result of his earlier involvement with canal business.[56] Back in 1767 Watt was disdainful of the political process, but in 1775 he acknowledged the necessity to play the game. He broke the news proudly to his father: "After a series of various and violent opposition I have at last got an Act of Parliament vesting the property of my new fire-engine in me and my assigns, throughout Great Britain and the Plantations, for twenty-five years to come, which I hope will be very beneficial to me, as there is already considerable demand for them."[57] Watt told his father that he would be in London for a few more days to make the final settlement transferring Roebuck's share in the engine partnership. Then he would return to Birmingham to start making engines that had already been ordered. Beyond that he was looking forward to "seeing you and the dear children."

Watt's long-term future was still not officially settled. The presumption in Watt's own mind was that it lay in Birmingham and the partnership with Boulton. But there were other possibilities open to him. He even received a very lucrative offer of £1,000 per annum from Russia (communicated by John Robison, himself then a professor at the Russian Naval School at Cronstadt). Erasmus Darwin wrote to Watt of his fright when he heard that "a Russian Bear had laid hold of you. . . . Pray don't go, if you can help it: Russia is like the Den of Cacus, you see the Footsteps of many beasts going thither but few returning." On the other side of the question was his old friend John Marr, the military engineer who was in Canada in connection with the troubles in North America. Marr considered the offer from Russia a very flattering one that Watt should accept, providing he could be assured of his salary being regularly paid and his being free to return when he pleased. Admittedly, the experience of others was not encouraging in that regard. Ironically, one reason for the Russian offer may have been Boulton's puffing of Watt's capabilities to the Russian Ambassador in London.[58] When it became clear that Watt was not going to Russia, close associates in Scotland presumed that Watt would be returning there to stay, rather than making a life in the English Midlands.

It was not until mid-August 1775 that Watt set off from Birmingham and traveled via Sheffield back to Glasgow. In Greenock he was reunited with his father, now in his mid-seventies, and with his children, James, then six years old, and Peggy, eight. His father had looked after the children as best he could, but Betty Millar and Gilbert Hamilton had also assisted.[59] The children were well cared for, but their emotional state must have been cause for concern, having lost their mother and then suffered a year-long absence from their father.

The children had at first lived in Greenock with Watt's father, and by September 1774 were in school there, but, as Gilbert Hamilton reported, the old man was "afraid for them for the Boats & Keys" where they were fond of playing.[60] Betty Millar, the unmarried sister of Watt's late wife Peggy, was keen to help with them but had been unwell. However, by early October she was much improved and returned to her late father's house in Calton, a district of Glasgow. The children joined her there, having been brought from Greenock by Miss Jean Cochrane, another cousin of Watt (the daughter of Watt Sr.'s sister), who was probably looking after his father already. Such caring duties were typically assumed to be the responsibility of the unmarried women of the family. Betty reported to Watt in November:

> the children is far from being any trouble to me all I wish is I could do more for them little dears they are both grown very big. Jamey has got Briches of which he is very proud they are to go to the school next week. . . . Nothing shall be wanting that I can do for them it is with pleasure that I see them obey and love me. O my Dear Jamey they prevent my Dwelling from being Solitery for I look around and am thankfull for my father's house to dwell in and to you for your consenting to let me have it and the dear children to look after.[61]

It seems that Daniel Millar's house was now, after the death of Peggy, either owned or controlled by Watt, who allowed Betty to live in her father's house with the children. Betty received funds, through Gilbert Hamilton, not only to help with the children's expenses but also to pay servants. There are signs that Betty was also left to deal with others who worked for Watt, perhaps in connection with the shop.

By February 1775 Betty reported that the children were "going on as well with there learning as can be expected from Chilldren so young." She found them "tender hearted" and "loving." They often spoke of their Papa, "and every good thing they get they are for keeping a part of it till you come."[62] Betty was reassuring Watt of the continuing devotion of his children and that his return to Glasgow was keenly anticipated, though he was not to arrive for some time.

Watt must have indicated already the likelihood of his making Birmingham his permanent home, for Betty expressed herself, with her erratic orthography, "vere sore [very sorry] that you have the thought of settling so far away," but "if you love the Country and find it more to your advantage" then, she said, so be it.[63] But Betty was afraid that Watt, in the search "to be rich" was launching into "too much Bisness" than was good for his health. Watt had also evidently signaled his intention to take young Jamie back to Birmingham after his forthcoming Glasgow visit. Betty pleaded that "little Jamy" was "too young to go among strangers," especially since Watt's business would take him "much abroad." Having said this, she was immediately afraid that she had overstepped the mark in expressing these opinions about Watt's career and Jamie's immediate future because she asked Watt to "excus this freedom," and conceded that he "will be the best judg" about his son.[64]

There was also worrying news about Watt's father. Robert Muirhead had been at Greenock in mid-February and found that, although in perfect health otherwise, the old man seemed to have "lost his Memory in a great degree." Jean Cochrane told Muirhead that she was afraid Watt Sr. had been "paying Acc^{ts} twice over," and there were many rascals in Greenock who would take advantage of him. Watt subsequently took a precautionary inventory of his father's goods and had his accounts examined.[65]

When Watt finally visited Glasgow in 1775, the pull of the various threads that bound him there would have been very apparent. Indeed, in the months before that visit he was inundated with Scottish business. Robert Muirhead pressed him about the Delftfield Pottery in which they were both involved. Watt had resolved to sell his share, but Muirhead was keen to keep him. The plan was to take out a patent on a fine white glaze that Watt had discovered. For revealing this secret, Watt would have "such an annuity during the continuance of said patent for your trouble & Expence and ingenuity as we can agree upon with you."[66] Watt did not rise to this bait, but neither did he sell his share.

In the same letter in which Gilbert Hamilton confirmed that Watt Sr.'s memory was failing, he emphasized that the old man wanted Watt to "come back again & desires I should inform you that the Greenock people are complaining that their Water Pipes are not laid yet, as they are waiting for you, & that a number of them had refused to pay the tax."[67] The son of Greenock, the latest in the Watt line of technically capable improvers of the town, was wanted back home. Not only was the path of improvement and civic order breaking down in Greenock because of Watt's absence, but also, on the other side of the Clyde, estimates for a steam engine were required for Lord Kilmaurs with prospects of more to be ordered. Hamilton had earlier, and repeatedly, forwarded news to Watt that Provost Buchanan wanted to know when

Watt was coming home since the proprietors of the Monkland Canal were now "very desirous" of carrying it fully into the city "and only wait for your coming here."[68] As Watt's 1775 visit approached, Hamilton told Watt that a number of people were daily inquiring after him "and I would fain hope that business might happen to keep you with us." There was the prospect of a rupture between the committee of the "Grand Canal" (the Forth and Clyde) and their imperious engineer, and then "they will naturally apply to you."[69] On another occasion Hamilton advised that Watt could get orders "for near a dozen of barometers," presumably thinking that his friend could still be tempted into instrument-making business.[70] Watt could have been forgiven for imagining a conspiracy was engaged to draw him back into his old patterns of business in Scotland. The threads that had bound him were indeed many, and Watt had to dampen down these expectations if he was to break free.

As his visit approached, Betty Millar told him that the children "danced and sang with joy" at the prospect of seeing him. Watt had either sent or promised the two of them "pretty little Bibles," a characteristically serious present from their father. Watt arrived back in Scotland intending to take young James back with him to Birmingham while leaving Peggy in Scotland a while longer. This, as much as anything else, indicates Watt's resolve to settle in Birmingham. There was, however, still no formal partnership with Boulton, though Watt wrote to him from Scotland in early July 1775 setting out a version of a contract of partnership and asking that an able lawyer be engaged to draw it up.[71] What appears to be a reply from Boulton rather strangely advises that he would be willing to buy Watt out of the partnership, as if he still believed that Watt might decide to stay in Scotland after all. Boulton was clear, however, that for his part "I had rather ware out our Act of Parl[t] ensemble than gain a greater fortune by being ye sole proprietor."[72] Boulton advised that he had been unable to have the article of partnership drawn up since the lawyer had been called to London. There the matter lapsed again. Boulton and Watt seemed committed to their partnership, but it was unofficial. It is not clear in fact that the contract was ever enacted, and it may have remained simply a mutual understanding. There is no clear record of it in the archive.

The question arose again the following year, when the father of Watt's intended second wife, Annie McGrigor, demanded to see the partnership agreement as part of arranging the marriage settlement: "the old Gentleman wishes to *see* the Contract of Partnership between you & I, and as that has never been formally executed, I must beg the favour of you to get a legal contract writen & signed by yourself—sent to me by return of post or as soon as may be."[73] Watt had been afraid that if McGrigor

Figure 3.3: *Mrs. James Watt* by John Graham-Gilbert (circa 1820). Even in old age Annie's countenance suggests a lively, commanding presence. © Science & Society Picture Library.

knew the truth—namely that there was no partnership agreement signed—then he would doubt Watt's prudence in business affairs. Consequently, Watt had given the old man the impression that such a document did exist. So what Boulton sent must masquerade as a duplicate of that nonexistent original document. Boulton obliged, and McGrigor was satisfied. Our truth-telling Presbyterian was not beyond a little duplicity if the stakes were high in business, or in love.

Watt married Annie McGrigor in the summer of 1776. It is not clear exactly when he formed his relationship with her. However, their courtship, and Watt's negotiations with her father, probably began during Watt's 1775 visit to Glasgow. A clue comes in a letter written to Watt by Betty Millar. Having expressed her happiness at the news received via Hamilton that Watt and her "sweet nephew" were well, Betty assured Watt that Peggy was not missing her brother too much. She continued: "I was a little uneasy at a long Conference that was held in your house betwixt a sertan [sic] Gentelman and you but I will still hope the Best. I have had 2 or 3 visits from sweet Miss. She helped me to put up your things. She would not let none of your old stocking nor breaches come . . . if she had her will but I told her that it would serve you while you remained a Batcheler [sic] upon which she smilled sweetly and I dropt the discourse."[74] Betty was recalling here, in all probability, a meeting between Watt and Mr. McGrigor at which he had asked for his daughter's hand. It also appears that the daughter in question was already taking a proprietary (or even commanding) interest in Watt's self-presentation!

There is no surviving correspondence between Watt and Annie from the period of their courtship,[75] but we can glean something of its progress from letters that Gilbert Hamilton wrote to Watt during this time. Hamilton, it will be recalled, was Watt's agent in Scotland and dealt with Watt's finances, business dealings of all sorts, and property, including the remaining stock of the shop. But Hamilton also dealt with more personal matters. As we have seen, he kept an eye on Watt's father and on other relations, including Betty Millar and the children. He also acted as a conduit for Watt and Annie's letters to each other between the summer of 1775 and their marriage on 29 July 1776. This was natural enough since Hamilton was married to one of Annie's sisters, Catherine McGrigor. It is likely that Watt and Annie first became interested in one another at family gatherings involving the Hamiltons. However, Hamilton experienced some sensitivity when his father-in-law, who also advised him on business matters, proved not to be entirely enthusiastic about "losing" Annie to Watt. Indeed the necessity for Hamilton to act as an intermediary between the two probably resulted from the fact that Annie and Watt wished to keep the extent of their communication—let alone its content—unknown to old man McGrigor for the time being.

A jocular letter from Hamilton to Watt of 18 October 1775 set out the situation: Hamilton's letter was just a cover to convey to Watt "the Jewel," a letter from Annie.[76] Watt is described as "mad"—a term often used to describe someone in the first flush of romance. Watt's intentions were made clear by Hamilton's remark that he would be giving up the pleasure of Annie's company altogether "unless your High Mightiness shall vouch safe to grant us a sight of her & for all your fine promises now . . . you may repent of visiting these cold Bleak northerly regions so often as you intended, especially as you will then have carried off what you reckon the soul & Sun of the Country, & there will be nobody to visit but a parcel of dull stupid Mortals." Watt had clearly expressed his intention of taking Annie to Birmingham and in doing so had promised they would return to Scotland regularly. (We will see that Annie did return quite regularly but Watt, obsessed and anxious about his work, much less frequently, and this was to remain a bone of contention between them). We do not know what "the Jewel" contained, but Hamilton let drop that Annie had told him "that the Contents are no ways displeasing to you."[77] Underlining the pact that this correspondence would remain secret to the three of them (or perhaps four, including Betty Millar) Hamilton gave security instructions: "Seal always with your Initials & there is no sort of chance of being opened as a letter of business in case I should be out of the way." One senses that much of the paranoia about being discovered in courtship may have lain with Watt.

In later correspondence the jocularity continued: Hamilton told Watt that for the comfort of a "certain person," he had told her that Watt had got "part of your winter store laid in & an ugly maid to attend you." It becomes apparent also that Annie was being a little coy in response to the ardor of her suitor: Hamilton reports her as saying that Watt should not expect an answer to every letter.[78] Beyond this we learn that Boulton was being kept in the dark about Watt's new attachment, and that James McGrigor was being uncommunicative. A promised letter from the father to Watt was not forthcoming, and Hamilton advised that McGrigor had "never spoken a word on the subject to me since you left, nor do I believe he has even to her."[79]

Over ensuing months the correspondence between the lovers continued apace; at one point Hamilton comments that he has received Watt's cover for "a Volume rather than a letter which I delivered & to which I now inclose you an answer which seems to be pretty tolerable likewise for length."[80] But repeatedly Hamilton inquired whether Watt had received a letter from McGrigor. Even in March 1776 Hamilton asked, "Have you ever heard from Mr McG or has he given his consent," and observed "I am shy of speaking to him about it."[81] Hamilton's discomfort was doubtless compounded by the fact that his father-in-law was currently advising him "in every step" of a venture into a calico printing business.[82] As late as 19 April 1776 a letter came from McGrigor, but Hamilton, who was aware of its contents, observed that Annie's father "seems to stick to his old opinions." The hope was that Watt's presence might change the old man's mind. But Watt himself was perhaps despairing of this since Hamilton informed him, "I told A. what you said, that you depended on her."[83] The only hope seemed to be that Annie might persuade her father to relent. Perhaps Watt would have said, as he did of business bargaining, that he would rather face a loaded cannon than try to bargain with a reluctant prospective father-in-law!

So Watt found himself in Glasgow in the summer of 1776 pleading with Boulton back in Birmingham to forge a contract of partnership to satisfy James McGrigor. Watt informed Boulton on 8 July that McGrigor, once satisfied of his daughter's wishes, "gave his consent with good grace."[84] In those strongly patriarchal times, typically the men would have sorted the matter out, but perhaps McGrigor, knowing his daughter to be a strong-willed woman, needed reassurance from her. Whatever the case, on Sunday 28 July Watt wrote to Boulton anticipating the marriage the next day: "Upon Monday the fatal noose is to be put about my neck, or more emphatically, I am to get a lick of the eternal gluepot. At present I am something in the state of a man waiting for his dinner after the cloth is laid, comforting myself for the delay by the certainty of a good meal. I wish my appetite may be as good—I dare say I have your prayers."[85] The record in the Old Parish Register for Kilpatrick, which was Annie's parish, has an entry for 27 July as follows:

> Mr James Watt Engineer in Birmingham in the County of Warwick & Miss Anne McGrigor in this Parish gave up their names for proclamation previous to Marriage
>
> NB The abovementioned Miss McGrigor is lawfull Daughter to Mr James McGrigor Mercht in Glasgow And were married the 29th July 1776.[86]

Watt had declared himself of Birmingham and so, finally and decisively, he was to be. He gathered up his daughter and his new bride and took them on Friday 2 August first to Edinburgh and then south, arriving in Birmingham on 9 August. James Jr. would then have met Annie and indeed probably became aware only on meeting her that he was to have a "new mama," for Watt had asked Boulton to keep that news from him.[87] No doubt Watt felt he was doing the right thing by his young son.

Watt now faced the issue of where his newly gathered family was to live. During his earliest time in Birmingham, Watt may have lived with Boulton at Soho House. We know that while he was in Scotland in the summer of 1775, Boulton was looking for a house for him and that when back in Birmingham Watt was furnishing a residence. Then in April 1776, perhaps anticipating returning with a new wife, he rented another residence at 1 Newhall Walk, where he moved on 11 April. This was the house in which Annie and Peggy came to be united with James Jr. after the marriage. But less than a year later, they moved to a house called Regent's Place at the top of James Street, Harpur (or Harper's) Hill. Both houses were in what is now the Jewellery Quarter of Birmingham, the location of the Birmingham Assay Office from 1773, but when Watt occupied Regent's Place it was the first of the villas being built on the then-rural slopes above Newhall by prosperous manufacturers and merchants seeking an increasingly fashionable suburban existence. This was perhaps more to Annie's liking. According to one report, Boulton leased Regent's Place for Watt.[88] He and Annie were to live there until 1790, when Watt built Heathfield Hall in Handsworth, a further and more substantial retreat from urban and industrial encroachment.

Watt had an office at Regent's Place from which he conducted his correspondence, drafting, and calculations. He also established a chemical workshop there, where much of his most significant chemical experimentation would be done.[89] Regent's Place was about a fifteen-minute walk from the Soho Manufactory, and it must have been there that Watt conducted his experiments on working engines. (This becomes significant for our understanding of his pattern of work and also his character as a natural philosopher. He presumably observed engines and had natural philosophical insights into them at the manufactory). But when living at Regent's Place, Watt would spend days on end in his home office without visiting the manufactory, a point reinforced by the fact that his assistants worked at his home also. This was the work pattern of a consulting engineer at the least and possibly, at times, the state of "ingenious indolence" that Watt had long craved. Such a work pattern was reflected in the partnership arrangement between Boulton and Watt. The draft of the intended partnership agreement that Watt sent to Boulton from Scotland in early July 1775 certainly set out the financial aspects of the arrangement: Boulton to have two-thirds of the property in the invention, Watt one-third; Boulton to pay all expenses of the Act of Parliament and of future experiments; Boulton to hold and own all stock in trade, and also to keep the books. One clause, however, referred directly to the work that Watt was to do: "I to make drawings, give directions, & make surveys, the company paying travelling expenses to either of us when upon engine business."[90] However much Watt might hanker after the life of the closet, he recognized, as did Boulton, that he would have to travel on engine business. Drawings, plans, and calculations were of no use without the experience of someone adept at putting engines together and making them go. Although Boulton was not without skills in that department, Watt's skills were nonpareil. They were to be in demand "on the road" until designs advanced, procedures became more predictable, and a new tribe of engine erectors was bred up. So, what of the early Boulton & Watt engines?

Early Engines and the Cornwall Business

Before they could enter fully into the engine business with Watt's new designs of the engine, Boulton and Watt needed a clear idea of how they were going to engage in business and, in particular, precisely how they were going to make money. They had been contemplating these questions individually and collectively at least since Small assisted Watt in drawing up the specification for the 1769 patent.[91] In fact, even in the early days of partnership with Roebuck, Watt had the idea that rather than earning money for the construction of an engine by charging for a license to build an engine of his design, he would link his own reward to the savings in fuel that were made as a result of the superior fuel efficiency of the Watt engine. Taking an annual premium rather than a one-off payment did not saddle the customer with initial expenses beyond the construction cost. By taking one-third of the annual saving and allowing the engine owner to take the other two-thirds, the customer continued to benefit greatly. If there was no saving to the customer, for whatever reason, there was no profit for Boulton & Watt. Where benefits would be most substantial, that is in places where the price of coal was very high, Boulton & Watt's premium would be proportionately higher.[92] Although this system of payment was Boulton & Watt's basic model, it was not always consistently applied. In situations where engines would work at a steady rate over long periods, Boulton & Watt allowed a one-off payment of a multiple of the annual premium. Customers who were close to Boulton & Watt were offered discounted single payments, often as a quid pro quo for allowing the engines to be used for tests or because Boulton & Watt had investments themselves in the company in question.[93]

Arrangements for building the engines were complicated. Recall that Boulton, when he was "courting" Watt, had painted a picture of a grand engine manufactory at Soho making engines "for all the world." This was not the business that they in fact set up. Only in the 1790s, with the establishment of Soho Foundry, did their company manufacture the main engine parts.[94] Their initial modus operandi was to provide plans for the engine and the building to house it.[95] This would include specifying all the iron engine parts. It was then up to the customer to obtain most of those parts and other materials, as well as workmen, to erect the engine under Boulton & Watt's strict supervision. There were some exceptions because of the importance of a high degree of accuracy in the manufacture of certain key parts of the engine. The accuracy of the cylinder was crucial, and Boulton & Watt usually recommended that it be obtained from Wilkinson. They also undertook to make at Soho the valves and other parts requiring high precision. This was essentially the way Watt had worked when he was erecting Newcomen engines in Scotland. Thus the business of finding materials, manufacturing parts, and procuring labor was left so far as possible to the customer. It is important to realize from the big picture of steam engine diffusion through British industry that the speed and breadth of that diffusion meant that Boulton and Watt could never have achieved the overpowering dominance of Boulton's grand vision. Nor could Boulton & Watt's impressive output of engines have been accomplished without the decentralized system they adopted.[96]

Boulton & Watt provided designs, drawings, some precision manufacture, and expert direction, supervision, and, for a time, maintenance. This, once again, was the consulting engineer's model. But it only went so far because Boulton & Watt wanted to retain control. They would not license others to build their engines, not least because their long-term success would depend upon the quality of those machines, the fuel savings that they could achieve, and the reputation thus established. They felt that they could only maximize their chances of success by retaining tight control. It seems likely, given what we know of their characters, that this carefully thought-through "business plan" originated more with Watt than with his flamboyant partner.

The first major Boulton & Watt engines that Watt worked upon were for Bloomfield colliery near Dudley, owned by Messrs. Bentley & Co., and for the New Willey ironworks near Broseley, owned by Wilkinson, which was to drive an air-blowing cylinder for a coke-fired blast furnace. These were both local ventures. The Bloomfield engine was built according to the announced modus operandi with Watt providing the drawings and supervising the engine erector while Wilkinson supplied the cylinder, and the colliery dealt directly for materials and labor. The machine was set to work on Friday 8 March 1776 before a party of the proprietors and curious local worthies. It was named the "Parliament Engine," presumably to mark the fact that it was the partnership's first completed order after the Act of Parliament extending Watt's patent. The report in *Aris's Birmingham Gazette*, which has all the hallmarks of a piece of Boulton's puffery, gave the dimensions of the engine and the

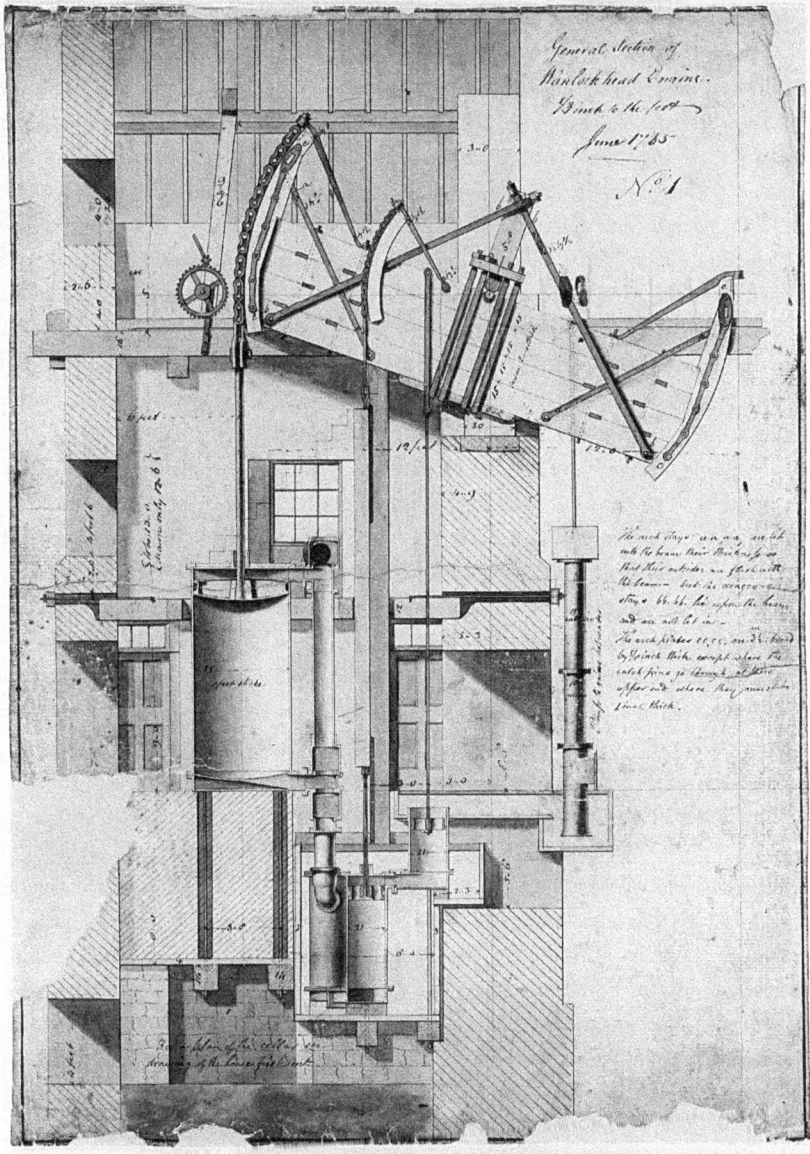

FIGURE 3.4: A Boulton & Watt pumping engine (1785). General section of engine for the Margaret Lead Mine at Wanlockhead, Scotland. Annotations are in Watt's hand. Courtesy of Museums Victoria, Melbourne, Australia, https://collections.museumvictoria.com.au/items/408010.

fact that it worked on principles "invented by Mr Watt (late of Glasgow) after many Years Study" and many experiments.[97] The report praised the proprietors for their "liberal spirit" in opting for this new machine rather than a common engine. This acknowledged that they were taking some risk in so deciding, in effect participating in an experiment, having been persuaded to do so by Boulton, who knew the colliery owners. The engine for the New Willey ironworks was probably in the same category of a friendly, not entirely commercial, transaction. Wilkinson could do very well from Boulton & Watt's activities in the future since they recommended that their customers obtain cylinders from him. Boulton had persuaded Wilkinson to apply his cannon-boring machine to produce accurately machined cylinders and then to design a boring machine especially for the purpose. So Wilkinson was likely to be willing to help Boulton and Watt in the "experimental" phase of early commercialization of their engines by purchasing this blowing engine.

Other early engines included a small water-raising machine for Cook & Company, a firm of distillers at Stratford-le-Bow near London. This caused some concern because it was likely to be seen by many people and it was important for Boulton & Watt's early reputation. A drunken engineer precipitated a disaster while John Smeaton was visiting to see the engine running. The broken engine was put right by the erector, so that its running became, and came to be seen as, satisfactory.[98] This lack of control was a headache for Boulton and Watt, and a weakness of their way of doing business to the extent that it relied on the efficiency and effectiveness of others in producing engines to which the Boulton & Watt name and reputation was attached.

Another early engine was at the colliery of Peter Colville, for whom Watt had earlier erected Newcomen engines at Torryburn in Fife. Watt visited the site in early July 1776 when in Scotland just prior to his marriage to Annie. Like other coal-mine engines, this was a large one, but already the partners were setting many design features as standard. There were hopes that this might be a demonstration engine in Scotland, leading to further orders there, but it was a disappointment in that regard. Boulton & Watt did not build many colliery engines, for good financial reasons. Such engines burned waste coal at virtually no cost, so the economy of an engine was of little consequence, and there was small incentive for collieries to purchase such engines or, given his mode of remuneration, for Watt to build them.

Metalliferous mines, especially in Cornwall, provided their major market for powerful pumping engines. Although Boulton & Watt prior to 1800 supplied significant numbers of blowing engines based on that constructed for Wilkinson, and also numerous low-lift pumping engines based on the Stratford-le-Bow machine, for waterworks and like applications, it was engines for Cornwall that formed the initial basis for their eventual success and prosperity.

One construction of Cornish history emphasizes the centrality of mining.[99] Certainly by the 1770s Cornwall was the largest copper producer in the world, and an important aspect of the county's visible industrial heritage consists of the engine houses that once accommodated the steam power used to drain the copper and tin mines. Deep mines in the hills of the Cornish peninsula presented severe drainage difficulties. Animal power was insufficient to pump to the necessary depths and waterpower was scarce, given the lack of streams in the region. This ought to have made steam power an attractive proposition once Newcomen engines became available. Indeed, by 1775 something like a quarter of all Newcomen engines at work in England were in Cornwall. But the use of the steam engine, while offering the technical means to drain the deeper mines, was often marginal economically. The nearest coal to fuel these engines was in South Wales and was very expensive on reaching a Cornish mine. The operation of Cornish Newcomen engines, and so the viability of mines, was very sensitive to the price of coal. In autumn 1775 Boulton reported to Watt that a Cornish informant had given him a list of engines in the county: "The whole number in . . . Cornwall are exactly 40 but there are only 18 of 'em in work on acct of ye high price of Coals."[100]

Here, then, was a major opportunity. But Cornwall was not an easy market to break into. There was a long history of engineers promising improvements to Newcomen engines and not delivering. Considerable skepticism resulted among mine adventurers about new technological claims. There was no particular reason to believe that Watt, as an improver of the Newcomen engine, was any different. Trust was not easy to establish. When a Cornish deputation visited Birmingham (while Watt was in Scotland getting married) to learn what Boulton & Watt had to offer, a drawing went missing from Soho. Boulton demanded its return: "We do not keep a school to teach fire-engine making, but profess the making of them ourselves," he wrote to Thomas Ennis, the head of the Cornish delegation.[101] The drawing was returned, pleading a "misunderstanding." But it was clear that sharp practice would have to be guarded against. There were also experienced steam engineers in Cornwall, notably the Hornblowers, who had erected many engines there. They and others were to be very useful to Boulton & Watt, but they also became competitors.

The first Watt engine was ordered from Cornwall in November 1776 for the Ting Tang mine. But the first to start up was at Wheal Bussy mine in September 1777.

Watt visited Cornwall for the first time to supervise the erection and start-up of the latter engine. Between that time and the end of April 1782, that is to say in a period of less than five years, Watt spent over two years in Cornwall, including a stay of eight months in 1778 and of ten months in 1781–1782. Any chance of that life of in-genious indolence after his move to Birmingham was long delayed, and though Annie accompanied him to Cornwall for some of that time, the children would have seen little of their father. During Watt's stay in Cornwall from September 1779 to Febru-ary 1780, Annie took the children to Scotland. Hamilton's prediction back in 1776 that her Scottish family and friends would be long deprived of her company when Watt whisked her away proved accurate. By the time Watt became more genuinely "settled" in Birmingham in 1782, Peggy was in her mid-teens and James Jr. about thirteen years old.

Watt's first visit to Cornwall was crucial since it was to erect, and ensure the successful start-up of, the engine at Wheal Bussy. A good start with this engine might overcome local skepticism. There were many spectators for the start-up of the engine including, Watt thought, "all the West-Country captains."[102] The mine captains were the men who were in charge of mines overall, so they were crucial targets for con-version to use of the new engines. The most influential among them was John Budge, who had earlier refused to be involved in work on any Boulton & Watt engine. Now that Budge had seen the Wheal Bussy machine operate, he had waited on Watt "and promises to read his recantation"—Budge had budged.[103]

Annie accompanied Watt on this first trip even though she was heavily pregnant; she gave birth to their first child, Gregory, on 11 October 1777, just over a week after their return to Birmingham. While they were in Cornwall, they lodged with Thomas Wilson, who was responsible for ordering the Chacewater mine engine, the third Boulton & Watt engine to be started up in Cornwall. Wilson had seen the Wheal Bussy engine in operation, and its performance convinced him not only to order the engine for Chacewater but also that Boulton & Watt had a great future in Cornwall. He signed up shortly thereafter to be the Birmingham partnership's commercial agent in the county, a position that he held until 1800.[104]

Watt arrived for his second visit to Cornwall on 28 May 1778, this time to attend to the Ting Tang engine, the assembly of which both Jonathan Hornblower, the el-der (1717–1780), and Jabez Carter Hornblower (1744–1814) were involved with. There had been a long delay in producing and shipping the cylinder. The engine ran for the first time on 30 July. During this same visit the Chacewater engine was con-structed and started on the same day as that at Ting Tang. This was another big test for Boulton & Watt in Cornwall. The volume of water to be "forked" from the Chace-water mine was enormous, and locals doubted that this single engine could drain the mine when a previous effort deploying two Newcomen engines had failed. Though it took three months, the mine was pumped dry. Watt left Cornwall on 31 December 1778 in triumph.

Thereafter, further orders flowed from Cornwall, and Boulton & Watt's activities there were consolidated. They took a long lease on a house—Cosgarne House in the Gwennap valley, near Truro and convenient for many of the mines where Boulton & Watt had business. It was well furnished and a comfortable base of operations, capable of accommodating the family on occasion.[105] Watt stayed there from June 1781 until April 1782, and the family joined him for six of those months. This was Watt's last visit to the county for a decade. Wilson was taking care of finances. As the engines themselves took on a more standard form and the volume of work increased, Watt produced a set of instructions to guide the engine erectors. *Directions for Erecting and Working the Newly-Invented Steam Engines* was written in the early part of 1779 and was available for distribution in printed form by that summer.[106] A great deal of effort also went into an attempt to rationalize the calculation of the premiums to be paid by those using Boulton & Watt engines.[107] Customers found many reasons not to pay, or to dispute the amount demanded. The standard against which savings of fuel made by Boulton & Watt engines were to be judged was, not surprisingly, a common bone of contention. Thanks to Watt's ingenuity, some progress was made in devising a system. The old engines at Poldice mine were monitored and agreed upon by a wide range of mine owners and captains as a standard. Watt devised a complex system to enable the calculation from the Poldice data of savings for engines of different sizes, loads, and stroke rates. A counter device was developed that provided a tamperproof measure of the number of strokes performed by an engine. These complex and ingenious methods of keeping account that Watt devised might be seen as another product of the culture of Presbyterianism that he imbibed in his youth. But, not surprisingly, given the ever-ambiguous nature of rules, Watt's stipulations shifted the terms of negotiation rather than avoided it. It seems likely also, given the impressive standardization of engine reporting by Cornish adventurers in the early nineteenth century (notably in *Lean's Engine Reporter*),[108] that Watt's methods were not moving into a vacuum but into competition with existing cultures of economy in Cornwall.

Watt's copying machine,[109] initially developed between 1778 and 1780, was another invention that was prompted at least partly by the exigencies of doing business in Cornwall. There was a large number of letters to be written, plans to be drawn, and instructions to be distributed, and all had to be copied. The only method of doing so then available was by hand. Copy clerks were used, but Watt, Boulton, and Thomas Wilson themselves worked long hours as copyists. Using copy clerks was risky, especially in the intensely competitive, and routinely underhand, commercial dealings of Cornish mining. But this business demand was not the only stimulus behind the copying machine venture.

Watt's attention had been drawn to the issue of copying by Erasmus Darwin, who developed a "polygrapher" or "bigrapher" in 1778. This was a machine that connected the writer's pen mechanically to a second pen so that two, or possibly more, copies of a document could be written at the same time. Darwin presented his machine at a Lunar Society meeting hoping that Boulton and Watt might take it on as a commercial venture. They were not tempted. Polygraphers were difficult to use, required very sensitive adjustment, and were not readily portable.[110] So, Watt came up with a copier using an entirely different process—as he saw it, chemical rather than mechanical. This involved pressing a dampened piece of thin, unsized paper against an original document that had been written in a specially developed ink. The press copy produced a reverse facsimile of the original; the image could be read through the other side of the thin paper employed.

Watt was informing friends about the technique by May 1779 when he was already using it to copy his own letters. His old friend John Marr, back from his North American exploits, considered it a "most astonishing Invention," surpassing in importance all Watt's former efforts, even his steam engine improvements. While Marr was probably exaggerating here, the fact that he put the invention above Watt's steam engine improvements reminds us that the latter had still to earn the status they would later acquire. Marr went on to tease Watt, whose staunch political caution was well known to him, that his copying machine would increase "the power of every villainous Incendiary to propagate sedition, Rebelion & high Treason without any possibility of discovery."[111] Marr was also not alone in worrying that Watt's "forgeing Roller press," as he called it, might be used by counterfeiters, but he was reassured when advised that the process would not "take on all papers."[112]

The patent that Watt enrolled on 31 May 1779 described a roller press and a screw press design, specially treated paper that would make good copies, and ink formulations directed to the same end. Watt, inveterate chemist that he was, experimented with preparations of ink for a number of years. Although much of the impetus for developing the copying machine derived from its value to Boulton & Watt's own commercial activities, it was decided to manufacture it for sale. If the machine was as transformative as people like Marr thought, then it might be a very valuable commercial proposition. A new partnership was formed as "James Watt & Co." Watt had a half share of the patent, Boulton a quarter share for financing it, and James Keir a quarter in exchange for managing the business. Copying machines were manu-factured at Soho, and the London stationer James Woodmason was appointed as sales agent. Gilbert Hamilton acted as agent for the machines in Scotland, and Jean Hya-cinthe Magellan (1723–1790), a ubiquitous "intelligencer" whose correspondence with Watt about heat we will also encounter, for overseas business.[113]

For five and a half guineas subscribers would receive a press, proper paper, and other materials and the "secret," that is the instructions for operating the copier. Subscriptions, however, were well below expectations even though Boulton had mar-keted the copying machine by demonstrations at the Royal Society, the Houses of Parliament, and London coffee houses. Deluxe machines were sold to the kings of England and Russia, and to a number of noblemen. While Boulton courted the elite, Watt fulminated about the need to direct efforts at merchants and the like. As he told Boulton, their money was just as good and there were more of them! There were steady sales throughout the 1780s, and business improved with the development of a portable press by Watt Jr. in the mid-1790s. But by then the patent had expired. As a commercial proposition, the copying machine was perhaps more valuable for promot-ing Watt's reputation for ingenuity than for boosting the bank balances of the part-ners, though it did earn *some* money at a crucial period. But its utility for Boulton & Watt's operations was significant.[114] Boulton & Watt certainly ensured that they had a copying machine wherever they were conducting regular business, and a machine was working hard in Cornwall.

For all the difficulties, the Cornwall steam engine business did begin to produce a flow of revenue, and Watt saw better financial times ahead, as reflected occasion-ally in his letters to Boulton from Cornwall. As we will see, Boulton perhaps knew otherwise, given his extensive and intricate pattern of borrowing. To compound the problems, the price of copper began to fall. When the price had been high in the early to mid-1770s, previously marginal or unworkable mines once again became of inter-est, and so, therefore, did the improved means of draining them that Boulton & Watt offered. Prices began to fall, thanks in part to the discovery of rich new deposits of copper in north Anglesey at Parys Mountain, accessible to open pit mining. As prices fell some mines were closed, and the owners of those still operating became very cost conscious. The premiums payable to Boulton & Watt were an obvious target, and hos-tility toward the company began to build. In many ways this was Watt's worst nightmare. While his move to Birmingham, his partnership with Boulton, and his marriage to Annie had lifted his spirits greatly, the frantic pace of business and incessant deal-making attendant on the time spent in Cornwall had seen the return of the frequent headaches and periods of despondency experienced during his civil engineering days in Scotland.[115] As difficult economic times hit Cornish mining, all of Watt's careful, rational calculations about their fair reward went by the wayside. Owners and captains argued for reduced one-off premiums, challenged Watt's complex calculations, questioned the veracity of the stroke counters on the engines, and even challenged the validity of Watt's patent. In these circumstances Watt called Boulton in to deal with the disputes, recognizing that his partner's "chearfull countenance and a good heart" would be more effective than his own reluctant efforts at bargaining.[116] Watt's sense of himself as honest and straightforward in his dealings, and his rational approach to everything, was greatly offended by the antics in Cornwall. He wrote to Boulton, "These disputes are so very disagreeable to me, that I am very sorry I ever bestowed so great a part of my time and money on the steam-engine. I can bear with the artifices of the designing part of mankind, but having myself no intention to deceive others, I cannot brook the suspicions of the honest part, which I am conscious I never merited even in intention, far less by any actual attempt to deceive."[117] In his worst moments Watt was ready to sell up his share in the patent, so outraged was he that his long labors to perfect the steam engine were regarded by some as worth nothing.[118]

Boulton's more accommodating view prevailed. He compromised on calculating premiums and negotiated reductions, even suspensions, of payments. He, and Watt with his encouragement, along with friends such as Josiah Wedgwood and John Wilkinson, made investments in several mines, helping to keep them afloat and gain inside knowledge of their plans and operations. In 1785 Boulton was even actively involved in an effort of the mines to combine, forming the Cornish Metal Company, in order to force the Anglesey company to cooperate in setting levels of production and price.[119] This was a type of business strategy at which Boulton excelled and of which Watt, for all his businesslike qualities in some other respects, was profoundly shy.

In the 1780s, as mines went to the wall and many workers lost their jobs, much anger was directed at Boulton & Watt. By 1792, when the Anglesey mines had been largely depleted and prices were recovering, the Cornish Metal Company was wound up. By this time, also, Boulton had become a large-scale purchaser of copper for his recently erected steam-powered coining mill at Soho. He no longer had reason to try to maintain the copper price—quite the reverse.[120] From that point, Boulton and Watt's primary interest in Cornwall lay in recovering the premiums that they felt they had been robbed of over the years by illegitimate refusals to pay and by the construction of "pirate" engines. It is no coincidence that Boulton and Watt finally swung into legal action against the Cornish "pirates" in that very same year, 1792, as their complex entanglements with Cornish mining were simplified and reduced.[121]

Boulton & Watt's Cornish venture was very important in the early history of the partnership and of considerable consequence financially, as we will see when we come to consider Boulton & Watt's earnings. But future, even larger, markets lay elsewhere, and with different engine designs that were intended to supply the need for power to drive machinery across a wide range of industries.

Powering the Machine: The Rotative Engine Realized

The pumping of water from mines was the major application of the Newcomen engine, and so it was natural for Boulton & Watt to proffer its improved engine to that market (at least where fuel was expensive). But they had long had other markets in mind. Even before they met, Boulton had been interested in Watt's steam-wheel invention because it seemed to offer the possibility of direct steam-powered rotary motion for the machinery at Soho.

The ultimate competitor for Watt engines, as for Newcomen engines, in many power applications was, as already noted, waterpower supplied by waterwheels located on rivers and streams and aided by waterworks that constructed dams and traces to control and direct flow. Such waterworks could go a long way in obviating the great problem of waterpower, its unavailability or reduced availability during dry weather and drought. Considered in the long run, waterpower, with its centuries of application and development, remained the most important prime mover in British industry until well into the nineteenth century. It continued to be a subject of technical interest and development. Indeed, for many decades after the appearance of more efficient steam engines, an important and common use of them was, as initially at Soho itself, to pump water up a gradient so that it could be reused in waterwheels. In that respect, for many industrial applications, steam power employed in such "returning" engines remained an adjunct to waterpower.[122]

In some circumstances, however, waterpower was seen as limiting because its economic use generally required industry to be located on fast-flowing rivers and streams. Hence, the location of much industry in the valleys of northern England was because of the availability of such waterpower there. Economical steam engines supplied with coal transported by canal gradually became one basis for the relocation and concentration of industry in growing towns. Boulton and Watt were not alone in anticipating the demand for rotary power, but Watt was to show characteristic ingenuity in making the inventions necessary to its realization. Those inventions were the subject of a second phase of patenting.

Economy was not the only desideratum in driving machinery by rotary power. Just as crucial was the constancy of the power—the smoothness of its application to the machinery. The use of the more economical Newcomen engines of the later eighteenth century as returning engines was a solution to the problem of the uneven output of reciprocating steam engines. Their combination with a waterwheel enabled steam to supply smooth rotary power. Where waterpower was available, a reciprocating Watt engine might be used in the same way to supply smooth power with even greater fuel efficiency. However, if smooth steam power was needed in the absence of the ready means of linking it to waterpower, then an engine had to be developed that provided smooth power directly. This was the task Watt took up in the early 1780s, encouraged by Boulton, who advised, "There is no other Cornwall to be found and the most likely line for the consumption of our engines is the application of them to mills, which is certainly an extensive field.[123]

Rotative power from steam was not an idea original to Boulton & Watt. Both its major elements—conversion of reciprocating motion to rotative and, equally important, ensuring a smooth delivery of power—had been attended to by others, notably Matthew Wasborough in a patent of 1779, and James Pickard in Birmingham, who in 1780 patented his engine using a crank to convert to rotary motion and a flywheel to smooth out the power.[124] At least part of the inspiration for these engines, however, had been provided by leaks from the Boulton & Watt organization, specifically from one of the pattern makers, Richard Cartwright. Watt was extremely angry, and his high dudgeon may have helped to fuel a purple patch of mechanical contrivance. Some of these competing engines were put into operation in the mill districts and had made significant inroads into the market in Manchester, for example. But they had their own problems and, legally at least, could not benefit from the fuel economies that Watt's patent engine achieved with its separate condenser.

So Watt, with Boulton's prodding, turned his attention to the problems of the rotative engine. Some of his solutions were incorporated in a patent specification of 25 October 1781 that was enrolled on 23 February 1782 (Patent No. 1306) "for certain new methods of applying the vibrating or reciprocating motion of steam or fire engines, to produce a continued rotative or circular motion round an axis on a centre, and thereby to give motion to the wheels of mills or other machines." Watt played the patent game here, providing five different methods for smoothing the motion, only one of which—the sun-and-planet motion—was to be used in his rotative engines as a way of avoiding Pickard's patent on the use of the crank.

Watt's next patent (No. 1321), of 1782, also addressed questions relevant to producing rotative engines, notably the problem of smoothing out the engine's cycle, but it did so disguisedly, by reference to the pumping engine and to a different way of working engines, so-called "expansive working." This method involved cutting off the injection of steam into the cylinder early, that is before the piston had moved fully to the end of the cylinder, allowing the expansive force of the steam to complete the job. It was possible to perform this early cut-off in Watt's engines because the separate condenser allowed the cylinder to remain hot throughout the cycle. In a Newcomen engine, steam had to be admitted for the full period because that steam had to heat the cylinder for the engine to work at all. Expansive working represented a further significant saving in fuel consumption, although at the cost of some power. However, while power was halved, consumption was reduced to a quarter of what it would otherwise be. This would render Watt's engines, whether pumping or rotative, more economical than ever. The presentation in the 1782 patent of mechanisms for equalizing the power stroke of the expansive engine applied to pumping were, again, a disguised way of claiming them for use in rotative engines. The 1782 patent also described a rack and sector device for connecting the beam of the engine to the piston.

Yet another key invention described in the 1782 patent was the use of steam to push the piston both up and down—the so-called double-acting engine. This was in fact an older idea that Watt had described at the time of the 1775 Patent Extension Act. The double-acting engine raised another mechanical problem, in that the usual way of connecting the beam to the piston rod could not be used. All Newcomen and Watt engines up to the building of double-acting engines had used a chain to pull down the beam during the power stroke and another chain to connect the beam to the pumping apparatus that pulled the other end of the beam down during the recovery stroke. The problem presented by double-acting was how to transmit force through the piston rod in both directions without allowing it to deviate from the vertical. Watt's solution was the "three bar" system, which after further amendment became known as "parallel motion." The three bar system was contained in his 1784 patent (No. 1432), but again was buried among other items including other ways of obtaining straight-line motions, another rotary engine design, and a steam-driven road carriage, which was an idea of William Murdoch's. The parallel motion, which was never itself patented, was a crucial part of the solutions that enabled Boulton & Watt to produce very economical, double-acting rotative engines that could operate smoothly with varying loads and speeds. This gave such engines a great advantage in use for flexible powering of machinery.

While these engine patents of the early 1780s were important ones, they were certainly an odd collection in many ways and could be seen as a complicated strategy to maximize Boulton & Watt's control of the engine business, both reciprocating and rotative, and as giving further protection for the central invention of the separate condenser. This was certainly how Watt's close friend Gilbert Hamilton saw the situation in 1782: "I hope that your new Patents, if they should not produce much money, will have the good effect to prevent every other person from making use of your Invention & I think the money that ought now to be coming in from the Engines already erected should soon make you very easy."[125] With the phrase "your Invention" here, Hamilton seems to refer not to the inventions protected by the new patents but to the prior central invention of the separate condenser. This process of surrounding a central patent by a range of other enabling or developmental ones would become a familiar strategy in the nineteenth century for keeping competitors at bay. When further developed by commercial licensing and secondary markets in patents, this produces what is now known as a "patent cluster" or "patent thicket."

The earliest Boulton & Watt rotative engines were one at Soho, set up in 1782, and one at Wilkinson's Bradley works, set up a year later. They relied on some of the key features buried in the 1781 and 1782 patents. Watt was skeptical about the profitability of rotative engines. He believed their small scale would limit profits in comparison with the large Cornish engines, since the small engines would still be expensive to design and erect. Boulton encouraged him with the point that the market for small rotative engines was virtually endless. All they had to do was ensure profitability by limiting costs, which could be done by only offering engines of standard design and size.[126]

It was dawning on Watt that another way to reduce the construction cost of rotative engines was to deliver more power by making the engines operate at faster speeds. Here the double-acting engine was the key. In late 1782 and early 1783, Boulton & Watt built an engine of this type at Soho, which proved very powerful. The company then proceeded to build an engine for a corn mill at Soho, which was operating very satisfactorily by the summer of 1783. Watt told Gilbert Hamilton that they had ironed out remaining difficulties and could now make "very perfect rotative engines," which would, however, still be rather expensive ones.[127] But he was now ready to take on orders for rotative engines from outside customers, and in 1784 he was working on a dozen orders for them. These engines were not of a standard type.

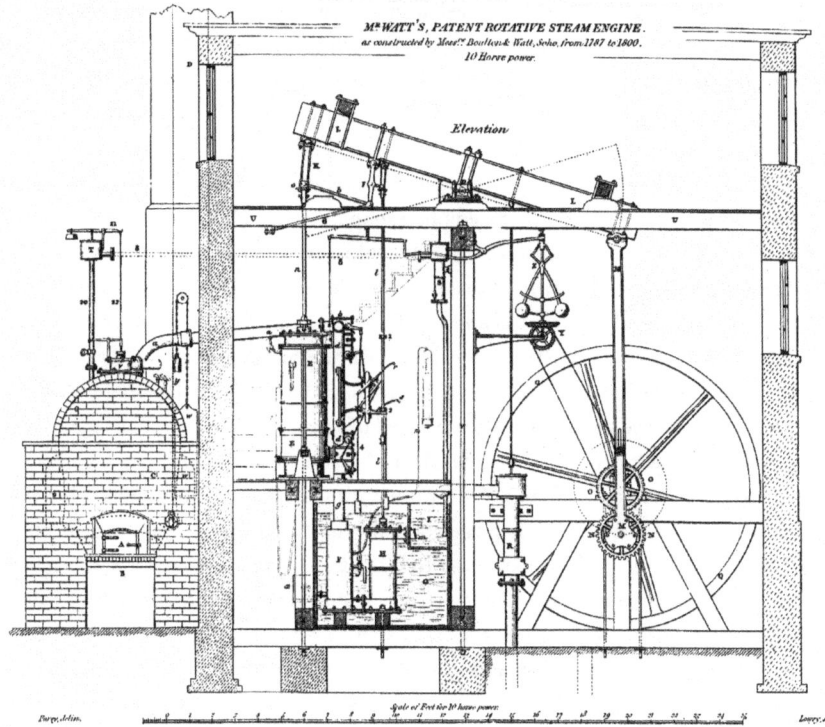

FIGURE 3.5: "Mr. Watt's Patent Rotative Steam Engine." This was the more or less standard engine sold by Boulton & Watt for driving machinery from 1787 to 1800, as represented in John Farey, *A Treatise on the Steam Engine* (London: Longman, 1827), I: plate XI. © Science & Society Picture Library.

Four of the engines were for forges (for Wilkinson and for Reynolds), three for breweries (Whitbread, Goodwyn, and Calvert), one for Wedgwood's pottery, one for an oil mill, one for a starch factory, and one other. It is notable that several of these customers were close associates of the firm. With the rotative engines, as with the early reciprocating ones, Boulton & Watt used commissions for trusted associates to further develop and test the designs of engines. Thus, for example, the engines designed early in 1784 did not use the parallel motion but the last two engines of that year did. By the late 1780s a more or less standard design had been arrived at (see figure 3.5) that would be used for rotative engines throughout the remaining term of the Boulton and Watt partnership.

One feature of the standard design, quite famous but not yet mentioned, is the centrifugal governor (also known as the "whirling regulator"), the triangular structure featured in figure 3.5 just above the sun-and-planet gear system. It consists of two pivoted arms carrying spherical weights at their ends. If the speed of the engine increased, then the weights would fly outward and upward, and, through a linkage to a valve, the engine throttle would be progressively closed, slowing the engine. The opposite process occurred if the engine ran more slowly. This meant that an engine could be self-acting in important ways. It could, for example, maintain a steady speed if boiler pressure varied or if the load of machinery being driven was changed.[128]

The centrifugal governor is another feature of the Boulton & Watt engine that, though often attributed to Watt alone, has a more complex history. In May 1788 Boulton spotted a device of this nature being used in the Albion Mill to regulate the distance between the millstones as their speed of rotation varied, and he reported it to Watt. It may have been John Rennie who had installed this device,[129] which had been patented by Thomas Mead in 1787 for use with millstones. It was quickly developed in the Boulton & Watt workshop and applied to use in opening and closing the throttle valve on their steam engines in order to make their speed and power self-regulating and render them steady and even.[130] In fact, Watt did not claim what is often referred to in later times as the "Watt governor" as his invention, but rather stated that he had improved the mechanism and adapted it to his engines.[131] Others have been less careful in their attributions.

There have been varied counts of the number of rotative and reciprocating engines that were supplied by Boulton & Watt. Counts for reciprocating engines vary between 180 and 188. Those for rotative engines are between 285 and 295.[132] The broad picture is clear: rotative engines were of great importance to success in the last fifteen years of the Boulton & Watt partnership. The single largest user of their rotative engines was the textile industry, with cotton accounting for most of these. The rest were spread over a great variety of industrial and mining concerns.

As with the pumping engines, so with the rotative ones much hung on how Boulton and Watt were to be rewarded.[133] They were to an increasing extent actually constructing and selling engines, a fact that led eventually to the establishment of Soho Foundry as a specialized engine fabrication plant in 1796. But they had to work out how they were to be rewarded for the use of Watt's patents. Rotative engines and their new market were not so readily charged for by a premium on fuel cost savings. Watt produced a scheme that relied on a more technical measure of the substitution of steam power for previous sources of power—animal and waterpower, most commonly the power of horses. This was a natural move in that mills and other machinery had often been driven by horses, and informal assessments had been made of equivalence between horses and other sources of power. On the basis of some experiments on the capabilities of horses used by the large London breweries, Watt settled upon one horsepower as equivalent to the raising of 33,000 pounds through one foot in one minute. This standard allowed comparison of the work done by horses, waterwheels, and engines. For their rotative engines, Boulton & Watt set an annual premium of £5 per horsepower *deliverable* by the engine. The charge was thus effectively for what would later be called a "standing reserve," rather than varying with the power actually delivered. In any case, until the mid-1790s when the steam indicator was developed, there was no way to measure variations in power delivered. The only concession Boulton & Watt made to the variable power requirements of industrial processes over time was in cases when an engine was not used for a year. In that case the payments were stopped. Otherwise, the annual payment was made on the horsepower deliverable by the engine. This caused some complaints. Very occasionally special deals were made, but Watt in particular was critically aware that these could undermine the payment system entirely. It would be all too easy for engine users to fabricate stories about the extent of their use. As with the payment system for reciprocating engines, almost as much ingenuity went into devising and operating payment systems, and ensuring accountability within them, as went into the engineering itself. Most of the time these systems worked at some level, although in the end the law might have to be the ultimate decider. But again, Boulton tended to be more flexible than his partner, whose Presbyterian outlook on accountability remained strong.

Rotative engines became more public machines than the Cornish reciprocating ones. The latter were located in the wilds and unlikely to be seen by many. But the location of rotary engines in major cities and towns meant that they readily became a spectacle. This was also relevant to Boulton & Watt's marketing strategy. Boulton & Watt's greatest demonstration project, but also a major headache and ultimately a major disaster, was the Albion Mill in London. The idea for a large-scale, steam-powered mill for grinding corn in the metropolis was floated in 1783. But because of considerable resistance from those who considered it too disruptive of the existing economy of flour supply, it was 1785 before a partnership began the process of building it and fitting the machinery at Southwark, near Blackfriars Bridge. Samuel Wyatt undertook the building of the mill, John Rennie was recruited to supply the grinding and dressing machinery, and Boulton & Watt was to supply the engines, which Rennie would fit. Parallel motion was employed in the engines, as was the sun-

and-planet gear system, the latter giving much trouble. Indeed, the engines required considerable development work over a period of time. The first engine was started in February 1786 and later attracted a crowd of spectators including Sir Joseph Banks, the president of the Royal Society. The second engine was not set up until 1789. The mill was a very large enterprise, with twenty pairs of stones in all, and was capable of producing prodigious quantities of flour. Steam had its advantages, some not so obvious. For example, an extreme frost in the winter of 1788 closed waterpowered mills all over England and further afield. Albion Mill sold a great quantity of flour as a result, including some to France. In one week in June 1790, sales of flour ground by the mill amounted to £6,800.[134] But the conduct of the mill itself and the financial side of the business were to cause great concern.

For the moment, though, Boulton was excited about it as a demonstration project to publicize the Boulton & Watt rotative engine business. Unsurprisingly, Watt saw the situation very differently. Writing to Boulton in 1786, he remonstrated with his partner about the number of people being shown the mill "as an object of curiosity." He was concerned that the teething troubles of the mill and its disarrangement were creating exactly the wrong impression. He wanted to avoid ostentation and advised that Rennie be warned against it on his own account: "Dukes & Lords and noble peers will not be his best customers," Watt suggested. He wanted the mill to be shut to all comers except under very strict controls. Watt believed that generating interest was better served "by making it a mystery to the many & by the external appearance of the business."[135]

The mill certainly generated valuable publicity and with time might have become profitable, though Watt doubted this, and he and Boulton were eager to unload the two-fifteenths share they had in the business. They believed that bad management was piling up losses. But in the early morning of 2 March 1791, a fire started and quickly became so intense that within half an hour or so the mill was destroyed, together with a great quantity of grain. A newspaper report of the time described the scene: "During the conflagration . . . the platform of cannon, which were kept loaded for the purpose of giving signals to the workmen, went off successively . . . and a barrel of gun-powder went off with a most tremendous explosion, and ascended in spiral streams of liquid fire to an amazing altitude, carrying immense quantities of burning wheat into the air, which were scattered profusely in the different streets of Westminster, and in St. James's Park."[136] House windows in surrounding streets cracked in the heat, and the pitch in river barges melted and ran out of the crevices. The watching crowds "seemed to rejoice at the misfortune," as they believed the mill to be an injurious monopoly affecting the trade of hundreds of millers around the country. The financial losses were very large, being calculated at £200,000, of which only £66,000 was covered by insurance. Boulton lost £6,000 and Watt £3,000, this at a time when income from the Cornwall engines had not yet amounted to much. Oddly, the fire was probably a relief for Watt, who was anticipating total ruination before they could extract themselves from the partnership.

Fortunately, other publicity was probably more productive. The engine at Samuel Whitbread's London brewery, among the early rotative engines that Boulton & Watt had installed, had attracted many eminent visitors. The king heard of it as a result and in 1787 expressed a wish to see it. Whitbread invited Watt to be present, and there was much hovering in London until a date was finally established. Watt reported to Annie on 26 May that he had been "most graciously received" that day "by the King, who has expressed himself most highly pleased with every thing he has seen here." Whitbread had organized a fine reception and "collation" for the royal party, which included three or four princesses, the Duchess of Ancaster, the Duke of Montague, and Lord Aylesford. Watt was particularly impressed by the queen, "a most agreeable personage, [who] also honoured me with much of her notice."[137] Despite his reservations about aristocracy, Watt was clearly taken with the monarchy and flattered by its attention. Newspaper reports of the visit, however, gave him no notice, mentioning the steam engine supplied by "the Birmingham Bolton" as taking up half an hour of the visit. The king, it was reported, "with becoming science," explained "the leading movements of the machinery" to the queen and princesses.[138] Farmer George was evidently up on the latest machinery. Watt was, perhaps, surplus to requirements!

Not long before this Watt had returned from a visit to the Continent with Boulton that was also concerned with marketing their engines further afield. Boulton already had considerable experience doing business overseas in the toy trade and had a network of agents, both salaried and on commission. Although the toy market and that for industrial machinery might be considered very different, Boulton tended to transpose to steam his established method of appealing to the apex of the social pyramid—particularly royalty and nobility—in the confident expectation that the rest of society would follow.[139]

In their overseas dealings Boulton & Watt were concerned not only to make sales but also to ensure that they could control future markets. The danger was that if their engines gained currency, pirates would copy them and crowd them out of the market. So Boulton & Watt were as concerned about receiving some sort of protection for their steam engine designs as about sales themselves. Without the former, the latter might backfire. In this respect also appeals to the rich and powerful were most likely to bring success. These and other difficulties with overseas dealings meant that during its first decade the Boulton & Watt partnership erected only four engines overseas, and during its second decade only twelve.[140] Interest in steam engines in the Caribbean sugar industry was perhaps anticipated, but there was little of substance in the eighteenth century. Samuel Whitbread on behalf of Wilberforce approached Boulton & Watt in 1789 about the potential use of steam in the West Indies plantations. Some thought that it might contribute to the abolition of slavery. The first Boulton & Watt engine was not sold to the West Indies until 1803, but from 1803 to 1825, 119 engines were sold there. Sugar was the largest single industry market for their engines after the cotton industry.[141] More generally, the significant expansion of overseas business came only after Watt had begun to wind down his involvement, as Soho Foundry was established and as control of the engine business passed to the next generation.

Nevertheless, these overseas contacts were important and revealing. Early attempts were made to sell engines and gain protection in France, but despite the granting of an *arête de conseil* in 1778, there was little business. The arête de conseil granted them an exclusive privilege for the construction, sale, and distribution of their engines for a period of fifteen years. But it was conditional on a demonstration of the superiority of the invention, and this for various reasons was never accomplished. Boulton and Watt themselves played "fast and loose," first making an agreement with one key official that he could be an exclusive agent for Watt engines in France, then finding excuses for breaking the agreement to supply the Paris waterworks with two large engines.

Business in France was one of the reasons for Watt's first trip to the Continent, taken with Boulton from mid-November 1786 to late January 1787. The trip involved all manner of business. Watt by then had established something of a reputation as a natural philosopher by publication in the *Philosophical Transactions* (the prestigious journal of the Royal Society of London) of work on the composition of water, which is discussed in the next chapter. In Paris he and Boulton attended a meeting of the Académie des Sciences and mixed with eminent natural philosophers. They also paid a visit to Madame Lavoisier's salon. Besides Antoine Lavoisier, Watt also held meetings with Pierre Simon de Laplace, Gaspard Monge, Gaspard de Prony, and Claude-Louis Berthollet that enabled subsequent ongoing correspondence with a number of them, which constituted an enduring and important part of Watt's rapidly expanding networks.

Watt described to Annie the hectic round of industrial, bureaucratic, and scientific meetings. Officially, the central purpose of their visit was to examine and consult on the large waterworks at Marly, where fourteen waterwheels supplied water from the Seine for the Palace of Versailles. They were being asked to advise on the best way to renew those waterworks. Importantly, however, meeting the academicians was not unrelated to their business concerns since in France the Académie des Sciences had a close relationship with the state. Watt told his former partner and old friend Roebuck: "We have contented ourselves with giving a general opinion [on the Marly works] until we come to closer terms, which cannot be before summer when the Academy shall have given their decision on the merits of the 400 proposals which are laid before them on the subject. The field being then clear, the minister can make a bargain with us if he shall then choose."[142] Thus the academicians were directly involved in the processes that would inform the minister's decision. Boulton and Watt hoped, apparently, to be brought in above the heads of the local proposers. By way of putting local competition on its mettle, Boulton and Watt also made a visit to the works of J. C. Périer in Paris, which Watt described to Roebuck as "a most magnificent and commodious manufactory for steam engines, where he executes all the parts uncommonly well." Périer was a competitor who Watt described as having assumed "the merit of our invention." But their surprise visit had set that matter straight, being as Périer described it "un coup soufflet diabolique pour lui." Boulton and Watt were being consulted by the French state and recognized as the most prominent authorities in the field. It was even mooted by the French that they might themselves move to France to serve the French state and industry. The seduction got nowhere. Neither, in the end, did Boulton and Watt's hope to secure a workable exclusive privilege in France.[143]

Attempted seductions of Boulton and Watt by overseas interests, often through scientific overtures, were a frustrating business. Boulton made it clear to Watt, who might have been more susceptible, that they were to be resisted. From Holland, Jan Daniël Huichelbos van Liender approached Watt about engines and linked them to the idea of scientific reward. Van Liender was director of the Batavian Society of Experimental Philosophy, a scientific society that had been established in the late 1760s in association with wealthy Rotterdam merchant Steven Hoogendijk (1698–1788).[144] Its central purpose was to promote the use of steam engines in the United

Provinces. A Newcomen engine was erected in Rotterdam in the early 1770s. As this was happening, van Liender was in contact with Watt and learned of his invention. Because their Newcomen engine was not faring well, the Batavian Society decided to draw more widely on expertise by setting a prize question seeking a more efficient arrangement of engine and pumping apparatus. As was customary in such a competition, a gold medal was to be awarded to the best entry.

Watt was among those to be notified about the competition, but Boulton received the correspondence because of Watt's absence in Cornwall erecting engines. Boulton made it clear to Watt how they should respond: "I believe the best answer that can be given . . . is to tell him that we are not anxious about the honour of acquiring gold medals nor of making an éclat in philosophical societies. We will not merely talk about a thing but will actually do it provided an exclusive priviledge [sic] could be obtained for us in the United Provinces."[145] Boulton replied in much these terms to van Liender on behalf of the partnership: "As Mr. Watt & I are engaged in the Fire Engine as a business or profession, we do not enter ourselves as candidates for honorary rewards, neither do we engage in any discussion upon theory of principles." But if van Liender was willing to assist them in gaining an exclusive privilege, then they were prepared to build a demonstration engine. They indeed obtained a patent in January 1786.[146]

Other agents and other privileges were sought elsewhere, in the Austrian Netherlands, the Austrian Empire, and Spain. Boulton and Watt even entered into negotiations that would have seen a joint enterprise with James Rumsey in the United States involving seeking a patent in all their names, promoting Boulton & Watt engine sales there and restricting others, with profits divided between them. But the negotiations came to nothing in the end, with Rumsey thinking that Boulton and Watt were driving too hard a bargain. Their insistence on very favorable terms derived from their conviction, almost certainly justified, that without them the difficulties of doing business overseas, particularly avoiding being swamped by "pirates," were so great that the venture would be scarcely worthwhile. They were, in any case, very busy at home both supplying engines and fighting pirates. Their decision in the late 1780s to intensify the battle with the latter meant that their interest in pursuing overseas business arrangements evaporated. Only in the years around the Treaty of Amiens (1802–1803) and after 1815 were significant numbers of Boulton & Watt engines sold overseas.[147]

Boulton & Watt and High-Pressure Steam

The Boulton & Watt partnership did not have the market for engines to itself, and others also made significant improvements in the design of steam engines. But Boulton & Watt could and did claim the credit for truly transformative invention. In one crucial respect, however, this was not the case—the development of high-pressure steam engines.

Hovering around, and sometimes acting against, Boulton & Watt were a number of characters who themselves are identified with substantial steam engine improvements. Those improvements, together with a large number of incremental ones, were to launch the high-pressure steam revolution of the nineteenth century. One of these characters was Jonathan Hornblower Jr., who began the use of higher-pressure steam in one of the cylinders of his compound engine. Even though Hornblower's machine remained a condensing engine, it began to associate high-pressure steam with opposition to Boulton & Watt. Another key figure was Richard Trevithick (1777–1833), who went into competition with Boulton & Watt along with Edward Bull. At the turn of the nineteenth century, Trevithick built the first noncondensing, high-pressure steam engines and made a number of other improvements to boilers that made such engines possible.[148] A third important character was Arthur Woolf (1766–1837), who developed compounding high-pressure engines and made many of the improvements in materials, construction, and design necessary to the viability of high-pressure engines.[149]

Watt himself had certainly considered the use of high-pressure steam to drive a piston directly, the steam being simply vented to the atmosphere rather than condensed. His patents both of 1769 and 1784 (in its accounts of the Murdoch-inspired steam carriage mechanism) discussed this possibility. But Watt was also quick to reject it. His announced reasoning was that high-pressure steam generation was a dangerous business because of the risk of explosion. But as he and Boulton developed, and sold large numbers of, their low-pressure engines, there emerged from their point of view very sound business reasons for opposing high-pressure steam. Even when the extended 1769 patent expired in 1800, Boulton, Watt & Co. , as the business was now called, retained some powerful "first mover" advantages, and Watt Jr. in his improvements of rotative engines and development of marine engines continued with the low-pressure paradigm.[150]

Very early in the nineteenth century, high-pressure engines began to appear as a genuine commercial proposition. They found their most attentive audience in Cornwall. The potentially large savings in fuel costs of high-pressure over low-pressure engines were very attractive there. Trevithick began serious development of high-pressure engines in Cornish mines in 1800, and soon the so-called "puffers" (because of the noise made by the vented steam) began to appear in the Cornish landscape. He also promoted high-pressure steam through his celebrated demonstrations. In 1801 he drove a road locomotive through the streets of Camborne; in 1808 he demonstrated a railway locomotive at Euston. He installed a high-pressure engine in a dredger in 1803 and used one to pump water from a Thames tunnel. The engine he first patented in 1802 was a small noncondensing, high-pressure engine intended for a locomotive. He had been working on this for some time.

Then on 8 September 1803 an event occurred that checked the high-pressure engine builders' momentum. The boiler of one of Trevithick's high-pressure stationary pumping engines at Greenwich exploded. The cause of the explosion was undoubtedly human stupidity rather than a failure of the machines as such: the boiler exploded because the engine was stopped without releasing the safety valve. But it was a golden opportunity for Boulton, Watt & Co., never a group to waste an opportunity to press home an advantage. Trevithick thought they would seize upon this accident and publicize it widely. He wrote to Davies Giddy in the aftermath of the explosion: "I believe that Mr B. & Watt is abt to do mee every engurey in their power for the[y] have don[e] their outemost to repoart the exploseion both in the newspapers and private letters very different to what it really is."[151]

It is in fact very difficult to find reports of this explosion in the daily press. There is a report in the *Philosophical Magazine* cited as an example by Dickinson and Titley, but it expresses confidence in the viability of high-pressure engines and boilers and the belief that they could be "worked with as much safety as those on the common principle." However, it also noted that "such an accident . . . cannot fail to intimidate some people from adopting them."[152] Periodically into the 1820s, items appeared in the press explicitly blaming high-pressure designs for accidents and also on occasion demanding that Parliament pass legislation to prohibit their use.[153]

Early in the new century, Trevithick continued to build and trial larger engines in Cornwall, but their performance was not good enough to make them a paying proposition at that point. Only later did he develop the idea that he could use high-pressure steam in a Watt-style engine and then expand it down to low pressure before condensation. As boiler design also improved, he was able by 1812 to erect what became known as the "Cornish" engine, the first at Wheal Prosper mine. But prior to this he failed financially, returning to Cornwall as a consulting engineer. Even without opposition and sniping from Boulton, Watt & Co., Trevithick's engines would not have been an instant success. Their development took some time, just as Watt's machines had.

The development of the high-pressure pumping engine from the basis provided by Trevithick and Woolf took place over a long period in the early nineteenth century, primarily in Cornwall, in what has been called a "collective invention" setting. Although they competed, the Cornish engineers did not generally patent their improvements. This, together with regular publication of engine performances, allowed for the transformation of the high-pressure engines through mutual emulation and incremental improvement.[154]

While Watt and more frequently Watt Jr. took opportunities to cast aspersions on their Cornish enemies, there was little direct competition in Cornwall since Boulton, Watt & Co. had effectively abandoned Cornish engine markets and been sealed out of them at the conclusion of the Boulton & Watt patent. We have seen that even in the 1790s their interest in Cornwall was not as a live market for new engines but as a place from which to extract what they considered owing from existing ones. Most of their engine production was of rotative engines for the manufacturing districts. That application of steam remained resistant to the use of high-pressure engines. There were no strong economic reasons for fuel saving, but instead a premium on reliable, smooth operation and little interest in investing in what could still on occasion be a dangerous type of machine. Low-pressure engines dominated in manufacturing until the mid-nineteenth century.

Some early steamboat developers used high-pressure steam engines, such as Oliver Evans in the United States and Trevithick himself, who put a high-pressure, non-condensing engine in a paddle steamer in 1812. But most early marine steam engines like those produced by Boulton, Watt & Co. were low-pressure, condensing engines. The explosion of a boiler in a boat at Norwich in 1817 led to an inquiry, which recommended that pressures be kept below six pounds per square inch.[155] Once again, the cause of the explosion had been improper use of the safety valve, but the incident led to caution about the use of high-pressure machines rather than acceptance with better policing of use. For a long time there was no reason for Watt and his successors to worry about high-pressure marine engines.

Only in the realm of railways did high-pressure engines necessarily dominate, and it was not until Robert and George Stephenson's famous *Rocket* in the 1820s, with its use of a new multi-tubular boiler, that the weight of the engine could be reduced far enough and the steam pressure raised high enough to produce a reliable machine capable of hauling a significant load. With the opening of the Liverpool & Manchester Railway on 15 September 1830, a new era of transport certainly began.[156] This was an era that, despite the overblown rhetoric that so often credited its inauguration to Watt, had little to do with his story or that of his son.

There has been considerable controversy about the relationship between Boulton & Watt and the development of high-pressure steam engines. Some of this focuses on the often harsh accusations and counter-accusations that each side threw at the other. Watt has been charged with great vehemence in his opinions against Trevithick. Not long before his death, Trevithick reflected on his life, his hard work, and the scant reward he had received. Worse than this he had been, he claimed, "branded with folly and madness for attempting what the world calls impossibilities; and even from the great engineer, the late Mr James Watt, who said to an eminent scientific character still living that I deserved hanging for bringing into use the high pressure engine; this so far has been my reward from the public."[157] The attribution of this violent expression is quite clearly directed at Watt himself since it is made to "the late Mr James Watt" and Watt Jr. lived long beyond Trevithick's demise. However, Dickinson and Titley suggested, confusingly given the dates but in other respects plausibly, that it was Watt Jr. rather than his father who made the remark. They elsewhere describe Watt Jr., with justice, as a "hot-head and given to immoderate language such as his father would not have used."[158] Indeed Dickinson was inclined to attribute to Watt Jr. most of the nastiness that went on between Soho and Cornwall from the 1790s onward. This view has merit, as we will see, but Watt was not so innocent of such dealings as Dickinson imagined.

Much of this personal animosity was driven by controversy about the patent system, then and now. Critics of that system argue that the extended patent monopoly granted to Boulton & Watt, and the thoroughness with which they defended it in the courts, acted as a brake on invention and innovation in the field. As evidence for this they point to the burst of activity in high-pressure steam immediately after 1800. But those engines were not "ready to go" until after 1800; their very viability depended upon a number of crucial developments, in boiler design especially, that only occurred in the new century. Thus, the Boulton & Watt patent did not block the implementation of high-pressure steam.[159] Boulton and Watt certainly did not encourage its development—Watt considered that he had good technical and safety reasons for his criticisms. Boulton and Watt, as we know, tangled with Jonathan Hornblower Jr., though by opposing the extension of his patent rather than blocking him with their own. They did sue Jabez Carter Hornblower, along with his partner Maberley, but that was as late as 1796. Jonathan Hornblower Jr.'s engine provided a design principle that was to be important to the full incarnation of high-pressure steam, but as a working engine before 1800 it failed, and from Boulton and Watt's perspective it did not need to be blocked.

There are, in fact, arguments on both sides of this controversy, and it is certainly true that Boulton & Watt and the new Soho Foundry–based partnership Boulton, Watt & Co. played a ruthless hand in defending their position. As we will see there were squabbles in the early nineteenth century over Watt's growing reputation between the Watt camp and the mainly Cornish opponents. The patent abolition debates of the mid-nineteenth century returned to these arguments, and, as we will also see, they have resurfaced in recent years. But the black-and-white cases presented by modern protagonists are not helpful historically. For example, it is not helpful to entirely dismiss Hornblower Jr.'s compound engine as a total failure. Nor is it helpful to paint it as a fully formed success that was obstructed by the Boulton & Watt patent alone. The truth, as even Watt's close friend Robison was to acknowledge in the first version of his article on the steam engine in the *Encyclopaedia Britannica*, was that it represented a significant and worthy improvement. In private, Watt in the early 1780s genuinely feared that Hornblower's engine would severely disrupt or destroy their business.[160] In public he dismissed its claims as derivative.

What is certain is that the partnership of Boulton & Watt proved remarkably effective—through a combination of acute invention, flamboyant marketing, ruthless business practices, and a good dose of luck—in developing newly economical low-pressure reciprocating engines and in their most enduring achievement, producing smaller, powerful rotatory engines that were readily applied to a wide range of manufacturing applications. The Boulton & Watt legacy continued to dominate in those applications through much of the nineteenth century. It was the inventions of others in high-pressure steam that produced that great disruptive nineteenth-century technology of the railways, which springs to most people's minds when steam engines are mentioned.

We backtrack now once more to the early days of Watt's engagement with steam and consider the experimental work that was to gain him some status as a natural philosopher. But as we will see, he was very much engaged in the business of natural philosophy.

Four

Watt as Natural Philosopher

The Experimental Life

In the 1780s, as the steam engine business began to bed down, Watt had some time to devote to other inventions and to the pursuit of his varied natural philosophical interests.[1] Watt himself sometimes depicted those interests as a refuge from the world of business. Undoubtedly, they did offer Watt some escape. However, we will see that Watt's pursuit of natural philosophical inquiries was driven by a range of practical projects. In this sense Watt's natural philosophy was not a refuge from the invention business but rather an adjunct and an aid to it, part of the business of natural philosophy as he had understood it since his Glasgow days. At times Watt did participate in the world of "high" natural philosophy, and yet there are strong indications that he and Boulton were not entirely at home in those circles and, indeed, regarded organized natural philosophy chiefly as another forum through which their business interests could be promoted.

During his lifetime and in the numerous accounts of his career written since his death in 1819, varied estimations have been offered of the importance of natural philosophical inquiry in Watt's life and technical achievements. There are many reasons why this question has been of interest. For Watt himself and his immediate circle, a philosophical persona and approach were claimed in part for commercial reasons, and in part for reputational and filial ones.[2] Ideological debates in the nineteenth and twentieth centuries about the process of invention and how it should best be encouraged (especially the issue of patent protection) saw the use of various characterizations of Watt as empiric or as philosopher by interested parties in those debates.[3] Major anniversaries have been occasion for the engineering profession to discuss characterizations of Watt, and in many cases to portray him in their own image.[4] Historical discussions about the driving forces of technological change, especially about the relationship between industrialization and Enlightenment, have been another forum in which the status of Watt as natural philosopher (or not) has been of interest and consequence.[5] Thus, Watt as natural philosopher is a much examined phenomenon.

In this chapter we will be reexamining periods of Watt's life that we have already traversed, especially the early period of his engine experiments during the 1760s and 1770s. We have so far treated the experiments of that period only in relation to the general strategy of Watt's path to improve his steam engine. In this chapter we will see that, in the course of these experiments, Watt arrived at more general philosophical propositions about heat, specifically about the chemistry of heat. These propositions were important to how Watt came to understand the operation of the steam engine. In the 1780s he conducted further experiments of this sort in connection with his ongoing engine innovations, especially the idea of expansive working. But these further experiments carried him, with the help of others, from the study of steam to the study of airs—the realm of pneumatic chemistry—to ideas about the composition of water, and thence into the mainstream debates of the so-called chemical revolution.

We will also see that in the 1780s and 1790s Watt pursued further experimental inquiries in relation to other practical projects that made use of his earlier chemistry of heat and further developed it. Understanding the deeper philosophical underpinnings and outcomes of this work shows us that Watt, although an episodic philosopher, did develop a much more consistent and coherent chemistry of heat than is usually realized.

The major transformation in chemistry of the late eighteenth century referred to as the chemical revolution was long characterized rather simply as an evidential struggle between the proponents of two chemical theories.[6] One of these was phlogiston theory, which explained chemical processes such as combustion and the production of metals as involving the loss or addition of the substance "phlogiston," and which had been developed by Georg Ernst Stahl (1659–1734). By the time of Watt's introduction to chemistry, phlogiston theory was very successful and taken to explain a great deal. The opposing theory, sometimes referred to as "oxygen theory" but actually involving a complete overhaul of how chemical processes were conceptualized and of claims about the centrality of quantitative analysis of those processes, was developed most famously by Antoine Lavoisier and other French chemists. It gained prominence and momentum in the 1780s with Lavoisier's 1786 anti-phlogistic essay "Réflexions sur le phlogistique," his *Méthode de Nomenclature Chimique* of 1787, and his 1789 text *Traité Élémentaire de Chemie*. Watt was caught in the middle of these disputes, in which his friend Priestley played a leading role as defender of phlogiston theory, not least by incorporating newly discovered airs into that theoretical framework, notably the air that Priestley called "dephlogisticated air." Along with Priestley, Watt became one of the last of the phlogistians, holding out against the new theory into the early years of the new century. Inevitably, during the period of the French Revolution and the French wars, chemical controversy was closely tied up with competing conceptions of how scientific inquiry should proceed and the extent to which it was a search for system or a more empirical form of inquiry.[7] Watt, like his mentor Black, was skeptical about grand systems, and for the purposes of developing his chemistry of heat he borrowed ideas eclectically from all sides in the great dispute, adopting in his chemistry, as in much else, a pragmatic approach. In fact, the story of the chemical "revolution" has itself in recent interpretations become more of a tale of hybridization and conceptual compromises than of a grand set battle in which the forces of modernity were decisively triumphant in victory.

Before I examine Watt's experimental inquiries in detail, I will recall some of the people, events, and associations that helped to construct his natural philosophical persona. I will also make a crucial point about how Watt's fame and the need for his successors in science and engineering to make him into the founding father of their ventures has distorted our picture of him as a natural philosopher.

The Making of Watt as Natural Philosopher

We have seen that Watt had no formal training in the sciences. But in the eighteenth century this was not unusual. Watt was an autodidact, but he had considerable help from circumstance and various people along the way in accumulating much natural philosophical knowledge and experimental facility.

We might start with his family, who, as we saw in chapter 1, provided an environment in which he imbibed a good deal of scientific and mathematical education. Recall that his grandfather, Thomas Watt, was a teacher of mathematics. Watt's uncle, John Watt, was also a teacher of mathematics, surveying, and navigation and also a practical surveyor. Watt, though he knew neither his uncle nor his grandfather, received a legacy of books, instruments, and professional possibility. We have seen that Watt's schooling was patchy, but his time in London with John Morgan in 1755–

1756 gave him opportunity to become aware of the metropolitan scientific community, important clients for the upper echelons of the instrument-making trades.

Back in Glasgow Watt's natural philosophical development was encouraged, as we have learned, by professors of the college, most notably Joseph Black and John Anderson, with whose lecture materials he became familiar in his role as instrument maker to the college. Black's famous discovery of "fixed air" (in modern terms, carbon dioxide) was a key foundation of the chemistry of airs that later preoccupied Watt. The chemistry of heat that Black pursued in the train of William Cullen was carried even further by Watt and formed a central motif of Watt's natural philosophy. Black and Watt were regular correspondents through much of Watt's life, and when Black died in 1799 Watt was thought of as a suitable person to act as editor of Black's chemistry lectures for publication, though he declined the task.

John Anderson is less often mentioned, except as the person who, as professor of natural philosophy from 1757, asked Watt to repair a model of a Newcomen engine used to demonstrate in his classes. Again we have seen that Anderson was important to Watt more generally as a source of books in natural philosophy, which Watt consulted during visits to the professor's home. Another influential figure during Watt's Glasgow College years was his friend John Robison. The two young men became firm companions in natural philosophy.

It was not only at the college that Watt was able to expand his philosophical, particularly chemical, understanding and experimental skills, but also through the practical ventures that engaged his experimental talents, including the scheme to produce alkali from sea salt engaged in with Black and with John Roebuck. The Delftfield Pottery business also drew upon and expanded Watt's talents as a chemical philosopher. Important above all, of course, were the events that prompted his experiments on steam made in the early to mid-1760s and then sporadically through his partnership with Roebuck and in later decades.

There was, however, an important change in Watt's character as natural philosopher as a result of his move to Birmingham in 1774. However well known Watt's philosophical knowledge and experimental talents might have been to his circle of friends and collaborators prior to his move to Birmingham, they were not in any significant sense "public." But with the move to Birmingham, they became more so. This happened in two related ways. First, Watt and the Birmingham group decided that it would be important and useful to present Watt's steam engine improvements as the product of a philosophical mind and of extensive experimentation. We have seen that there was a long "courtship" between Watt and some members of the Birmingham group beginning in the mid-1760s, and William Small became a friend and close advisor on the patent for the separate condenser in 1769. The rather abstract form that the patent specification took in emphasizing the principle behind Watt's improvements was an early indicator of the use of a philosophical persona for Watt with commercial purposes in mind. This became a routine aspect of the commercial rubric used to promote Boulton & Watt engines. While not untrue, this depiction was selective in laying to one side a wide range of practical skills that Watt exercised, and the complex measures he took, in producing his efficient engines.

Broader and at the same time more specialized recognition of Watt's natural philosophical talents occurred through his participation in the meetings of the Lunar Society. There are no systematic records of the activities at Lunar Society meetings, but Peter Jones has compiled a calendar from the evidence of primary sources encountered in his extensive archival research in Lunar Society–related archives. He finds that the society's wide-ranging discussions, and sometimes experimentation, were largely devoted to matters of natural philosophical interest to those present. They appear to have pursued natural knowledge rather than dealt directly with its relations with the industrial arts, or those arts themselves.[8] Certainly some of the "Lunatics" (Joseph Priestley after his arrival in Birmingham in 1780 is the prime example) were engaged in trains of research that became a focus of interest for the society and supplied it with much material for philosophical rumination. The same might be said in lesser degree of Watt, who enjoyed a close relationship with Priestley on natural philosophical matters in the 1780s. Watt's correspondence with Priestley provides the most extensive discussion of a clearly natural philosophical nature, much of it to do with airs and the nature of water. These topics were regularly before the Lunar Society from 1781 to 1784 and again from 1788 to 1791. Other Lunar members were also involved in this, including Erasmus Darwin. Boulton and Watt undoubtedly shared natural philosophical interests even outside Lunar Society meetings. James Keir shared research with Watt on alkali and also on inks for Watt's copying machine venture. With such encouragement, in which Priestley was of central importance, Watt began in the 1780s to aspire to contribute to "high" natural philosophy, specifically to the chemistry of airs.

Watt recognized the extent to which natural philosophy could be a *refuge* from the cares of business, as from the disagreements of politics. He put it this way in a letter to Jean André De Luc (1727–1817) in 1786, at a time when business pressures and associated travel meant that Lunar Society meetings were less frequent and he was suffering from one of his bouts of "low spirits": "I know that disease too well having been the victim of it almost all my life. The irresolution, the procrastination, the diffidence, the pusillanimity, and dread of mankind which it inspires would have totally ruined me, if I had not been saved by an order & enthusiasm in the pursuits of science, that at times forced their way through the languor to which I was subjected & commanded exertion. Lucky circumstances seconded this favourable disposition otherwise I would have sunk under my afflictions."[9] De Luc was one of the natural philosophers from further afield into whose orbit Watt moved in the early Birmingham years. He was a prominent Genevan citizen, a businessman who pursued scientific investigations in the Alps and built a large collection of minerals and fossils. After some business troubles De Luc moved to England in 1773, already well known for a six-volume geological work, and was elected a Fellow of the Royal Society in that year. On his appointment as reader to Queen Charlotte in 1775, he became a significant figure in the Royal Court at Windsor, where he served for the rest of his life.[10] De Luc's other great interest was meteorology, and it was through reading De Luc's work on that topic that Watt first encountered him. De Luc also used Black's and Watt's work on heat in his meteorology, and he became an important sounding board for Watt in his ongoing experiments on airs and the compound nature of water. In the early 1780s they worked in the laboratory together on occasion, and the De Luc and Watt families became close. James Watt Jr.'s education on the Continent was partially overseen by De Luc. Of particular importance is the fact that De Luc encouraged Watt to publish his work on airs and water and to assert his claims to priority of discovery.

Links with the likes of Priestley and De Luc fostered Watt's contact with the world of "high" natural philosophy and its ethos of the free exchange and publication of ideas. But the relationship between philosophy and business was a complex one. It is not always easy, for example, to conclude from the topic of a discussion at the Lunar Society whether or not it had industrial interest. Even Watt's experiments with Priestley on the composition of water, which may well have had therapeutic value for him and philosophical interest alone for others, also had links for him with ideas about the nature of steam and heat and their operation in steam engines. Watt's tendency was to be prompted to philosophical investigation by some practical problem and then pursue it well beyond the requirements of practicality. Losing himself in experimental inquiry was a way of getting away from the cares of the business that often prompted him to the inquiry in the first place.

Another collaborator of great importance to Watt's natural philosophical inquiries into airs was the medical doctor, former reader in chemistry at Oxford University, and radical democrat Thomas Beddoes. In the 1790s Beddoes's pneumatic medicine, which enabled people to respire airs for therapeutic purposes, became very important to Watt. The two men published together, and their correspondence is another important source in trying to reconstruct Watt's natural philosophy.

Watt also made important philosophical contacts overseas. After the trip to France made with Boulton in 1786–1787, during which he met a number of the leading French philosophers, Watt struck up an extensive correspondence with the celebrated chemist Claude Louis Berthollet (1748–1822), who played an important role in the study of chemical reactions and chemical nomenclature. Their correspondence was practically about bleaching methods—Berthollet famously developed what we call "chlorine bleaching"—but also of a more theoretical nature.

Like many autodidacts, then, Watt built and honed his natural philosophical interests and skills through interaction over many years with friends and colleagues, through his reading, and through his incessant experimentation. He gained recognition among his contemporaries as a philosopher by his election as Fellow of the Royal Society in 1785. Many years later he became a corresponding member of the French Institute, and then one of the eight Foreign Members, a rare honor indeed. But what manner of natural philosopher was Watt?

Watt as Chemist of Heat

Writings about Watt as natural philosopher were for a long time obscure. One major reason for this was that Watt's actual natural philosophical inquiries from, say, the 1760s to the 1810s were separated from the major mid-nineteenth century accounts of his life and work by both the chemical revolution (or at least transformation in chemistry) and the beginnings of a thermodynamic revolution.[11] From a mid-nineteenth century vantage point, Watt's chemical work looked confused and wrongheaded, conducted as it was under a modified version of phlogistic chemistry, that chemistry being on the losing side of history in its struggle with Lavoisier's system of chemistry. For this reason, the details of Watt's natural philosophical ideas were rarely examined. In the controversy about the history of the discovery of the compound nature of water, for example, Watt's advocates simplified and streamlined his chemistry in order to avoid embarrassment. His critics, still respectful of the "Great Steamer," were also disinclined to dwell on the details more than was necessary to question Watt's credibility as a chemist, preferring to pay tribute to him as a philosophical engineer.[12] One is reminded to some extent of the horrified response of David Brewster in the nineteenth century to the discovery of some of Isaac Newton's alchemical and hermetical writings.[13] In both cases "opening the trunk" on what were regarded as a great man's "unfortunate" intellectual peccadilloes saw it swiftly closed again and a sanitized, retrospectively reconstructed, account of his "true" achievements rolled out.

The accounts of Watt's "true" achievements as a natural philosopher took two main forms. One emphasized the connection with Joseph Black and Watt's adoption of the concept of latent heat. The erroneous impression was often given that Watt's status as a natural philosopher depended on whether or not this concept was mobilized by Watt in his improvements of the steam engine. The second version of Watt as natural philosopher also drew on Watt's work on heat but linked it directly to his and John Southern's development of a device called the "steam indicator" (and its graphic output, the "indicator diagram") in the 1790s, and so treated Watt as an early actor in the thermodynamic revolution as a kind of proto-thermodynamicist. This dubious move was possible because the indicator diagram produced by Watt's steam indicator became, in the hands of W. J. M. Rankine (1820–1872), identified with the "diagram of energy." Rankine and others during the mid-nineteenth century thus linked Watt directly with a major representation within thermodynamics. This led many to believe that Watt had somehow intuited the central truths of that science.[14] The fact of Watt's already remarked upon concern with accountability and prevention of waste both in everyday affairs and in relation to engines makes it very tempting to align his cultural Presbyterianism with its counterpart in mid-nineteenth-century Scotland. Crosbie Smith, Norton Wise, and Ben Marsden have made a convincing case for seeing such cultural concerns as allied with and indeed structurally part of the conceptual framework of thermodynamics.[15] But continuities in cultural Presbyterianism do not allow the reading back of continuities in natural philosophy. Indeed, Wise and Smith have made the case themselves that there was a major discontinuity between the guiding metaphor of late-eighteenth-century natural philosophy (the balance) and that deriving from the steam engine during the thermodynamic era.[16] Other less reflective historical work on Watt's steam engines has often teetered on the edge of treating those engines as if they were considered by Watt to be "heat engines" in that thermodynamic sense.[17]

Against these retrospective reconstructions, then, we need to explore Watt's chemistry of heat on its own terms and in its own context. Before doing that, however, I need to mention another feature of the traditional accounts of Watt's natural philosophy. They saw the *extent* of his philosophical activity as quite limited. This was a result, I think, of a failure to pursue Watt's experimental work in any but the most obvious sources. The only record of much of Watt's philosophizing is in his corre-spondence. Inspired in part by Richard Hills's work, I see Watt as an episodic natural philosopher.[18] Some episodes of work were motivated by Watt's contact with partic-ular individuals or groups, notably with a string of teachers and collaborators. Other episodes were initiated by particular projects: the Delftfield Pottery, steam engines, the letterpress copier, and pneumatic medicine. Once we read Watt's correspondence carefully in and around these projects, we find a much richer vein of natural philo-sophical activity than had previously been appreciated.

Although Watt's philosophical investigations were episodic, Watt's chemistry of heat was an ongoing natural philosophical venture pursued through and across proj-ects, and further developed through those projects. Watt was certainly not a concert-ed natural philosopher to the same extent as his friend Priestley, for example (and Priestley himself was not exactly a paragon in that sense!). But Watt was a consistent and "cumulative" scientific worker, and there was a system to his work that was sometimes hard to find in the efforts of his more prolific friend. Indeed, Watt once re-marked on some of Priestley's experiments that he had conducted them in "his usual way of groping about."[19] Some might judge Watt also as "groping about," but there was continuity in Watt's thinking from project to project. Later projects also further developed ideas originating in earlier ones. So, there was in this sense a "train" of philosophical research.[20]

　　　　※　　　※　　　※

We will see that Watt first developed his ideas about the nature of water and its relation to heat and airs by means of his experiments with steam between the 1760s and the early 1780s. He had previously learned about the tradition of chemistry of heat that had been developed at Glasgow by William Cullen and then by Black. With-in this tradition heat was regarded as a substance that engaged in chemical combina-tion with other substances. Steam was, Watt believed, a combination of water and the substance of heat. Moreover, the strength of a given body of steam (or its elastic-ity) depended upon the *amount* of heat that had gone into combination. Watt arrived independently at a concept previously developed by Black, which we call "latent heat." This was the additional heat that had to be supplied to convert water at a given temperature and pressure into steam, or more generally the heat required to produce, or released during, what we call a "change of state." Watt extended the chemistry of heat into ideas that the quantities of latent and sensible heat (heat detectable, unlike the latent variety, by the thermometer) that were combined in substances changed their chemical nature, and that, in particular, water, steam, and air were related in this way. Subsequently, and through later projects, Watt applied these ideas to the understanding of other chemical changes.

Watt's Steam Experiments

The earliest of Watt's laboratory experiments with steam of which we have documentary evidence were performed in the early 1760s as a result of puzzling over the model Newcomen engine that Watt repaired for John Anderson. These experiments may have been conducted initially at his workshop at Glasgow College, though the necessary apparatus was quite simple—a source of heat, a balance for weighing, a thermometer, and a barometer that could also be deployed as a manometer to measure steam pressures. The workshop where the steam experiments were done has been variously identified as located at Delftfield, Anderston Walk (this is perhaps a reference to Watt's first marital home on Delftfield Lane), and at or near King Street "in a little court at the north end of the beef market."[21] Watt was assisted in these experiments by John Gardner, who worked for him from about 1760 until 1772 and is sometimes described as apprenticed to him.[22]

Watt had conceived an overall approach to improving steam engine efficiency that involved increasing the production of steam and utilizing it as fully as possible to generate a vacuum and operate the piston. This meant minimizing the waste of steam in the cylinder and elsewhere in the engine. The notebook recording these 1760s experiments contains a variety of subject matter entered over an extended period.[23] When Watt returned to steam experiments in later years, he sometimes returned to this notebook, so it includes entries concerning later experiments in the 1770s, 1780s, and beyond, the last entries concerning the work that Watt did (with the help of his assistant John Southern) in the late 1790s and early 1800s to improve the results. This was with a view to Watt's contribution of notes to an intended volume, John Robison's *System of Mechanical Philosophy*, then being edited by David Brewster, which was finally published in 1822.[24] The notebook was thus a working document over many decades. Its historical reliability is problematic since there is evidence that Watt annotated, and possibly modified, its early contents in the light of later work. Nevertheless, it does give us an indication of the range of experiments that Watt undertook directed toward maximizing production of steam, minimizing its waste, and generally understanding its properties, knowledge of which was conducive to these objectives. The first experiment in the section of the notebook headed "1765" was to investigate ways to make "the boyler perform with as little heat as possible."[25]

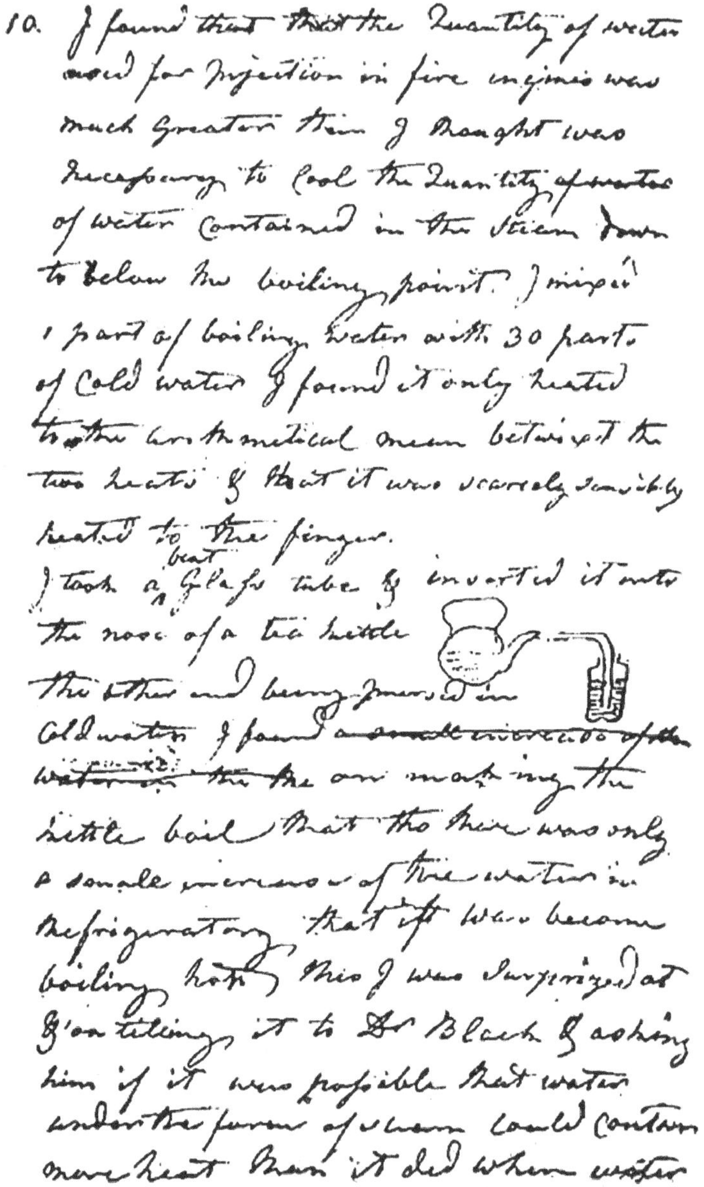

10. I found that that the quantity of water used for injection in five engines was much greater than I thought was necessary to cool the quantity of water of water contained in the steam down to below the boiling point. I mixed 1 part of boiling water with 30 parts of cold water I found it only heated to the arithmetical mean betwixt the two heats & that it was scarcely sensibly heated to the finger.

I took a Glass tube & inverted it into the nose of a tea kettle the other end being immersed in Cold water I found a small increase of the water the air making the kettle boil that tho there was only a small increase of the water in Refrigeratory, that it was become boiling hot. This I was surprized at & on telling it to Dr Black & asking him if it was possible that water under the form of steam could contain more heat than it did when water

FIGURE 4.1: A page from Watt's "1765" notebook, in which he recorded his experiments on steam. This page indicates that Watt did experiment with kettles in his maturity. By permission of Archives of Soho, Birmingham Archives and Heritage, Birmingham Libraries.

Watt conducted a simple experiment in which he compared the time taken to evaporate a given quantity of water using two different ways of heating it. The first way was to simply heat the given quantity of water in bulk in a pan. He found that it took twenty minutes for the water to evaporate completely. The second way was to deliver the same quantity of water a drop at a time to the bottom of the pan, allowing each drop to evaporate before the next was added. There had been earlier expectations that this would lead to quicker evaporation, but Watt found that the time taken to evaporate the same quantity of water by the two methods was little different. He concluded that the rate of evaporation of water varied not with the surface area exposed to heat or to the quantity of it, but simply with the quantity of heat that "enters it."

Although this is a neat experiment, the investigation of this question of boiler design was not original to Watt but had been discussed for some time by others. Ironically, those others included Matthew Boulton, Erasmus Darwin, and William Small, who were, quite independently, at the very same time addressing the same question and arriving at the same conclusion.[26] Watt concluded that the basic "tea kitchen" as used on the domestic hearth was as good a boiler as any. One constructed of thin iron and clad in wood would be ideal since very little heat would "be able to penetrate" the wood.

According to the notebook and to a retrospective account Watt wrote in 1809,[27] at this point in his early experiments, Watt noticed phenomena in his wooden-cylindered model Newcomen engine that led him to deeper investigations into the operation of steam within the engine and then to further laboratory experiments. Specifically, Watt had noticed that his wooden cylinder model consumed many full cylinders' worth of steam for each stroke of the piston, much more than might be expected, given Desaguliers's figures for the steam consumption of large colliery engines. He also found that producing a good vacuum in the cylinder by the introduction of cooling water to condense the steam was more difficult than expected. At lower pressures water boiled at lower temperatures, with the result that the cooling water itself turned to steam—a phenomenon that came to be known as "back pressure."

Watt needed to determine more accurately what was going on, but first he required a better method of measuring steam consumption. He experimented to establish how much steam a given volume of water produced by obtaining the ratio of their volumes at boiling point. Using a "Florence flask" he inserted a glass tube into it and sealed it at the neck, the bottom of the tube being a little distance above the surface of an ounce of water that he placed in the bottom of the flask.[28] He then put the flask in an oven "before the fire" until all the water had evaporated. The form of the flask and tube ensured that air was driven out of the flask as well as all the steam produced by the water except for a flask full of steam. Allowing the flask to cool, Watt weighed it and the condensed steam. Then he thoroughly dried and reweighed the flask, finding that the difference between the two weights was four grains. He now knew that this was the weight of water that had produced a flask full of steam. Knowing the weight of a flask full of water, Watt determined the volume equivalent of four grains and so found that the water, in being converted to steam, had expanded to over 1,600 times its bulk. This experiment enabled Watt to determine how much steam an engine used by simply measuring the volume of water evaporated from the boiler, providing in effect a yardstick with which the steam consumption of different designs of engine could be compared.

In trying to understand why his model engine was consuming so many cylinders of steam for each stroke of the piston, Watt's thoughts turned to the material from which the cylinder was made. The question was: How much steam would be needed to heat a cylinder made of a given mass of a particular metal to the point where the steam being introduced into it was no longer vitiated by its surroundings, so that it could fill the cylinder, be condensed, and so produce a high quality vacuum? So, Watt set about investigating the capacity of various metals to absorb heat.

There is in fact only a short entry in the notebook concerning these experiments: "A piece of iron being heated to 120° & then plunged into water at 60° was found to have heated the water as much as an equall bulk of [water] heated to 120°. Copper was found the [sic] to do the same tin is said by Dr Black only to hold the half of the heat of an equall bulk of water."[29] But we learn from other sources that Watt's experiments of this sort were probably more extensive, and that he and Black cooperated in them, with Watt perhaps taking the lead.[30] These were experiments to establish what we would call the "heat capacity" or "specific heat" of various substances. Knowing the specific heat and the weight of an engine cylinder, Watt was then able to calculate how much cold injection water would be needed to cool it to a given temperature, and to determine how much steam would be used up in raising its temperature.

Watt's experimental attention also turned to the injection water and the condensate that was drawn out of the cylinder via a pipe known as the eduction pipe. He was surprised at the quantity of injection water needed to condense the steam. It far exceeded the amount needed to cool the quantity of water in the steam. At this point he undertook the famous "kettle experiment": "I took a bent Glass tube & inverted it into the nose of a tea kettle the other end being Immersed in Cold water I found on making the kettle boil that tho there was only a small increase in the water in the frigeratory that it was become boiling hot this I was surprised at & on telling it to Dr Black & asking him if it was possible that water under the form of steam could contain more heat than it did when water He told me that had long been a tenet of his."[31] Black had earlier done extensive experimentation on heat along lines first suggested by Cullen. Focusing on the creation and melting of ice, Black had found that considerable quantities of heat not visible to the thermometer were involved in those processes. When Watt told Black that he had found similar heat invisible to temperature involved when steam changed to water or water to steam, this answered a question that Black had posed but never answered as to whether this invisible (or latent) heat was involved at 212°F as well as at 32°F. Black began a systematic rep-etition of Watt's experiments and set his abler students onto them too. Thus William Irvine and Black performed from October 1764 a series of experiments to determine what we would call the latent heat of vaporization of steam.[32] John Robison had been set to work on determining the boiling points in vacuo of various liquids. This related to another set of Watt's experiments prompted by the tendency of the injection water to boil at low temperatures in the cylinder when the pressures there were low, and so prevent the formation of a good vacuum (referred to earlier as "back pressure").

I leave for later the thorny question of precisely how, if at all, these experiments related to Watt's steam engine innovations. For the moment it is important to consider the milieu and style in which Watt conducted these early experiments. He worked within a tradition of work on heat, and with a laboratory setup that dated back through Black to William Cullen. His focus on heat experiments prompted directly by the working of the steam engine gave new impetus to Black. It must be remembered that the connection of natural philosophy with improvement was of constant concern to the Glasgow professors and their patrons, and Watt had opened a door on a new arena of research that promised practical utility. The work was collaborative. It also did not see the light of day in philosophical circles through conventional publication. If Watt had published his work at the time, then there would have been no need for us to engage in elaborate reconstructions of it from a variety of evanescent later ac-counts. But in not being primarily concerned with publication, Watt was behaving perfectly naturally within his community in the early 1760s. Black did not publish much in the conventional way; rather, he published by discussion and presentation in his university lectures. He probably began at this time to mention Watt's work in those lectures, and the important philosophical and practical issues that it raised. He certainly did so in later years. Given the large numbers of people gaining medical training at Glasgow and later Edinburgh who attended Black's lectures, and given the significant number of them who later populated the medical and scientific communities of Scotland and England, it is likely that Watt's experiments became rather well known. To the extent that the natural philosophical circle in Scotland of which Watt became a part was insular and guarded, it was for proprietary reasons. They were engaged in the business of natural philosophy.

The breakup of this little group at Glasgow College when Black left for Edinburgh and Robison left in 1770 for St. Petersburg, then Edinburgh in 1773, would have deprived Watt of immediate experimental company. His increasing volume of civil engineering work also absorbed his time. But he continued to conduct experiments, especially on latent heat. These began to acquire broader significance for him—that is to become a general study of the heat content of solids, liquids, and airs. Signs of such generality appeared as early as 1765, when he developed what became known as "Watt's Law." This concerned the quantities of latent and sensible heat in steam at different pressures. As the pressure and temperature of steam rose, so latent heat in the steam decreased while the quantity of sensible heat (detectable by the thermometer) increased. Watt's Law stated that the sum of the latent and sensible heat was always constant.[33] This was later to lead him to consider what might happen when all the latent heat was converted to sensible heat. His speculation would be that the steam would turn into air. In this way Watt's steam experiments led him into the territory of pneumatic chemistry, a field of great importance in high natural philosophical inquiry in ensuing decades.

In late 1773 and early 1774, Watt did further experiments on latent heat to examine its variation with pressure. His wife Peggy died in September 1773, and this renewed bout of experimentation was in part a refuge for Watt in his despair. But even before his great loss, on 17 August 1773 he had told William Small that he had been reading De Luc lately "and I have tried a curious experiment to determine the heats at which water boils at every inch of mercury from vacuo to air. De Luc's observations and mine agree; but his rule is false."[34] Watt had been reading De Luc's *Recherches sur les Modifications de l'Atmosphère*, published in 1772, and was evidently repeating an experiment that he had read about in that volume. The immediate context for this experiment was Watt's efforts to construct barometers, but it also helped to develop a deeper understanding of the steam engine. He informed Small: "I have some thoughts of writing a book, the 'Elements of the Theory of Steam-engines,' in which, however, I shall only give the enunciation of the perfect engine. This book might do me and the scheme good, and would still leave the world in the dark as to the true construction of the engine. Something of this kind is necessary, as Smeaton is laboring hard at the subject; and if I can make no profit, I ought not to lose the honor of my experiments."[35] This passage gives us insight into Watt's attitude to his experimental philosophy. It is important to his theory of the steam engine, though it can, and must, be expounded publicly in a way that only deals with the principles of the engine's operation and construction. Showing his command of the theory of the steam engine would bolster his reputation, and that of the engine venture, without enabling others to pirate the actual engine since the latter involved many crucial engineering details untouched by such an abstract treatment. Watt's concern with the "honor" of his experiments shows his awareness that those experiments had a general value and importance beyond their relevance to the engine. The study of chemical variability on the basis of changing heat content of substances was an area in which Watt could see the possibility of a reputation to be earned.

By the late 1770s and early 1780s, Watt's intercourse with the Lunar Society group, especially with Joseph Priestley but also with James Keir, as well as a closer relationship with De Luc further reinforced this sense in Watt. He was clearly on an expansive natural philosophical path that would extend his chemistry of heat into the heart of pneumatic chemistry, or the chemistry of airs, one of the "hot topics" of the time. In the early 1780s he was to experiment extensively on the chemistry of airs, supported from a distance by Black but involved most intensively with his new friends and collaborators Priestley and De Luc.

Watt had been puzzling about air for some time before he began serious investigations in the early 1780s. Thus, in 1774, writing to Small about his experiments on the "heats of water boiling from the open air to vacuo under different pressures," he observed that he had "still to try the expansion from 212° to 230° when the steam is not pressed, and also the quantity of air in water, when I think our theory will be tolerably complete."[36] Watt was curious about what happened when water and steam were subjected to ever-higher degrees of heat. In 1782, in a letter to Boulton he revealed his thinking about such matters: "You may remember that I have often said that if water could be heated red hot, or something more, it would probably be converted into some kind of air, because steam would in that case have lost all its latent heat, and that it would have been turned wholly into sensible heat, and probably a total change of the nature of the fluid would ensue."[37] This recollection makes sense of scattered earlier links that Watt had made between water and the production of air. Thus in March 1770, when he was designing alternative condensers for the Kinneil engine, the issue of air being mixed with the steam was a concern. There was no doubt that air was present, although Watt had changed his mind about in what quantities.[38] That Watt paid attention to the presence of air along with steam in his early engines suggests that he considered them both to be products or modifications of water.[39] The fact that he drew philosophical lessons from manifestations in his engines as well as from laboratory experiments is also noteworthy.

The excited letter to Boulton was prompted by Watt learning about, and witnessing, experiments by Priestley, who claimed to have converted water into air, thus "proving" Watt's theory. Watt reported the experiment thus:

> He [Priestley] took lime and chased out all the fixed air, and made it exceedingly caustic by long continued and violent heat. He added to it 2 ounces of water, and as expeditiously as possible subjected it again to strong heat, and he obtained 2 ounces weight of air; and what is most surprising, a balloon which he interposed between the retort and the receiver was not sensibly moistened, nor at all heated that could be observed. The air produced was but very little worse than common air, and contained scarce any fixed air. So here is a plain account of where atmospheric air comes from. The Doctor does me justice as to the theory.[40]

Watt wrote similarly excited letters to Black, De Luc, and his friend Joseph Fry at about this time. The letter to Black makes it clear how the theory for which Priestley gave him credit had emerged from Watt's experiments on the values of latent heat in vacuo and from Watt's Law regarding the changing relationship between the amount of latent and sensible heat in steam: "As steam parts with its latent heat as it acquires sensible heat or is more compressed, that when it arrives at a certain point it will have no latent heat and may under proper compression be an elastic fluid nearly as specifically heavy as water and at which point I conceive it will again change its state and become something else than steam or water. My opinion has been that it would then become air, which many things had led me to conclude and which is confirmed by an experiment which Dr Priestley made the other day."[41] These experiments of, and with, Priestley were a conceptual bridge between Watt's steam engine work and the philosophy of airs and water. They drew Watt, rather haltingly, into the world of high natural philosophy and also into controversy.

Watt, Priestley, and the *Philosophical Transactions* Papers

The most famous outcome of Watt's chemical work was his only publication to achieve a high profile in the scientific world of the time: "Thoughts on the Constituent Parts of Water and of Dephlogisticated Air," which was published in the *Phil-*

osophical Transactions of the Royal Society of London in 1784. This became the basis for Watt's claim, against those of the eccentric aristocrat Henry Cavendish and Antoine Lavoisier, to be the discoverer of the compound nature of water. The Aristotelian view of water as one of the four elements had survived well into the eighteenth century, and so to show that it was not an element was of great significance. Watt, as we have seen, was not indifferent to "literary fame"; indeed, by the early 1780s he felt he was due for such recognition. Watt thought that in his ideas about the compound nature of water he had hit the jackpot. He was embittered by the fact that he was unable to claim credit unequivocally, especially the fact that, as he believed, his ideas had been stolen. Such controversy as there was at the time was not pursued very far, but in the nineteenth century, urged on by Watt Jr., the so-called water controversy became for a time a cause célèbre. One consequence of this is that a distorted picture of Watt's ideas about water was created. In what follows I will seek to recreate those ideas and trace their sources.[42]

At 10 Priestley Road in Birmingham, a plaque commemorates the site of the residence that Priestley occupied at what was known then as Fair Hill. This was where he settled when he arrived in Birmingham as minister of the new meeting house in September or October 1780. Priestley already had a very substantial reputation as a natural philosopher, being the author of major works on electricity, optics, and pneumatic chemistry and credited with a number of important discoveries. He had been elected a Fellow of the Royal Society in 1766 and was awarded its prestigious Copley Medal in 1773. But his religious and political views, Unitarian and radical that he was, made him a divisive figure. Nevertheless, the Lunar Society welcomed him and even changed the night of their meetings from Sunday to Monday nearest the full moon in order to accommodate him. Boulton and Erasmus Darwin both assisted Priestley financially to help him build and equip a new laboratory. Although his own investigations were generally not directly concerned with the industrial arts and commercial benefit, Priestley was willing to assist those of his Lunar Society friends who were more preoccupied with such things.[43] Thus, soon after his arrival in Birmingham, he helped Boulton and Watt assess the likelihood, which Watt much feared, that some of the new airs that Priestley was discovering might substitute for steam in the production of mechanical power. This was considered a serious enough possibility for Boulton to draft up a patent specification concerning it, but Priestley offered reassurance.[44]

Despite their divergent politics, Watt and Priestley appear to have got on very well. Watt had long been interested in Priestley's work, and Priestley, less than en-

XXV. *Thoughts on the conſtituent Parts of Water and of De-phlogiſticated Air; with an Account of ſome Experiments on that Subjeĉt. In a Letter from Mr.* James Watt, *Engineer, to Mr.* De Luc, *F. R. S.*

Read April 29, 1784.

DEAR SIR,   Birmingham,
November 26, 1784.

IN compliance with your deſire, I ſend you an account of the hypotheſis I have ventured to form on the probable cauſes of the produĉtion of water from the deflagration of a mixture of dephlogiſticated and inflammable airs, in ſome of our friend Dr. PRIESTLEY's experiments.

FIGURE 4.2: Part of the title page of Watt's major excursion into public natural philosophy, in *Philosophical Transactions of the Royal Society of London*, 1784. Hathi Trust Digital Library, Internet Archive.

tirely confident in his own chemical knowledge and abilities, welcomed the scrutiny of his work by those among his new circle whom he regarded as proficient chemists, namely James Keir but also Watt.[45] Priestley became part of the Watt family circle in the early 1780s, and the two men were to experiment together on numerous occasions. Priestley collaborated with Watt (and De Luc) in experiments on heat, steam, and evaporation in 1783.[46] The two men commented closely on each other's ideas. In volume two of his *Experiments and Observations on Natural Philosophy*, the preface of which was dated March 1781, Priestley acknowledged Watt's valuable comments on his work.[47] It was in this context that Watt's paper on the compound nature of water was first produced.

The paper started life in April 1783 as a letter from Watt to Priestley intended to accompany Priestley's submission to the Royal Society of his original experiments on the direct production of air from water in an earthenware retort. Watt's letter circulated in London before Priestley discovered problems with his experiment: it appeared that the air produced in that experiment was an artifact of the apparatus being sucked into the porous earthenware retort. Priestley withdrew his paper and Watt his letter. But Watt remained confident that his ideas about the production of air from water were sound. He reworked his paper, this time in the form of a letter to De Luc. He removed reference to the retort experiments and used instead a range of other experiments by Priestley, De Luc, and others, as well as the products of his own work in the laboratory. In fact, in May, June, and July 1783, after his letter to Priestley was withdrawn, Watt conducted an extensive series of experiments on dephlogisticated air (called "oxygene" by Lavoisier), which were resumed in November of that year. These experiments are recorded in his Common Place Book, and have been little noticed.[48]

In the *Philosophical Transactions* paper of 1784, Watt most famously offered an interpretation of experiments in which dephlogisticated air and inflammable air (produced by methods that we would see as generating hydrogen) were sparked together. He described what he saw: "The first effect was the appearance of red heat or inflammation of the airs, which was soon followed by the glass vessel becoming hot . . . and as the glass grew cool, a mist or visible vapour appeared in it, which was condensed on the glass in the form of moisture or dew."[49] He found that this moisture was water, the weight of which was equal to that of the airs that had been sparked together. What had happened, said Watt, was that the two kinds of air had united violently, become red hot, and upon cooling had totally disappeared—this was demonstrated by immersing the neck of the sparking vessel under water, upon which it almost completely filled with fluid. The only products of the process were water, heat, and light. Watt offered his interpretation of what had happened:

Are we not then authorized to conclude, that water is composed of dephlogisticated air and phlogiston, deprived of part of their latent or elementary heat; that dephlogisticated or pure air is composed of water deprived of its phlogiston, and united to elementary heat and light; and that the latter are contained in it in a latent state, so as not to be sensible to the thermometer or to the eye; and if light be only a modification of heat, or a circumstance attending it, or a component part of the inflammable air, then pure or dephlogisticated air is composed of water deprived of its phlogiston and united to elementary heat.[50]

To a modern, or a nineteenth-century, reader, if we put aside the apparently unfath-
omable complications involving latent or elementary heat, it becomes easy to interpret Watt as saying that water is composed of two airs, dephlogisticated and inflammable air. Once we identify those airs in modern post-Lavoisierian terms as oxygen
and hydrogen, then we might conclude that Watt had, in effect, discovered that water
is made up from oxygen and hydrogen. Watt had, in effect, done no such thing. To
understand this we need to take the rest of Watt's paper seriously, something that
few if any accounts of the water controversy have done. Doing so enables us to make
sense of that extensive series of experiments on dephlogisticated air recorded in his
Common Place Book in 1783.

The very next page of Watt's 1784 paper offers an account that startles us out
of the above easy equation of his understanding and ours. We are reminded that his
chemical conceptual world is not one of chemical elements and compounds but rather
an eighteenth-century one of principles, affinities, and substances uniting in varying
degrees. He explains:

> It appears that dephlogisticated water, or, what may be a better name for the
> basis of water and air, the element you [he is referring here to his correspon-
> dent De Luc] call humor, has a more powerful attraction for phlogiston than it
> has for latent heat, but that it cannot unite with it, at least not to the point of
> saturation, or to the total expulsion of the heat, unless it first be made red-hot,
> or nearly so. The electric spark heats a portion of it red-hot, the attraction
> between the humor and the phlogiston takes place, and the heat which is let
> loose from the first portion heats a second, which operates in like manner on
> the adjoining particles, and so continually until the whole is heated red-hot
> and decomposed.[51]

What we learn here is that Watt considers water and air to have a common base.
"Bases," like "principles" in chemistry, are the elementary materials of the world. So
Watt is saying that water and air share an elementary principle, which he calls "humor." In using this term he is adopting the terminology that De Luc used in his writings about meteorological processes. The key point is that Watt believes that beneath
the chemical diversity of water and the different forms of air there are fundamental
unities, so that each is a variety of the other. Moreover, he notes that, depending on
the circumstances, dephlogisticated air and inflammable air (or phlogiston) can unite
in varying degrees. They can, for example, unite without forming water. He noted an

experiment of Priestley's in which they could be interpreted as forming fixed air. In this experiment iron filings and mercury calx were heated together, and fixed air was formed. If the iron filings were heated separately, inflammable air was produced, and if the mercury calx was heated separately, dephlogisticated air was the result. But when the substances were heated together, neither of these airs resulted and instead fixed air was found. Watt endorsed Priestley's suggestion that when heated together these two substances produced inflammable air and dephlogisticated air in a *nascent* state, in which condition their combination produced fixed air, not water. Watt also noted that a mixture of dephlogisticated air and inflammable air could remain inert without any combination occurring.

Watt's idea, then, was that a range of different airs and also water were products of the uniting of dephlogisticated air and phlogiston (or inflammable air) in differing degrees and by different means. Focusing solely on the passage in which Watt appears to anticipate modern ideas about the composition of water blinds us to a much more general proposition that underlies his chemistry of heat and, as we will see, unifies what otherwise appear to be quite different chemical projects.

In light of these experiments, Watt reconsidered the process of the production of air, the explanation of which, building from Watt's Law, had led him into those experiments. Now rather than seeing the production of air as being the change of state culminating from the transfer of all the latent heat in steam into sensible heat, he thought the transformation could be accounted for on better principles, namely the differential affinities of substances for humor and phlogiston: "In every case, wherein dephlogisticated air has been produced, substances have been employed, some of whose constituent parts have a strong attraction for phlogiston, and, as it would appear, a stronger for that substance than humor has; they should, therefore dephlogisticate the water or fixed air, and the humor thus set free should unite to the matter of fire and light and become pure air."[52] We can now understand why Watt spends so much of the 1784 paper dealing with the great variety of ways in which dephlogisticated air can be produced. His concern is to show that air is always produced from water via humor and never from some other constituent. He is particularly intent on showing that air is "not a child of acids," as he put it in a letter to Priestley,[53] by showing that reactions involving acids that do produce air contain as much acid at their end as at their beginning. This is why he recounts in detail a large number of chemical reactions, many of them conducted himself, with a range of chemical substances including niter, magnesia alba, calcareous earth, and minium (the red calx of lead) with mercury, sulfur, and phosphorus. To a casual reader these seem out of place and quite extraneous to Watt's purpose. But they are in fact central because he sees them as backing his claim that airs are always produced from water via humor. Watt also published a second, short paper immediately following the first one in the *Philosophical Transactions*, which essentially provided advice to anyone seeking to repeat his experiments, pointing to their delicacy and the various causes of varia- tion in the results he had observed.[54]

Those who sought to minimize Watt's contribution and achievement in these pa- pers, either immediately subsequent to their publication or in the nineteenth-century water controversy, often characterized what he had produced as a lucky speculation sparked initially by a spurious experiment of Priestley's. Rather, we see it as a sus- tained, reasoned piece of research underwritten certainly by some of Priestley's experimental work and that of others, but more particularly by a long sequence of experimental inquiry that Watt conducted himself and recorded in his Common Place Book. The scope of Watt's claims is also much greater than is usually perceived. He was offering not just a claim about the composition of water but also a general one about the production of airs. Moreover, the role of heat remained central, the content of elementary heat in reactions being a crucial determinant of what was produced. Watt was still engaged in a chemistry of heat.

We need to note here the wider body of work and controversy within which Watt's *Philosophical Transactions* papers were embroiled in order to realize that Watt's excursion into the public world of high natural philosophy left a bitter taste. The water controversy was conducted mostly in the nineteenth century, quite remote- ly from the eighteenth-century claims to discovery with which it was concerned. We will consider it when we examine the "afterlife" of James Watt. But for the moment we need to understand the rival claims that were made at the time, particularly the relationship between Watt's work and that of Henry Cavendish.

The Honorable Henry Cavendish (1731–1810) was the grandson of William Cav- endish, 2nd Duke of Devonshire. He was a notoriously shy and eccentric character, but by the 1780s already a highly distinguished natural philosopher and Fellow of the Royal Society with the discovery of what he named "inflammable air" among a number of scientific achievements to his credit. He was, thus, a key member of what we might call the scientific establishment of the metropolis.[55]

Cavendish had begun a series of experiments in 1781 upon which his claims to discovery of the composition of water were to be based. He published his "Exper- iments on Air" in the *Philosophical Transactions* of 1784 and 1785. Cavendish's announced aim was to study the "phlogistication" of common air and specifically the fate of the air that was lost or condensed in that process. His experiments dealt with a variety of instances of the phlogistication of air, including the calcination of metals, the burning of sulfur and phosphorus, and the respiration of animals. The experiments he conducted on phlogistication of common air by exploding it with inflammable air were just one other instance of the process. Cavendish was not the first to explode inflammable air. Others, like John Warltire, had found a loss of weight in the process but Cavendish did not, or at least only a very small one if any. Cavendish had previously published experiments in which he had determined the point at which complete phlogistication of common air by inflammable air occurred, finding it happened when 423 measures of inflammable air were exploded with 1,000 measures of common air. In his new experiments he found that when a mixture of these proportions was exploded, "almost all the inflammable air and about one-fifth of the common air, lose their elasticity, and are condensed into the dew which lines the glass."[56] Cavendish then scaled up the experiment so that he could produce a more substantial amount of the dew and test its nature. He found that the dew was plain water. As he put it, the inflammable air and about one-fifth of the common air "are turned into pure water," but at this stage he almost certainly meant that a process of condensation, rather than one of compounding, was happening to produce the water.

Cavendish's next set of experiments involved exploding various mixtures of inflammable and dephlogisticated airs, and he found that at a proportion of two to one the explosion produced a weight of water equal to the weight of the airs before the explosion. Once again, there are great temptations to see a straightforward modern interpretation in what Cavendish had done, but in various ways this would be mistaken.[57]

Cavendish's experiments had begun in 1781, but it was not until early 1783 that Priestley learned of Cavendish's experiment sparking inflammable and dephlogisticated airs to produce an equal weight of water. Priestley set out to repeat this experiment and claimed to have done so. He reported that he too "always found, as nearly as I could judge, the weight of the decomposed air in the moisture acquired by the paper."[58] On learning all this through Priestley, Watt ventured his hypothesis concerning the composition of water in a letter to Joseph Black on 21 April 1783: "Water is composed of dephlogisticated and inflammable air, or phlogiston, deprived of part of their latent heat." Letters to Gilbert Hamilton and De Luc carried the same news at about the same time.[59] At this time also, Watt wrote his letter to Priestley that was to accompany Priestley's paper to the Royal Society, but that letter was withdrawn, as we have seen, along with Priestley's paper because of the problem with the earthenware retort experiment.

As Watt launched into his experiments on dephlogisticated air, now with knowledge of both Priestley's and Cavendish's active work in the area, other developments were occurring overseas. Specifically, Antoine Lavoisier and Pierre Simon Laplace performed an experiment on 24 June 1783 in which they burned inflammable and dephlogisticated air in a closed glass vessel and obtained a quantity of water. A brief account of this rather hasty experiment was published in December 1783, and Lavoisier and his collaborators subsequently conducted further experiments, this time quantitative ones.[60]

Cavendish reacted to these developments by having a paper on his experiments on air read to the Royal Society on 15 January 1784. This was when the trouble between Watt and Cavendish began. When De Luc obtained a copy of Cavendish's paper, he found, as he told Watt, that Cavendish "expose et prouve votre système, mot pour mot, et on ne dit rien *de vous.*"[61] De Luc urged Watt to submit to the Royal Society his original letter to Priestley as well as a new one written to De Luc. These could be read to a meeting of the Royal Society and would amount to a prior claim by Watt on the discovery of the composition of water. While De Luc was encouraging Watt to stand up for himself in the face of apparent plagiarism of his ideas, he also advised him to be cautious since he was dealing with influential people and should not jeopardize his steam projects. Watt's response is worth quoting at length since it captures the mixture of defiance and diffidence in his reaction to his treatment, as he saw it, by the denizens of high natural philosophy:

On the slight glance I have been able to give your extract of the paper [Cavendish's], I think his theory very different from mine; which of the two is right I cannot say; his is more likely to be so, as he has made many more experiments, and, consequently has more facts to argue upon.

I by no means wish to make any illiberal attack on Mr. C. It is *barely* possible he may have heard nothing of my theory; but as the Frenchman said when he found a man in bed with his wife, "*I suspect something.*"

As to what you say of making myself "des jaloux," that idea would weigh little; for were I convinced I had had foul play, if I did not assert my right, it would either be from a contempt of the modicum of reputation which could result from such a theory; from the conviction in my own mind that I was their superior; or from an indolence, that makes it easier for me to bear wrongs, than to seek redress. In point of interest, in so far as connected with money, that would be no bar; for, though I am dependent on the favour of the public, I am not on Mr. C. or his friends; and could despise the united power of the *illustrious house of Cavendish*, as Mr Fox calls them.[62]

From what we know of Watt's character, he was capable of all the responses that he offers as alternative reactions, that is the contempt, the conviction, and the indolence. He certainly betrayed a proud independence of aristocratic power on this and other occasions. There is also a clear reference here to Watt's political sympathies. Charles James Fox's East India Bill of November 1783 had created a political storm in which eventually a new ministry was formed led by the younger Pitt. Fox had tendered his resignation but then stood as "the man of the People" in the election for the City of Westminster against the ministerial candidate. Georgiana, Duchess of Devonshire, prominently supported Fox in the election, even attending the hustings with him. It was in this context that Fox had referred to the "illustrious house of Cavendish."[63] Thus Watt's scientific and commercial defiance of that House and its representatives was tinged by his political antipathy to the radical Whig reformist cause with which it was then associated.[64] One cannot escape the view that this also showed his ongoing sense of being an outsider to the world of high natural philosophy in which Cavendish was such a major figure.

Nevertheless, Watt took De Luc's advice and trod cautiously. Having resolved to pursue the course that De Luc suggested for resubmitting his ideas to the Society, Watt had an interview with the president of the Royal Society, Sir Joseph Banks. He was concerned that the president might misinterpret his previous withdrawal of his paper and his current reaction to Cavendish's paper as tied up with the dissensions then disrupting the Royal Society. Accusations against Banks included the claim by some Fellows that he discriminated against certain classes of person, especially provincial dissenters.[65] Priestley had earlier threatened a boycott of the Society's publications, and Watt was clearly a close associate of the radical divine, though anxious not to be tarred with the same brush. The likelihood that the reading of Cavendish's paper in January was, as McCormmach puts it,[66] a "power move" in the struggle within the Society might lead to Banks seeing Watt's response to Cavendish as motivated by an oppositionist stance. Watt wanted to avoid such imputations. Besides the fate within the Society of his philosophical contribution, there were other reasons (to do with Watt's schemes for modifications of the patent system and his and Boulton's predilection for using the Society for discrete marketing) to remain on good terms with the president.

When he wrote to Banks to submit his letter to De Luc for reading at the Royal Society, Watt adopted a deferential tone and apologized for his defective writing style, which, as he put it, "must savour more of the mechanic than of the philosopher."[67] Watt's paper was duly read to the Society over more than one meeting in late April and early May 1784, and Banks relayed a message that it had met with approval. The Society's Committee of Papers ordered that the paper be printed in the *Philosophical Transactions*, and Watt negotiated its final form with Charles Blagden, who had recently become secretary of the Society with responsibility for the *Transactions*.[68] However, by the time this was done, the deadline had passed for publication of the first half-volume of the *Transactions* for 1784, in which Cavendish's paper was to appear. So, Watt's paper appeared in the second half-volume, adding to the impression that Cavendish's work was prior to Watt's. In dealing with Blagden, who was also Cavendish's assistant, Watt must have found it hard to contain his private feelings on all this, because he knew, or thought he knew, that Blagden had been active in ensuring that his ideas were plagiarized by both Cavendish and Lavoisier. This is how he put it to his close friend Joseph Fry in Bristol:

> I have had the honour, like other great men, to have had my ideas pirated. Soon after I wrote my first paper on the subject, Dr Blagden explained my theory to M. Lavoisier at Paris; and soon after that M. Lavoisier invented it himself and read a paper on the subject to the Royal Academy of Sciences. Since that, Mr Cavendish has read a paper to the Royal Society on the same idea, without making the least mention of me. The one is a French Financier; and the other a member of the illustrious house of Cavendish, worth £100,000, and does not spend £1000 per year. Rich men may do mean actions, May you and I always persevere in our integrity, and despise such doings.[69]

Apart from such complaints to friends, Watt seems to have done little to pursue the question and remained on friendly terms with Cavendish. Indeed, both Cavendish and Blagden, on one of their habitual summer tours together around Britain, visited the Soho works in early August 1785 and were cheerfully shown Watt's latest experiments on the steam engine.[70]

This initial phase of the water controversy was thus a rather mild affair. Watt did not pursue publicly the private complaints about his treatment. Perhaps De Luc was right that cultivating friends for the engine business meant avoiding making enemies

in high places. Perhaps also "literary fame," as Watt called it, had only limited appeal. We have seen him express the view that men who lived by their ideas should do
what they could to profit from them and were under no obligation to publish them to
the world. A corollary of this view would be that literary fame was itself not worth
fighting over since there was little direct profit in it. However, Watt's appearance
in public as a serious natural philosopher did have many positive effects for him. It
would certainly have assisted in ensuring his election as a Fellow of the Royal Society
in 1785, and it allowed him a respected, if shadowy, presence in the international
debates about pneumatic chemistry and the fate of phlogistic chemistry over the next
fifteen years. Only in the nineteenth century did the supposed injustice done to Watt
become a matter of extensive public debate.

Berthollet, Bleaching, and Watt's Chemistry of Heat

The next episode that contributed significantly to Watt's construction of his
chemistry of heat and airs grew out of a friendship that he formed during his visit
to Paris in late 1786 and early 1787, primarily on engine business. This was Watt's
first trip overseas and proved to be quite a heady affair. Although the engine business
bore little direct fruit, the visitors were feted by many of the leading scientific men
of Paris, including Lavoisier, Laplace, Monge, de Prony, Fourcroy, Hassenfratz, Delessert, and Berthollet. Watt famously described himself during this visit as being
"drunk from morning till night with Burgundy and undeserved praise."[71] This visit
established many lasting associations. Among the closest was that with the chemist
Claude-Louis Berthollet.

Berthollet demonstrated to his guests a method he had developed for bleaching
cloth using dephlogisticated muriatic acid air—what we know as chlorine. Watt was
very interested in this because his father-in-law, James McGrigor, was a proprietor
of bleachfields, open areas of land on which cloth was laid out so that the sun could
play its part in the traditional method of bleaching. So, Watt took careful notes of the
process and immediately wrote an account of it to Annie. On Christmas Day 1786 she
replied, with an air of skepticism: "I have wrote to my Father about the Bleaching
it is very wonderful but is your Bleacher not like the searcher after the philosopher
stone put in gold to produce gold."[72]

In this area, as in so many others, Watt was tackling a problem that had long concerned a number of his Scottish compatriots, including Cullen and Black—finding
quicker chemical alternatives to the use of repeated treatment of cloth with alkali
and its exposure to sunlight in the bleachfields. On his return Watt quickly launched

into a series of experiments, carefully recorded in his Common Place Book. They concerned methods of production of the gas and the best method of applying it, all the
time having a close eye on the economics of the process to be used. They constitute in
many ways a model of practically directed investigation.[73] Watt sent samples of liquor
and instructions for trials to his father-in-law. At the same time he sought Berthollet's permission to use his ideas, which was readily granted. Initially, Watt may have
been considering going into business with his father-in-law using the process he had
developed from Berthollet's basic approach. Watt sought to keep the process secret,
particularly his improvements, at least until a patent could be taken out. Berthollet
declined a share of the prospective patent. He was not interested in exploiting his
discoveries in business, something that Watt found puzzling. Berthollet's openness
meant that others had already learned of the process and began to offer the "secret"
for sale. Later in the year, on one of her regular visits to her father, Annie, under
Watt's instruction, undertook a number of trials. Though she was pleased with her
efforts, there were accidents with retorts and the exercise was fraught with danger.
In the end Watt managed to develop a large-scale process that could be used. But, as
Annie had anticipated, the economics of it were not transformative, and Watt was too
busy with other things to pursue it much further.

Watt's excursion into practical chlorine bleaching did, however, have other, as
yet unnoticed consequences—another burst of chemical philosophizing, much of
it recorded in correspondence with Berthollet. Once again a practical project set
Watt off on a chain of investigation that went well beyond the immediate demands of
that project. Consideration of dephlogisticated muriatic acid air further stimulated
Watt's thinking about dephlogisticated airs in general, continuing on from his earlier
investigations that had led to the Royal Society papers. Probably through Berthollet
and Jean-Henri Hassenfratz, Watt learned of the new chemical nomenclature being
developed in France, and he wrote a substantial chemical essay on the subject in response specifically to a scheme of nomenclature devised by Hassenfratz and Adet,
which was noted some years ago by David Larder. Watt shared his essay with Berthollet, and it sparked a long discussion between them.[74] While many of Watt's suggestions about changes to the Hassenfratz-Adet scheme were pragmatic ones, others
were chemically important.

Crucially, Watt wanted to extend the symbolism to emphasize the physical states
of compounds.[75] This was consistent with his long-standing belief that changes of
state were in fact chemical processes, a central tenet of his chemistry of heat. Watt
suggested a restatement of the definition of "caloric": "Caloric. Latent heat or mat-

ter of heat meaning that heat which enters into combination with substances & con-
tributes to their assuming new forms, & which is not shown by the thermometer except when a substance changes its state."[76] He offered for the symbol denoting "gas"
the following: "that modification of matter . . . not condensable by cold alone, which
state seems to be produced by the union of heat & water & perhaps light with the
respective bases, in some manner not yet perfectly explained." And "oxygene" Watt
defined as the basis, or base, of dephlogisticated air. Putting these ideas together,
we arrive at the view that dephlogisticated air is oxygene in combination with heat,
water, and perhaps light.

Watt is here maintaining a position between the old phlogistic chemistry and the
new chemistry of Lavoisier. He is accepting oxygene up to a point but wants to treat
it as a base, not as an element, and he will not abandon dephlogisticated air. He links
the two by means of his chemistry of heat. More broadly, the chemical role of heat
remains for him crucial since the combination of different quantities of heat with the
same base will produce different substances. This is an approach that Watt carried
into yet another practical project that produced a further burst of philosophizing,
which once again has attracted little attention.

Watt, Thomas Beddoes, and Pneumatic Medicine

In 1794 in Bristol, one of the rare works that bore James Watt's name as author
was published. Titled *Considerations on the Medicinal Use of Factitious Airs and the
Manner of Obtaining Them in Large Quantities*, its first part was credited to Thomas Beddoes, MD, and the second part to "James Watt Esq."[77] This publication gave the
impression that Watt's contribution was that of an engineer who had designed an apparatus for producing airs, collecting and storing them, and then administering them
to patients. This impression was perhaps reinforced by the fact that the company of
Boulton & Watt produced the apparatus at Soho Manufactory and sold it to interested physicians as well as to others keen to treat themselves. And it is certainly true
that Watt's design was ingenious, effective, and of lasting utility. Watt's design drew
upon existing scientific instruments, and his pneumatic apparatus was in turn to contribute to the basic model of the gas plants later developed by Boulton & Watt for gas
lighting.[78] But there were two senses in which Watt's interest and contribution went
beyond such engineering. First, he had an acute personal interest in the therapeutic
value of airs. His daughter Jessy had been ill with tuberculosis, an illness to which
she succumbed before the Beddoes-Watt collaboration could try to save her but not
before she had been subjected to various heroic therapies, including, at the last min-

ute, Beddoes's administration of fixed air.[79] Richard Hills sees this personal interest as Watt's chief motivation, and it was certainly important.[80] Watt was convinced that pneumatic medicine could be effective if only it could be perfected, and his use of it on his close family continued. Watt's son Gregory was to die of the same disease as his sister a decade later, in 1804, despite receiving airs as treatment. Watt also treated himself with oxygene for asthmatic symptoms and in 1796 deputed Gregory while in Glasgow to prepare airs to treat his dying grandfather James McGrigor, a treatment resisted by the old man's doctors there, and also to treat Gilbert Hamilton's asthma.[81]

In addition to his engineering and medical interest in airs, however, Watt saw the pneumatic project as an occasion to pursue more chemical experiments on the nature of airs. Such experiments might help in perfecting the medical uses of airs, but as so often happened with his research, they went beyond immediate practical concerns. As on other occasions, also, Watt did not publish this work in a way that drew attention to his novel ideas. We find it scattered through his other incidental contributions to *Considerations* and in his correspondence and private papers. As a result, this episode in which Watt exercised his talents as a natural philosopher long went unremarked and unstudied.[82]

Watt's and Beddoes's ideas concerning the nature of airs and their therapeutic activity were rather different, not surprisingly since Beddoes became a convert to the new chemistry of Lavoisier while Watt long retained his adherence to a version of phlogiston theory. This may explain Watt's enduring faith in pneumatic medicine almost to his dying day.

There were important differences in the approaches of the collaborators. While Watt's interest in pneumatic medicine was humanitarian and personal, for Beddoes it was also part of his politically radical outlook and activity. Beddoes had studied chemistry with Joseph Black as a medical student in Edinburgh in the early to mid-1780s. Beddoes maintained contact with Black as he began to lecture in chemistry at Oxford University. In the late 1780s Beddoes oscillated between support for phlogiston theory and for Lavoisier's chemistry as Priestley and the French chemists fought over the issues. But long before his collaboration with Watt, he had settled upon his own version of the French doctrine. Also long before, Beddoes had identified himself through public pamphleteering as a radical democrat, a stance that curtailed his tenure at Oxford.[83] For Beddoes, and for his political opponents, the plan of the Pneumatic Institution was deemed to be democratic in its tendencies, echoing views on the liberating potential of natural philosophy that had caused so much disquiet. Sir Joseph Banks, the president of the Royal Society, pointedly refused to support

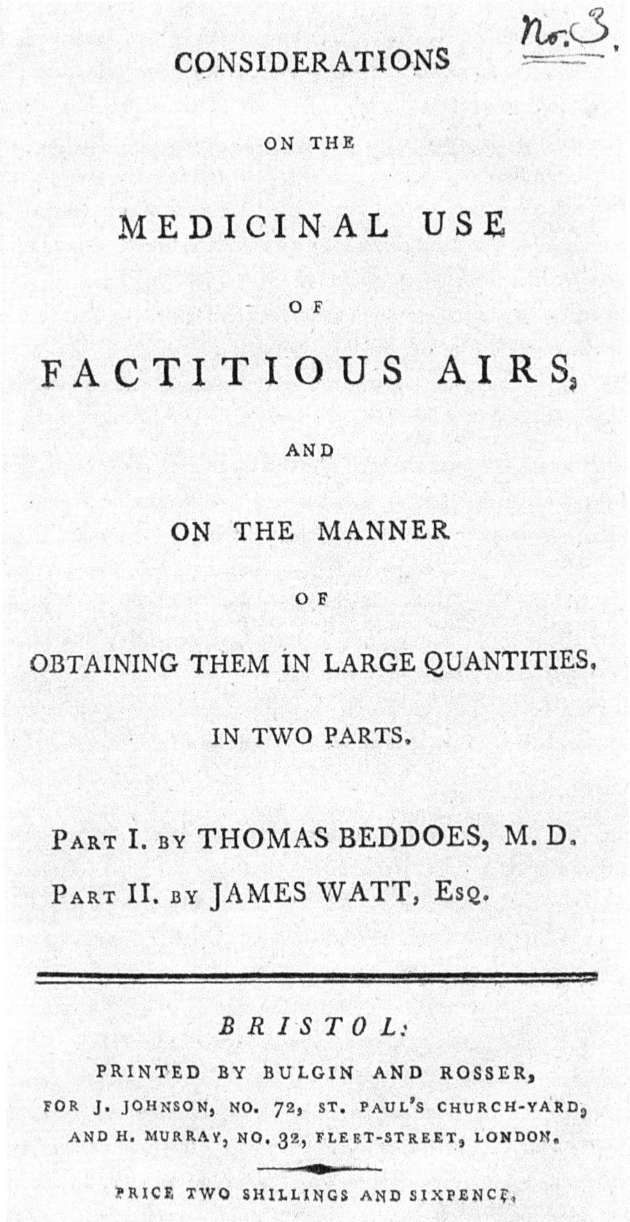

No. 3.

CONSIDERATIONS

ON THE

MEDICINAL USE

OF

FACTITIOUS AIRS,

AND

ON THE MANNER

OF

OBTAINING THEM IN LARGE QUANTITIES,

IN TWO PARTS.

PART I. BY THOMAS BEDDOES, M. D.

PART II. BY JAMES WATT, ESQ.

BRISTOL:

PRINTED BY BULGIN AND ROSSER,

FOR J. JOHNSON, NO. 72, ST. PAUL'S CHURCH-YARD,

AND H. MURRAY, NO. 32, FLEET-STREET, LONDON.

PRICE TWO SHILLINGS AND SIXPENCE,

FIGURE 4.3: Title page of Thomas Beddoes and James Watt, *Considerations on the Medicinal Use of Factitious Airs . . .*, 1794, the first of several editions. Courtesy of HathiTrust Digital Library.

the institution when approached to do so by, among others, Watt himself.[84] Watt was only just recovering from his son's long sympathetic association with French Jacobinism—Watt Jr. had returned in prodigal mode to England only a few months before Jessy's death. Watt Jr. was indignant at Banks's refusal: "The fact is I suppose he [Banks] has seen Beddoes's cloven *Jacobin* foot and it is the order of the day to suppress or oppose all *Jacobin innovations* such as this is already called. It is said to be the same spirit operating in a different way. Even the purity of my father's principles cannot absolve him from the contagion of the connection."[85]

Prominent medical friends such as James Lind apprised Watt of their opposition to pneumatic medicine because of its associations with French chemistry and politics, while at the same time celebrating Watt's experiments on airs, most lately, and ironically, stimulated by his collaboration with Beddoes. Lind wrote:

> He [their mutual friend De Luc] is much pleased to be informed by you of your having made the different Airs, without either Water, or Acids, He being no friend to Theories, either Philosophical or Political that are founded on falsehood, and are propagated by force, and such he thinks French Chemistry and Politics to be, Indeed there seems to have been a wonderful coincidence of sentiments, of some of the modern Chemists in this country that I have declined taking a part on the Pneumatical practice of Medicine from a detestation of having any connexion with such a set of miscreants, notwithstanding I am induced to believe from several reasons that in many Diseases, the practice will be of use.[86]

Watt found the situation very frustrating. His long bout of experimentation on airs in the months prior to Lind's communication was designed to uncover the nature of various airs, their differences, and how those differences might relate to their take-up in the human system. The experiments extended over the summer months of 1794, and Watt's younger son Gregory helped him much of the time, even in self-experimentation. These intense investigations ended in late October, when Watt remarked to Beddoes, "Gregory went off [for Glasgow College] last night, & my laboratory is full of potatoes, so no more airs just now."[87]

Watt's experiments were a contribution to what he and Beddoes had agreed should be an empirical project, or at the very least presented as such, precisely to try to avoid the alienation of medical men and potential patrons. But Beddoes persisted not only in invoking French chemical theory and nomenclature but also in linking the pneumatic project in peoples' minds to radical politics by pamphleteering against the

Pitt administration.[88] During a visit to London in early 1795 on patent trial business, Watt was able to assess the reaction of medical men there to pneumatic medicine and, it seems, make an effort to counter their concerns:

> Doctors in London in General condemn the practice in toto & some other people are sure it must be bad 1st because you believe in Lavoisier's theory, 2^{d} because you have the character of a Jacobin 3dly because they have found out from some expressions in your tracts on air that you are a Materialist 4^{thly} because in trying to do good some animals may be suffocated & some men get some new and unheard of diseases 5^{thly} 6^{thly} 7^{thly} for various equally good reasons best understood by themselves. To this I have answered *fas est et ab hoste doceri* that I never am too scrupulous in inquiring into the theories either in religion politicks or chemistry of those who are able & willing to do me or society any good, that I am no Jacobin nor Materialist, nor believe in Lavoisier's theory; but that I have an unaccountable tendency to believe in *facts* which pass under my own observation that though I think part of your theory false yet I see no reason to doubt that the lungs can absorb oxygene or carbone & that I am sure the vital & carbonic airs are powerful stimulants of the whole nervous system. I wish however you would observe the above maxim, ab hoste doceri, your republicanism may do more hurt by preventing the Pneumatic practice than it can ever do good, leave chemical or medical theory as much out of play as possible but ply them hard with facts—these are understood by every body & all the Doctors in London cannot overcome them. If they write against you answer them with more facts.[89]

Watt's avowed empiricism inevitably privileged his own theory-loaded facts. Yet his claim that he lacked interest in theories of religion, politics, or chemistry, focusing instead on facts and outcomes, does capture something crucial about his pragmatic approach. Watt combined, on all these fronts, a profound temperamental conservatism with a distrust of elites. Even as he chided Beddoes for his radical entanglement of the pneumatic project with politics, Watt was defiant toward medical elites. He told his old friend John Robison in 1798 that he was often accused of quackery, "by which appellation I am honoured by the enemies of innovation."[90]

So much for the context in which Watt's chemistry of heat underwent further development in the 1790s. But what were his ideas? While Beddoes concentrated on the effects of oxygene by "heightening" or "lowering" atmospheric air, Watt was more interested in substances "that may be supposed to act chemically," particularly charcoal. This was related to his occupation of a halfway house between the old and the new chemistry. Crucial was the fact—as we saw in his steam, water, and bleaching experiments—that Watt still believed in the inter-convertibility of airs. Watt's pre-occupation with the nature and role of charcoal is closely connected with his belief that charcoal either was, or was akin to, phlogiston. At one point Watt suggests that "hydrogene" is nothing more than charcoal "perfectly unified" with water.

Watt contributed to the *Considerations* in two ways. First there was his account of the construction and use of his apparatus for producing airs and delivering them to the patient. This was in part II of *Considerations*.[91] But it was not lacking remarks by Watt on the nature of airs. Watt's second contribution was a sequence of four letters written to Beddoes between June and October 1794 and included in the *Considerations* as "Mr Watt's hints on the operation of different airs."[92] In both contributions there are strong hints that Watt still considered different airs to be modifications of common air by its combination with different quantities of heat, as Partington long ago noticed.[93] In the course of discussing his apparatus, Watt described the production of "inflammable or hydrogen airs." The production of inflammable air from zinc he characterized as "the purest, or at least the lightest species of this air," while that produced from iron was of somewhat higher specific gravity. "Heavy inflammable air" was produced from charcoal, water, and heat. Watt called this "carbonated Hydrogene, or Hydro-Carbonate."[94] This air is produced by applying water to red-hot charcoal in a closed vessel, and it contains, Watt says, "inflammable air, properly so called, fixed air, separable by water or by alkalis, and some other substance, which, when the inflammable air is deflagrated with oxygen air, produced fixed air. This substance I consider as charcoal in a state of solution."[95] While here Watt treats "inflammable air, properly so called" and "charcoal in a state of solution" as different, elsewhere he speculates that light inflammable air (hydrogene) is in fact a solution of charcoal in water.

The term "solution" as used here is unusual. A clue to what he means by it comes when Watt is discussing the delivery of substances to the lungs. Some physicians suggested that powdered substances like Peruvian bark or the calces of lead or zinc might have therapeutic value in treating the lungs. Watt was skeptical but saw other possibilities: "To the use of powders, however finely *mechanically* divided, I think there are some objections; particularly I doubt whether they could enter the minute vesicles of the lungs, but if such substances can be *chemically* divided and obtained in the same state of solution in air of some congenial species, they might have their full effect."[96] The key idea here is "chemical division." This implies an association between an air and a substance that is intermediate between mechanically separable and chemically compounded. Watt is not entirely consistent in how he expresses this idea. Sometimes he uses the term "suspension" as synonymous with "chemically divided." For example, inflammable air produced by iron and vitriolic acid "always carries with it, even through water, a large quantity of iron; some of which afterwards deposits, but *very probably some part still remains suspended.*"[97] For Watt, this form of combination intermediate between mechanical and chemical was a different form of compounding. Thus, when referring to inflammable air produced from zinc, he distinguished between suspension and solution: "This air carries with it a large quantity of the flowers of zinc in suspension, which it deposits by standing at rest; it probably also contains another quantity in a state of solution, which seems to form a part of its substance, and on which some of its virtues may depend."[98] What interested Watt most were solutions of charcoal in this sense. Heavy inflammable air, or hydrocarbonate, could be stripped of its components progressively: the fixed air deflagrated with oxygene air. But when the latter was done, "some other substance" in the hydrocarbonate (charcoal in a state of solution) led to production of fixed air. Watt concluded that the charcoal in a state of solution was incompletely formed fixed air. The dissolution of charcoal or other substances in airs was conceived as a process of chemical union that could exhibit degrees of intimacy.

If the charcoal was more or less tightly "bound," then in some forms it might be more readily released. This idea was, Watt thought, perhaps crucial to the therapeutic possibilities of airs. To have therapeutic value charcoal's power of correcting putridity had to be enabled to operate in the lungs. Watt considered fixed air unpromising because in it the charcoal was tightly bound and the lungs would not be able to decompose it. But that other substance in hydrocarbonate (being charcoal in a state of solution) was more likely to be absorbable.

The larger point to be made here stands independently of whether we have managed to decipher correctly Watt's thinking about therapeutic airs. This is that Watt's chemistry of airs as he continued to develop it through this new practical project was still grounded in his earlier ideas about the inter-convertibility of airs as modified by the debates on phlogiston theory associated with Priestley. But now another set of ideas about airs as solutions has been incorporated. The origin of these new ideas is a matter for speculation. Watt's statements about degrees of intimacy of chemical union, and on processes of solution as creating compounds, are reminiscent of Berthollet's ideas in his major works on affinity. But these major works did not appear until the early nineteenth century.[99] Nevertheless, historians have suggested that the origins of Berthollet's ideas on affinity lay in his involvement in the practical projects of niter production and chemical dyeing.[100] This would make them potentially available to Watt during the course of his continuing correspondence with Berthollet in the late 1780s and early 1790s. Once again, Watt's chemistry is evolving as a consequence of his magpie-like collection of new ideas and his attempts to integrate them into natural philosophical explanations of practical experimentation.

Watt certainly saw his experiments for the pneumatic project as directly relevant to the questions at issue between the new antiphlogistic chemistry and the modified phlogistic views to which he adhered. He linked the specific project and larger theoretical issues when discussing pneumatic medicine with friends.[101] Most tantalizingly, there are signs that Watt produced a memoir on airs at this time. Just as they were finalizing the proofs for the first edition of the *Considerations*, Watt wrote to Beddoes mentioning that he had enclosed a "2d memoir upon airs."[102] The enclosure has not been located, but Watt's immediately previous letter to Beddoes, itself imperfectly preserved, contains a disquisition on airs that may have been Watt's first attempt at that memoir.[103] It seems likely that these memoirs were additional to his contributions to the *Considerations* and intended for a later edition of the work or possibly for separate publication. However, as so often before, a publication did not appear. Watt probably decided against a public disquisition so closely identified with the views of his friend Priestley, then at his most controversial and on the verge of leaving for the United States, a caution perhaps related to the fact that Boulton & Watt was at that time engaged in a long struggle for public support as it prosecuted engine pirates and defended Watt's patent in the courts.

The fragment of what is probably the first memoir on airs refers to "another proof that no air can be made without water & that it forms the principal part of its gravitating matter. . . . Airs, then, consist of water or steam united to some other substances, all in themselves unknown, except charcoal."[104] Shortly afterward, Watt clarified his point: "My opinion I have hinted at before that water in specie forms the great part of all airs, & I think the heavy inflammable air should be named Hydrocarbonate for if it can be produced in so small a mixture as 1/12 of fixd air as Dr Priestley affirms what becomes of the oxygen if the water is supposed to be decomposed[?]"[105] Watt refers here to the experiment of dripping water onto red-hot charcoal, which produces what he wants to call hydrocarbonate. If the water was decomposed in the process, then one would expect a significant proportion of fixed air to be produced. Since this was not always the case, the water must not be decomposed but rather enter into airs in specie, that is in the actual form of water. It is evident that Watt was contemplating in this putative memoir, stimulated by the pneumatic project, a contribution to the chemical debates on airs in which he was aligned with Priestley in a rearguard action against the new French chemistry.

Priestley had fled Birmingham after the Bastille Day 1791 riots, in which his house was attacked and burned, his laboratory destroyed, and his effigy consigned to the flames. He retreated to London, and in 1793 he published a work in which he argued "that water, as I have lately advanced, is only the *basis* of those kinds of air [dephlogisticated and inflammable] as well as of every other."[106] In that work, which was dedicated to his Lunar Society friends, Priestley anticipated his imminent departure for the United States as a secure haven. He and Watt remained in touch infrequently. Watt sent him a pneumatic apparatus and Priestley saw value in the medicinal airs. Priestley's *The Doctrine of Phlogiston Established* noted that Watt, "whose accuracy no person will call in question," had confirmed that the slow passage of steam over red-hot charcoal produced only inflammable air.[107] Watt may have been unwilling to enter the lists with his memoir on airs, but Priestley was happy to implicate his old friend in the final rearguard defense of phlogiston theory. In this case, as on other occasions, Watt proved a reluctant public natural philosopher.

Not many historians have taken the trouble to try to reconstruct Watt's natural philosophy. The task is difficult, and given gaps in the historical record, a deal of surmise and speculation is involved. However, we can certainly say that natural philosophical inquiry was, if not a constant preoccupation of Watt's, at least a regular, episodic, and enduring one. That kind of inquiry was generally stimulated by particular projects but often expanded way beyond their immediate needs. Our investigation of Watt's steam experiments, bleaching inquiries, and pneumatic medicine research reveals a coherence across projects that has not been noticed or at least emphasized before. He was certainly engaged in a "train of research" that for those of his contemporaries in the know would be taken as a sign of the work of a serious natural philosopher. For most people, however, Watt is of interest for, and only for, his steam engine improvements. So, I turn now to the vexed question of what role, if any, natural philosophical inquiry played in those innovations.

Natural Philosophy and Steam Engine Improvement

Whether Watt's celebrated improvements to Newcomen's steam engine depended upon science has been a long-debated issue. The question gained much impetus from broader debates from the 1960s onward about the place of science-based technological innovation in the step-change in industrial activity that occurred in Britain in the late eighteenth and early nineteenth centuries. What was the role of science in the first Industrial Revolution? A number of traditions of historical work have arisen in relation to this question.

Some historians who posed the question "What did the Industrial Revolution owe to science?" answered, "Not much"![108] A. Rupert Hall, for example, insisted that we should only grant a debt of industrial development to science if we could show cutting-edge science being applied directly to produce new technologies of economic importance. This did not happen, he contended, until the later nineteenth century in the electrical and chemical industries. Some economic historians who maintained an opposing position, especially Edward Musson and Eric Robinson, rejected this formulation and enumerated endlessly the *associations* between natural philosophical activity and industrialists. The Lunar Society was, of course, taken as a prime example of this. And within the Lunar group, Watt preeminently, but also Wedgwood, personified the connection.[109]

More recently, historians have developed the notion of "Industrial Enlightenment," following in some respects the lead of Joel Mokyr.[110] This shifts the focus from individual applications of scientific research to depiction of a broad cultural movement in which accumulated knowledge and technique together with cultures of communication (education, publications, and associations) enabled the codification of useful knowledge and its rapid diffusion through the community. From this perspective the details of Watt's, or anybody else's, natural philosophy have been less important than the experimental approach or attitude that was embodied in the individual practitioner. In characterizing the individual practitioner, Peter Jones' use of the term "savant-fabricant" as applied to Watt and some other members of the Lunar Society group, is important and perceptive.[111] I will return to this when I examine the sites of natural philosophy.

I acknowledge this approach's importance as an explanation of the dynamic of industrial change. I also see great value in the related effort to rescue the work and role of "ordinary people" in the laboratories and workshops of the Industrial Revolution, as Larry Stewart has done.[112] Nevertheless, it is important for other reasons to probe deeply into natural philosophical understandings. Mokyr's units of knowledge and technique, through which he builds an insightful economics of useful knowledge,[113] are disconnected from the living and breathing understandings of the natural and manufactured worlds of the eighteenth century. The "ordinary" natural philosophers rarely leave records sufficient for us to reconstruct their work and understandings. So, it remains important to question as fully as we can the rich documentary resources left by "major" figures. Indeed, if we understand natural philosophy properly, we can interrogate those resources more fully and begin to see the full extent to which the worlds of knowledge and technique interpenetrated in the eighteenth century.

Now to the question: To what extent, if any, did Watt's invention of the separate condenser depend upon his natural philosophy and more narrowly on his chemistry of heat? It has to be said that there was no direct connection. The idea that Watt's rediscovery through his steam experiments of Black's concept of latent heat had anything *directly* to do with the separate condenser is quite mistaken, though a connection was maintained by some early historical treatments.[114] Many who have recognized this lack of connection have nevertheless maintained that the earlier-established concept of specific heat was important, and indeed I have done this myself. The argument here was that Watt reasoned that the separation of the condenser and the main cylinder of the engine meant that less heat was wasted because the process of condensation of steam by injection of cold water did not affect the temperature of the cylinder. So, it was no longer necessary for that temperature to be restored in the next cycle. If indeed reasoning about specific heats *was* important to the invention of the separate condenser, then at least a natural philosophical concept was involved. However, it seems that the importance of the concept of specific heat has also been overestimated. Jim Andrew has apparently pushed any explicit theorizing about heat out of the picture in his treatment of the origins of the separate condenser.[115] Andrew shows that, given the size of the cylinders in both Newcomen and Watt engines, and given the engines' speed of operation, it is not convincing to find the main cause of inefficiency in the Newcomen engine in the continual heating and reheating of the cylinder. He points out that in that sense the model Newcomen engine that Watt was supposedly fiddling with leading up to his inspiration for the separate condenser would have given a greatly exaggerated impression of the importance of that particular source of waste in a full-scale engine. Rather, there were many other causes of the "waste of steam."

The danger in Andrew's analysis, from the perspective of historical method, is that he uses modern thermodynamic understandings to establish the "real" efficiencies of the early Watt engines and then, by assessing what efficiencies might be expected (again using modern insights) from various aspects of the engines' operation, identifies the "real" sources of efficiency improvement. To say that the real source is the fact that the Watt engine operated over a wider temperature range between steam and condensate than Newcomen engines did is to give a modern thermodynamic interpretation. Of course, this was not and could not be known to Watt *as a thermodynamic explanation.* As Andrew notes, Watt and Boulton did make systematic measurements of the temperature of the condensate in their own and Newcomen engines. This would indicate to them variations in the degree to which steam was being wasted and hence the relative overall economy of the engines, but it did not in any sense *explain* those variations. Any such explanation lay in the chemistry of heat.

The question then becomes: What did they know about the possible sources of wastage? Boulton and Watt were undoubtedly aware of a number of sources of wastage. In particular they were aware that a variety of happenings in the cylinder were responsible for such waste, of which the injection of cold water into the cylinder was only one. Others, as Andrew explains, were the presence of condensate in the cylinder, exposure of the cylinder and piston to cooler ambient temperatures at certain parts of the cycle, and leakage into the cylinder of the standing water used above the piston to seal it. These were problems that would apply to the Watt engine also if it were distinguished from the Newcomen only by the separate condenser. Knowing these sources of wastage of steam, a variety of detailed engineering solutions were implemented to reduce them. The claim is that these engineering solutions collectively, rather than simply the separate condenser itself, were crucial to the advantage of the Boulton & Watt engines.

What do we make of all this so far as natural philosophy and the steam engine are concerned? It is clear that any stories told by historians in which heat theory directly prompted the improvements responsible for the Watt engine efficiencies are mistaken. The range and variety of improvement undertaken smack more of the empirical tinkering account given by many economic historians, although it is also clear that Boulton and Watt, with their systematic measurements of pressure and temperature in assessing the performance of engines, were taking a "scientific" approach to that tinkering and giving themselves a crucial yardstick by which to measure engine performance. What about the stories of the invention told by Watt himself?

The main account that we have from the horse's mouth, Watt's so-called Plain Story of his invention, is obviously an important document. The Plain Story's emphasis on dealing with steam wastage, and the concept of the "perfect engine" that it invokes, can be safely taken on board, providing always that we avoid eliding Watt's "perfect engine" into nineteenth-century notions of "ideal heat engines." (Such avoidance is necessary because that term too easily carries with it a misleading attribution to Watt of the basic frameworks of thermodynamics.)[116] But in other respects the Plain Story is either based on delusion on Watt's part or on deliberate "simplification" of the story. Placing the repair of the model Newcomen engine at center stage may be an example of not allowing the facts to get in the way of a good story. Watt told the Plain Story not in some dispassionate context but in the course of patent contests during which an effort was made in various ways to wrap the Boulton & Watt innovations in a cloak of scientific respectability. In 1794, in the midst of the patent contests, an innocent inquiry from Lord Lansdowne to Matthew Boulton about domestic heating stimulated a remarkable reply explaining that "Mr Watt & myself have been persuing a constant series of Expts upon Fire, Water & Steam for 30 years past," which had resulted in various original discoveries. Boulton then claimed: "Upon the foundation of this knowledge we have made many important improvements on steam engines."[117] Boulton was engaged in marketing here, taking the opportunity to impress upon an important personage a view of the Boulton & Watt innovations as deriving directly from philosophical inquiry. There are plenty of other examples of "team Watt" treating natural philosophy as a marketing tool. It would not be surprising if in the Plain Story Watt chose not to elaborate the great variety of improvements in response to numerous ways that steam was wasted but instead concentrate on a single idea, the route to the separate condenser, that provided an easily digested picture of experimentation on the model engine and the inspiration of the separate condenser. This story supported the image of Watt that distinguished him as an engineer guided by natural philosophical inquiry, and it disguised the importance of the more mundane reality of engineering improvement that Watt and his steam engine competitors shared.

Even though the springs of Watt's invention may be only indirectly informed by his chemistry of heat, the latter remains important to understanding how Watt thought about the motive force of the steam engine. This would be true even if his natural philosophical understandings had nothing whatever to do with his efficiency improvements. In that sense, the chemistry of heat is vital. It underlies Watt's understanding of the origins of the elasticity of steam and of variations in that elasticity. If we were able to question Watt about *why* the steam in an engine does what it does, then the fact that it was a compound of water and heat would figure centrally in his response. If we could interrogate him on the case of the expansive working of a steam engine, then there too his chemistry of heat would be crucial, including in that case the concept of latent heat.

The Sites of Natural Philosophy

The laboratory is clearly an important site of natural philosophy for Watt, as for other practitioners. In Glasgow in the early days, Watt went to some lengths to ensure that he had a laboratory at home as well as at the college and, after 1763, at other separate premises in central Glasgow. In Birmingham he maintained a chemical laboratory in Regent's Place and also from 1790 at Heathfield House. But what about other sites? One is the "out of doors." In the eighteenth century there is a long tradition of linking steam in engines with the role of steam in meteorological phenomena. I will not pursue this in detail here but just give an indication of a tradition of inquiry that Watt was almost certainly part of.[118]

Various authors in the eighteenth century, including Bernard Annely and J. T. Desaguliers, had discussed meteorological phenomena and the operation of steam engines in analogical ways, usually taking insights from steam engines and applying them to meteorological understanding. Watt and De Luc experimented together on steam in the early 1780s, and De Luc used the Watt/Black chemistry of heat and analogies with steam engines to explain processes of evaporation and consequent meteorological phenomena. The famous geological ideas of Watt's friend James Hutton were also partly inspired in all probability by analogy between steam engine processes and the Earth machine. When Watt repeated many of his steam experiments in 1803 with the assistance of John Southern, he was keen to develop as full an account as possible of the elasticities of steam. In the course of this work they investigated whether the powers of steam might be the cause of earthquakes.[119] Steam in its out-of-doors operations was an integral part of the way that Watt thought about steam. The apparent success of his chemistry of heat in accounting for other phenomena out of doors would have boosted his confidence in its value in the laboratory.

Engines aside, out-of-doors phenomena were important to Watt's theorizing about heat more generally. There was a great craze for ballooning in the early 1780s, and the Lunar Society was keen to learn of experiments conducted abroad and in Britain, as well as conducting experiments themselves. Characteristically, Watt was dismissive of the sensationalist aspects of human flight in balloons but keen to understand and exploit the art for scientific purposes. Writing to De Luc on 2 February 1784, for example, Watt discussed the burning of straw to generate the hot air for Montgolfier's large balloon. The reported quantities involved had strange implications either for the heat that could be generated from straw or for the value of the heat capacity of air. Perhaps the reports were wrong, he conceded, but,

> If the thing be literally fact, it entirely demolishes the origin of heat from capacity and will I believe furnish the means of making an Engine in which heated air would be the acting power, and which would be at least on a par

with steam. I therefore wish that these observations of mine may remain locked up in your own heart until I shall get some experiments made on the subject and see whether we have anything to hope from it. . . . Making all possible allowances, I cannot see that the capacity of air for heat cannot be greater than that of water weight for weight so thus the Doctrine of capacity must fall to the ground and the expts. which have been made on it appear to be mere chimeras. But that of Latent or elementary heat shall be exalted and glorified.[120]

In this case Watt runs off excitedly from reports of a full-scale balloon flight, through calculations to conclusions, admittedly tentative, about basic theoretical issues in the theory of heat. Watt was always looking to find philosophical significance in large-scale practical ventures. Having informed De Luc of some experiments that he was trying, Watt asked for De Luc's opinion "as soon as you can, it seems to me to open a new peeping hole into the mysteries of nature."[121]

There was a third site of natural philosophical activity, another "peeping hole into the mysteries of nature" so far as Watt's natural philosophical practice was concerned; this was the manufactory itself. Early natural philosophizing about steam hopped promiscuously between laboratory work, observations of natural phenomena "out of doors," and the prosecution of the practical arts. There is work still to be done on Watt's natural philosophy of steam—at the very least we need to reassess the possible occasions for insight that Watt experienced and allow that such occasions may not have been limited to the laboratory.[122]

Let me give a specific example of natural philosophical "traffic" apparently moving from industrial site to laboratory. In early 1783 Watt was considering that set of experiments undertaken by Priestley which were described as "turning water into air." Watt explained these experiments in a letter to Black: "Water being put in a gun barrel and distilled over slowly gives no air but on being confined by a cock and let out by puffs it produces much air, which agrees with my theory and also coincides with what I have observed in S[team] Engines—In some cases I have seen [the] 10th of the bulk of the water of Air extricated or made from it."[123] Watt must have communicated these observations to Priestley, too, since the Reverend philosopher recalled, in a published account, a conversation with Watt about the possibility of converting water into "permanent air" in which Watt had observed that "some appearances in the working of his fire engine had led him to expect this."[124] It is possible that those "appearances" were the same as the great amount of air that Watt had noticed in his condensing apparatus in 1770. In Watt's early engines, Richard Hills notes, Watt made "ample provision for the removal of air and water with triple pumps."[125]

Exactly what was being said here about the processes of the supposed production of air both in the laboratory and in real engines is puzzling, and Priestley's experiments were certainly problematic.[126] But the key point to be drawn from this example is that we should not look only to the laboratory to find Watt's natural philosophical practice. It appears from this example that Watt had stored away in his mind his observation of a strange and suggestive phenomenon in the operation of a full-scale engine, and that subsequent experimental findings caused him to recall it as relevant evidence in relation to a natural philosophical question. More generally, we should treat Watt's work in devising, erecting, and operating steam engines as "experimental" in this sense and at least potentially informing his chemical thinking. There is likely much of interest to discover from manufactory, or workshop, sites about natural philosophical thinking.

Larry Stewart, in rescuing the unknown technicians of laboratory and workshop, has observed: "ordinary workers assisted manufacturers, unknown technicians made possible experiments, and common craftsmen brought to the bench their daily experience to illuminate the confusions of philosophers."[127] But it was not only the "common craftsmen" who did this. As Ben Russell has emphasized, Watt too was a craftsman, if a rather "uncommon" one.[128] In fact, he had multiple technical personalities and embodiments, and they were not strangers to each other. It was not just the common craftsmen who drew on daily experience of other sites. Surely it is plausible that a philosopher like Watt illuminated his *own* confusions and those of his philosophical friends by bringing daily experience to the laboratory bench, and also, importantly, by carrying the philosophical problems and agenda of the bench with him into his daily practical experience.

Watt did not draw a sharp line between laboratory and other spaces within which natural philosophical insight might be had. We will see now that Watt did not treat the social worlds of business and natural philosophy as separate spaces either.

Watt and Organized Natural Philosophy

In his own lifetime Watt certainly gained recognition as a natural philosopher. That recognition came, for example, by his election as a Fellow of the Royal Society of London in 1785 following closely upon publication of his experiments on the composition of water in the *Philosophical Transactions*. But for Watt, other considerations were perhaps at the front of his mind. A number of the Lunatics were, in fact, elected FRS at about the same time. In 1785, every member of the Lunar Society circle (except one) who was still alive, still active among the group, and not already a Fellow was elected FRS.[129] Their election as a phalanx came when the Royal Society seemed likely to take on a central role in adjudicating patents of invention. Watt was deeply involved in this affair and building strength within the Society to support his plans. In 1785 Watt drew up a document titled "Heads of a Bill to Explain and Amend the Laws Relative to the Letters Patent." In this and his "Thoughts on Patents," Watt proposed to remove the decision on the adequacy of a specification from the legal arena and assign it to expert committees consisting of three FRSs and two artisans.[130] This was deeply self-interested activity given Boulton & Watt's heavy reliance on patents and the anticipation that the Society would play a central role in arbitrating patent disputes. Here, plainly, was positioning by Watt and friends for potential use of the Royal Society for trade. Boulton & Watt's reliance on witnesses who were FRS in its defense of Watt's engine patent in the 1790s is another, related, example of the use of the authority of science for business purposes.[131]

From Watt's and Boulton's points of view, the Royal Society represented a marketing opportunity, as did the publication of philosophical work. We have seen that as early as 1773 Watt told his Lunar friend William Small that he was contemplating writing a book titled "The Elements of the Theory of Steam-Engines": "This book might do me and the scheme good. And would still leave the world in the dark as to the true construction of the engine. Something of this kind is necessary, as Smeaton is labouring hard at the subject."[132] The intention was clearly to garner reputation without sharing his knowledge of the subject so fully that it would be bad for business. The same idea lay behind discussion of another possible use of the Royal Society in 1776, when Boulton wrote to Watt:

> Perhaps it might not be impolitical if you were to publish a paper in the Philosophical Transact[ion]s . . . intimating that We have a variety of Engines invented very diff[eren]t in their construction some where the piston is press[e]d upwards & without a great Beam others where there is a constant vacuum under ye piston &c &c &c; & that you have annexed a drawing of one (. . . . .) w[hic]h is erected at Bedworth & that it doth so & so with such a quantity of Coals. . . . The Curve of Boiling points under diff[eren]t pressures will do you honour if you think it prudent to publish it. I w[oul]d explain ye Engine & things but little further than most Philosophers may do by inspecting an Engine Intimate that great Mechanical difficulties have occur[re]d but that

we have now conquer[e]d them & render[e]d the Engine less liable to be out of order than a Com[mo]n one. I think the best & most reputable advertisement w[oul]d be a Paper in the Ph[i]lo[sophica]l Transactions.[133]

In 1780, after Watt had developed his letter press copying machine, Boulton market-ed it fervently among high society. In May that year he demonstrated the machine to a meeting of the Royal Society, with Banks in the chair. Boulton reported to Watt on the occasion, betraying the knowledge that he was engaged in a ticklish business. The machine, he said, "afforded much satisfaction to a crowded audience" but he "did not show the list of subscribers and proposals, nor dishonour philosophy by trade in that room."[134] I suspect that he had the proposals ready for viewing, and the subscrip-tion list ready for signing, right outside the door!

When challenges or inquiries came from outside his immediate circle concerning publication, Watt showed his attitude to such questions. He was contacted by the philosophical "intelligencer" Jean Hyacinthe Magellan, who was writing a work dealing with the history of discoveries about heat.[135] Watt not only pressed Black's claims to be the original "inventor" of the theory of latent heat but also defended the latter's decision not to publish except through his lectures: "You seem to consider it as wrong in Dr. B not to have published his discoveries as he made them—in that I cannot agree with you. Every man who is obliged to live by his profession ought to keep the secrets of it to himself so far as is consistent with the use of them: It is only People of Independ‍ᵗ fortunes who have the right to give away their Inventions without attempting to turn them to their own advantage."[136] The way Watt puts this is interesting. He is not saying that keeping the secrets is *justified* but that it is in the nature of a moral *obligation*. This suggests that Watt's views were informed, in a way we have yet to appreciate, by moral and even religious considerations about how society and commerce should operate. Watt advised Magellan, however, that Black was not motivated in his noncommunication of discoveries by any prospect of gain but more by "the love of peace and quietness" and by a concern to perfect his discov-eries before publishing them. For his own part Watt observed modestly, "My own experiments in that way . . . are few and unconnected and I have at present no violent desire to shine forth in the page of Literary fame."[137] The desire may not have been violent, but it was real enough. Watt betrayed a different attitude, one of growing impatience, to Black himself. During the to and fro regarding Magellan's book, Black had said to Watt, "My present intention is to publish myself next Summer."[138] Watt's response now was firm: "If you do not publish this Summer, I will, as I will not longer run the risk of having the little pretension I have to literary fame run away with and I think I have completed the theory of the causes of the different heats at which [water] boils under various pressures and have found the true series of the Curve."[139] Though Watt was clearly resolved to publish, this no more came to pass than did his earlier promised book on the theory of the steam engine. Presumably, more pressing business issues got in the way.

What are we to make of these exchanges with Magellan and Black? Watt's response to Magellan is revealing, first because Watt does not seem to distinguish between scientific discoveries and inventions; he maintains a view that the inventor of either, in the absence of independent fortune, has a clear moral right, even obligation, to exploit his invention without necessarily publishing it to the world. Black is a natural philosopher, but he is also by trade a professor, and the goods of his trade are his lectures, which students within the Scottish system paid a fee to attend. Thus, Watt argued, it was entirely proper for Black to release his discoveries only through his lectures. Of course, "literary fame" could also be profitable, though in natural philosophy usually in a more indirect fashion. Watt put it to Magellan that for men like himself the love of fame was a failing: "The man of Ingenuity from that weakness which is called the love of fame frequently publishes what he ought to conceal; for which the world is obliged to him; but the man of fortune seldom bestows his money with so liberal a hand."[140] The key point is that Watt did not see anything wrong with the use of natural philosophy as a means of trade.

Indeed, closeness between philosophy and trade was a positive thing for Watt. Back in 1769 when still in Glasgow, Watt had met "a truly chemical Swiss dyer" by the name of Chaillet, who was helping Watt with his German so that he could read a book he had just acquired about the mines of the Upper Hartz region. Watt described Chaillet to William Small: "He is, according to the custom of philosophers, *ennuye* to a great degree, but seems to be more modest than is usual with them; and, what is still more unusual, is attached only to his dyeing, though he has a tolerable knowledge of the rest of chemistry."[141] Although Watt was to claim the philosophical mantle for business purposes, he clearly regarded the effete, impractical philosopher with some contempt, being impressed by Chaillet for his relative lack of such characteristics.

There is another, quite startling, much later episode in which the Watt circle used the Royal Society and its journal, the *Philosophical Transactions,* to promote trade. The chief protagonist here was James Watt Jr., but Watt was entirely complicit. Watt Jr. had diversified the company's activities into marine steam engines and into the production of town gas for illumination. Leslie Tomory has shown that the firm of

Boulton & Watt was the lone success story among those who were attempting to commercialize gaslight in the first decade of the nineteenth century.[142] Part of the reason for their success, Tomory shows, was their ability to publicize their work. Boulton & Watt needed to promote its venture because competition had emerged in the form of Frederick Winsor, who was applying for patents for gaslighting apparatus and in the form of a company that intended to use Winsor's inventions to commercialize gaslight. Faced with this challenge Watt Jr. hatched a stunning plan. A paper on the industrial art and economics of gaslight would be written, nominally by William Murdoch. (We have met Murdoch as an engine erector for Boulton & Watt in Cornwall and an inventive Watt collaborator. We will later consider their relationship in detail. But for now we need note only that Murdoch participated in the development of gaslight). The paper supposedly by Murdoch would actually be produced by Watt Jr. himself. Through the representations of his father and Matthew Boulton to Sir Joseph Banks, the paper was to be read by Banks at the Society, published in the *Philosophical Transactions*; then, to top it all, Murdoch was to be awarded the Society's Rumford Medal for his invention. Winsor had been trying to claim credit for the invention himself. Watt Jr. intended that gaining the Rumford Medal for Murdoch would make Winsor appear as a pirate trying to steal Murdoch's true desserts, and so ensure that Winsor's patent application and the related business venture would be stymied. The award to Murdoch would also clothe the whole venture with scientific credibility.

Even more remarkable than this plan is the fact that it all came to pass exactly as intended. Watt Jr. wrote the paper, it was read by Banks at the Royal Society as if it was a paper by Murdoch, it was published in the *Philosophical Transactions* in July 1808 under his name, and Murdoch was awarded the Rumford Medal at the Society's anniversary meeting in November. Watt himself confirmed the importance of the *Philosophical Transactions* article to the business plan in 1809 when the opposition were seeking to promote their bill before parliament. He wrote to his lawyer and friend Ambrose Weston: "It seems to me that the best way to oppose this Bill would be for B&W to advertise that they furnish apparatus for lighting Manufactories &c with gas lights, under the direction of Mr Murdoch the original Inventor & refer to the Philosophical Transactions."[143]

There are many intriguing aspects of this saga,[144] but most notable are the audacity of the trading on the Royal Society's authority that it reveals, and the fact that Banks was complicit. That complicity is made clear in a remarkable letter from James Watt Jr to his father: "My time has been a good deal occupied in ascertaining what sort of paper would be most likely to answer Sir Joseph Banks's wishes, and secure

Mr. Murdock the Rumford Medal. And I have . . . drawn up one which fully met Sir Joseph's approbation and by his interest, was last night read at the Royal Society, and I suppose will be published in the next volume of the Transactions, it being the desire of Sir Joseph, that it should appear before the public, as early as possible."[145] The gaslight example makes one wonder whether Banks knew all along what the Lunar Society group were up to in *all* their dealings with the Society. Why did they not meet with his censure rather than enjoy his complicity? It seems to me that what Banks tried to protect the Society from was exploitation of its reputation for what he judged to be narrow *personal* interest. If he could be persuaded that the public interest was being served, then he would be less worried about private benefits. Certainly by the 1790s Boulton & Watt ventures were seen as having profound public benefits, and the partners were careful to emphasize this at every opportunity.[146] Not only was the improved Watt steam engine seen as a harbinger of a new industrial era, but Boulton's manufacturing enterprise at Soho and especially his collaboration with Banks in developing and supplying new machinery for the mint clearly placed Boulton & Watt on the side of public virtue so far as the president was concerned.[147] One must assume that in the very blatant case of Murdoch and the Rumford Medal, Banks would have justified his complicity in terms of the public interest served, as he saw it, by giving the lead in gaslight provision to a company so demonstrably concerned with promoting development and public utility.

The institutions of organized science were of interest to Watt for their philosophical productions, though it is notable that he rarely attended meetings of the Royal Society between his election in 1785 and his death. But such organizations were certainly of interest as a business opportunity. In March 1792, when Watt and Boulton were in London drumming up support for their fight in Parliament against the Hornblowers, Watt casually reported to Annie that they would "go to the Royal Society in the evening in hopes of meeting some friends who can be of use to us in Parlᵗ."[148] The world of natural philosophy might well be a refuge from business for Watt in some respects, but it could also be a vehicle for doing it. A tightrope was being walked here, as also when dealing with visitors to Soho.[149] The niceties of the Republic of Letters might be observed but industrial and commercial sculduggery had to be guarded against and, as occasion allowed, engaged in. Robinson Boulton captured the attitude of the Soho group very clearly in his reaction to the plans by the Royal Institution, newly founded in 1799 with considerable aristocratic patronage, and designed to spread knowledge of useful inventions: "It may be a very pleasant amusement for the Nobility and other idle Loungers who have never added an Iota to the Purse of the

Nation by the Sweat of their Brow, to diffuse the Inventions & Advantages acquired by the Perseverance & painfull Study of the grovelling mechanics, but how will it be relished by the Inventor himself? Will he think himself sufficiently rewarded for his Pains & Perseverance in struggling against Misery and Want, by a Vote of thanks for the liberal Communication of his Discovery?"[150]

As someone who lived by exploiting his ideas, Watt found it difficult to reconcile himself to giving those ideas away and found such behavior strange in others like Berthollet. This was why, even though he engaged in repeated and extensive trains of philosophical experimentation, he published little and had a rather distant relationship with the clubbable world of organized science except so far as it could be commercially useful to him. His experiences during the water controversy no doubt reinforced his view that he was as likely to find a "pirate" in the meeting room of the Royal Society at Somerset House as in the wilds of Cornwall. Natural philosophy was a business that echoed with many meanings.

Five

Team Watt

.ʃ   ʃʃ   ʃʃ

Collective Genius

*O*ne way of looking at Watt's life is as an evolving and progressive accumulation of collaborators and supporters. From the beginning this was so, whether we consider his family; his supporters at Glasgow College; his partners John Craig,
John Roebuck, and Matthew Boulton; or those associated more informally with him
in collaborative friendships, such as Joseph Black, Gilbert Hamilton, William Small,
Erasmus Darwin, Joseph Priestley, J. A. De Luc, and Thomas Beddoes. From the
1790s onward, the second generation took a greater part in business affairs. James
Watt Jr. and Matthew Robinson Boulton took on a prominent role in the engine business. Watt Jr. also played a major part in further developing the copying machine enterprise and in the legal action against the engine "pirates" in that decade. Although
these collaborations are not always emphasized, they are usually acknowledged. Usually unacknowledged, and certainly not stressed enough, are Watt's collaborations
with, and reliance upon, a range of assistants and workers who took on much of
the detailed work in the engine and other businesses and also, sometimes, in experimental inquiry.[1] Even during his struggling Glasgow years, Watt had had a number
of employees and assistants who did crucial work for him, notably John Gardner,
who made instruments and assisted in Watt's surveying activities and in his steam
experiments at Glasgow and Kinneil.[2] This mode of operation continued into Watt's
Birmingham years.

It was from a complex network of supporters and collaborators, built up particularly from the 1780s, that there emerged what I call Team Watt. Acknowledging the existence of this group and understanding what its members did is important in two respects. First, the picture of Watt as lone genius has undoubtedly been overdrawn. Patents require the nomination of an individual patentee or inventor, and those patents then reinforce that misleading individualism.[3] Historical treatments of Watt have reflected a similarly blinkered view. At the very least, many of Watt's supposedly single-handed accomplishments were in fact a collective achievement. Some authors have taken a stronger position and argued that Watt claimed, or was allowed to claim, some inventions that belonged to others. The second respect in which Team Watt is important is that some of its members played an important role—by their testimony, their silence, or in other ways—in actively generating the view of Watt as heroic individual inventor. His son James Watt Jr. was preeminent among these in the nineteenth century. But, as Melling has argued, pushing Watt forward as the central technical genius of the enterprise, even against the legitimate claims of the higher echelons of their workforce, was seen by Matthew Boulton as a managerial and commercial necessity.[4] Thus Boulton's much-vaunted lobbying and diplomatic skills were sometimes turned inward to the growing Boulton & Watt organization and directed toward maintaining and harnessing the artisanal and inventive capacities of their leading workers while at the same time ensuring that all paid due respect to Watt. That respect was both a business and a psychological necessity as Boulton saw it. It offered a balm for Watt's hypochondria and occasional lack of self-confidence. Encouraging Watt's ego was part of what being a member of Team Watt meant.

The creation and defense of Watt's reputation as a superior steam engineer by virtue of his philosophical approach had imbued the Boulton & Watt partnership from the beginning. It informed the early decision about how to specify the invention of the separate condenser, that is to present it as the outcome of implementation of a scientific principle. It informed the marketing strategy that the partnership developed. As we will see, the issue hung over labor relations in the engine business. It was also a central reason for the pursuit of pirates, as well as a crucial tool in the legal struggles against them in the 1790s.

Some who became faithful members of the inner group began their association as regular workers within the Soho organization. One such we have met already—Joseph Harrison, the smith who gave evidence before the House of Commons committee on the extension of Watt's patent. There he had testified to the superior performance of Watt engines over Newcomen ones, drawing on firsthand experience because he had assisted Watt's examination of a number of engines in London by way of comparison. As the engine business expanded through the early successes in Cornwall and the Soho staff grew accordingly, Harrison emerged as the leading man in the engine yard there. He also worked as an erector, and there are signs that his modest and calm approach was useful in placating the grievances that tended to arise in Cornwall. Harrison's love for the drink and his lack of education in the end limited his role, but as other engine erectors were trained up he became a permanent fixture at Soho, where he worked for the rest of his life.[5] As happened in a number of families, a son also entered the Soho service, William Harrison, who became an engine erector and then foreman of a small engine workshop.

Joseph Harrison was one of an early group of workmen of very variable abilities and utility who worked mainly on the erection of engines. While a man such as James Law proved very reliable, working in Cornwall for Boulton & Watt for three years from 1779 and having a long career as an engine erector into the 1790s, others proved a liability. Richard Cartwright, who assisted with the Bedworth engine and the engine at Chelsea Waterworks in the late 1770s, notoriously leaked the details of Watt's intended use of the crank to achieve rotary motion—necessitating the pursuit of alternative solutions, notably the sun-and-planet gear system. Watt, perhaps understandably, had no time for Cartwright's demands for more wages and was eager to let him go. But Boulton was more sympathetic, or more circumspect.[6]

Men such as Cartwright gained detailed insight into the operations of the partnership, and if they moved on to other employers could take that information with them. Industrial espionage was a perpetual problem, and not just with eminent visitors from overseas.[7] Contracts with engine erectors explicitly mentioned their obligation not to disclose what they learned about the business of Boulton & Watt.[8] But inevitably some men moved on, and sometimes they even became direct competitors in the engine business. Such was the case with Jabez Hornblower, the son of Jonathan Hornblower the elder.[9] Jabez was employed at Soho as an engine erector and worked on a number of machines for Boulton & Watt. Later, however, he set himself up as a rival, was treated as a pirate by his former employer, and was targeted in the patent litigation of the 1790s. Edward Bull and Robert Cameron, who had been recruited among a number of others in 1781 to work for Boulton & Watt in Cornwall, later set themselves up in business as rivals. Cameron's ambitions within the firm had been frustrated. He was engaged in the early 1780s in experiments on rotative engine design to some extent in internal competition with Watt. While Boulton gave some encouragement to this, he let key people know that in the end Watt's designs would, as a matter of principle and organization, take precedence: "I . . . can never consent to any person in our employ to say 'no—Mr Watt's plans shall not be followed but mine shall take place.'"[10]

Another engine erector who was with Boulton & Watt from around 1780 was the colorful character Isaac Perrins. His father, also named Isaac, had been involved in the erection of one of Boulton & Watt's earliest engines, at Bloomfield Colliery, and subsequently worked on other engines for them. Young Isaac appears to have been a good engine erector, but he also pursued a career as a prizefighter, a fact that caused some consternation among the Soho principals. At one point Watt Jr. was intent upon sending Perrins off on a job in order to try to divert him from taking up an offer to fight "Big Ben" Bryan, a matchup that evidently interested the "Birmingham gam- blers." There were perhaps good reasons to have a man of Perrins' stature (he was 6'2" tall and weighed close to 17 stones), strength (he reputedly could lift a lump of iron weighing over 800 pounds into the back of a wagon), and capabilities (he was literate and indeed quite well read) involved in engine work. But at one point Perrins, who had been erecting and tending engines in Manchester, where he also became a publican, was dismissed by Boulton & Watt. He had quarrelled with James Lawson and John Southern, other employees, about the condition of the engines he tended. In a parting shot in a letter to Boulton & Watt, he compared his own willingness to dirty his hands on engine business favorably with the attitude of others (meaning almost certainly Southern and Lawson) "that you send here with ruffles at hands and pow- dered heads, more like some Lord than an engineer."[11] One suspects that Watt, given his own often timid demeanor, might also have been seen by Perrins as a powdered head. But Perrins lived to fight another day for Boulton & Watt, subsequently offer- ing intelligence concerning patent infringements in Manchester and again erecting engines on their behalf. Fraught and colorful as employee relations could be, by the early nineteenth century Soho was recognized as having been an important institu- tional training ground for engineers, though few as interesting as Perrins.

James Lawson (1767–1818), the son of a clergyman in Dumfries, joined Boulton & Watt at Soho in about 1779 and in subsequent years operated in a number of roles, including in the central office at Soho and as an engine erector in Cornwall. Like oth- er Boulton & Watt employees in Cornwall, Lawson did work for the mine adventurers and in the 1780s spent some time surveying the mines and making plans and sections of them. He also worked a great deal in Yorkshire and Lancashire from his base in Soho. In the 1790s he added a role in Boulton's mint operations as well as taking part in the preparations for lawsuits. He also worked as Boulton & Watt's agent in Leeds,

Manchester, and Scotland at various times. Lawson became particularly close to Watt Jr. The quality of his work and his standing, maybe even his powdered head, were reflected in the fact that in 1812 he was elected a Fellow of the Royal Society of London. His election certificate described him as "of His Majesty's Mint a Gentleman well versed in Mechanics and other branches of Philosophy."[12]

Within this growing organization of workers assembled in the Soho-based steam engine business and arcing across the country, Watt's role began to change. In the earliest days he had been something of a one-man band, doing drawings, supervising erection of engines, and troubleshooting afterward.[13] This was frantic activity and drove him more than once close to breakdown. In that regard, the growing body of assistants was a boon. But in other respects it, and the processes of continual improvement, created problems for Watt. He was increasingly cast in a supervisory role, which he found just as troubling as being a lone technician, in some ways more so. In the mass of business, mistakes in drawings might go undetected and sometimes did. Watt's worries were thus transposed to all the possibilities for such mistakes and the consequences they might have for the business. While the involvement of other personnel was helpful, their tendency to have ideas and to innovate on their own account also caused Watt loss of sleep. After one particular mistake in a drawing that Watt had failed to detect, he wrote in a deeply troubled letter to Boulton:

> At the time that and other drawings were dispatched my head was so confused
> with various matters that the omission passed unnoticed, and in general in
> new and intricate contrivances it requires more attention and genius to obviate
> every difficulty than I am possessed of, especially where there is no *candid*
> reviewer to come after me & point out what seems to be mistakes. Such good
> office is hardly to be expected from those who consider their own inventions
> better than mine, and may think themselves crampt by having any rules laid
> down to them to work by. Such men are more likely to enjoy the gratification
> of their vanity in detecting one in error (which it is next to impossible that
> the man who projects upon paper, can intirely avoid) than to contribute any
> assistance in supplying the defects which may occur. This evil is augmented by
> my inability to lay down these drawings with my own hands as I used to do,
> consequently errors may escape me in a cursory looking over a drawing which
> would not have done so if it had been the work of my own hands.[14]

Watt was not coping well either with the complexities of organization or with not being the sole center of inventive attention. After this exculpatory account of the

FIGURE 5.1: *William Murdoch*, one of Watt's collaborators, by John Graham Gilbert. Reproduced by permission of Birmingham Museums Trust.

difficulties of delegation, Watt offered a fascinating insight into his thinking about invention, one that captures the deep conservative strain in the longer arc of his career: "On the whole I find it now full time to cease attempting to invent new things, or to attempt anything which is attended with any risk of not succeeding, or of creating trouble in the execution. Let us go on executing the things we understand and leave the rest to younger men, who have neither money nor character to lose."[15] Although this was written at a low point for Watt—he was ill and again plagued by headaches—and was sparked by his discomfort with the immediate problems of new working arrangements, it does suggest that Watt, despite his fecundity as an inventor, was prepared to call "Enough!" especially when money or reputation were at risk. This reinforces the sense that these anxieties activated Watt's gradual withdrawal from "frontline" exposure to the vagaries of the invention business from the late 1780s, long before his official retirement in 1800.

Occasionally, an exceptionally competent, challenging, but ultimately loyal and trustworthy individual emerged from the body of workmen to become a key member of Team Watt. One such was William Murdoch (1754–1839), even though he was probably the source of the irritation we have just witnessed with the "vanity" of employees who dared to invent for themselves. It has been claimed for Murdoch that he not only aided Watt but authored some of the inventions attributed to his employer, a fact probably deliberately concealed by those who became the self-appointed custodians of the great man's reputation. John Griffiths has suggested that the Boulton & Watt archive was deliberately weeded of correspondence with Murdoch that would have given a clearer picture of his pivotal role and accomplishments. This claim is difficult to substantiate, but it is certainly true that Watt Jr., to whom the bulk of the Watt archive passed on his father's death, had no compunction in massaging the historical record.[16]

Murdoch was born in Ayrshire, and his father was a miller and millwright who also turned his hand to various engineering projects. Young William was hired by Boulton & Watt through family connections in August 1777 and was based initially at Soho Manufactory as a pattern maker. Murdoch distinguished himself early through his hard work, his mechanical and inventive ability, and, unlike many of the Boulton & Watt workforce, his sobriety. He was soon engaged in repairing and altering the Bedworth engine (near Coventry) and then went to Scotland to erect a machine for Meason's lead mine at Wanlockhead. Watt described him most approvingly as "a very sober, ingenious young man," and Boulton considered him the best engine erector he ever saw.[17]

In 1779 Murdoch was sent to Cornwall as an engine erector and lived there, remaining throughout in the service of Boulton & Watt until 1798. Murdoch's ability to deal with the Cornish miners and adventurers whom Watt found so difficult made him a valuable asset. This, together with his technical abilities, ensured that Murdoch had some leverage in his dealings with Boulton & Watt. He was employed by a number of mine owners to attend to their engines and was well rewarded for this. So far as these were Boulton & Watt's engines, Boulton considered the arrangement to tie Murdoch more closely to them. But it also showed that Murdoch could exercise independence. He could also exercise ingenuity and often did so, taking it upon himself to depart on occasion from the drawings that were supplied from Soho, no doubt causing Watt considerable anxiety.

Griffiths makes the case that Murdoch, rather than Watt, was responsible for the sun-and-planet gear system to convert reciprocal into rotative motion, which proved so important to the firm's expansion into the supply of engines for direct driving of industrial machinery. If this is so then Murdoch is an example of an employee whose inventions were claimed by his employers, probably in this case with his consent.[18] In fact, the various inventions claimed in Watt's patents of 1782 and 1784 likely contained more than one that owed a great deal to Murdoch. As noted earlier, in 1784 Murdoch was working on his own account on a steam-powered carriage, which was a great enthusiasm of his. The partners were conscious of the need to manage their enthusiastic employee very carefully. Murdoch was evidently pressuring for his device to be included in the patent that Watt was working on. Boulton encouraged Watt to agree in order to retain Murdoch's invaluable services in Cornwall: "I verily believe he would sooner give up all his Cornish business & interest than be deprived of carrying the thing into execution." Murdoch regularly hinted that he might go it alone: "He doth not directly say that he hath a right to work pistons by steam, but he says he thinks or fears that the Trumpeters [code within Boulton & Watt for the Hornblowers] or somebody else will & then take out a patent for w[hee]l car[ria]g[e]s wch is the thing he says makes him so solicitous [sic] to have it secured." Boulton continued: "Now as you are going to specify sundry new applications of steam engines Q[uer]y if it may not be prudent to specify the application of it to w[hee]l car[ria]g[e]s without making any drawing & only describe ye application of the elastic force to steam to act upon a piston or pistons in cylinders & that force applyd to the turning of the wheels by means either of cranks or by one w[hee]l revolving round another or within another. . . . I propose this by way of taking possession and saveing [sic] the expense of a patent."[19] Watt did incorporate a steam carriage application in the 1784 patent specification, which, as we saw, also covered the parallel motion and other inventions. Boulton and Watt were skeptical about the feasibility of wheeled carriages powered by steam, but they humored Murdoch because he was such a valuable employee and probably also as a quid pro quo for his contribution to the sun-and-planet gear invention. Despite this, two years later Murdoch was on his way to London with a model of a steam carriage and the intent to patent it when he was literally met on the road by Boulton and persuaded to return to Cornwall instead.[20] There were two occasions when Murdoch did take out a patent in his own name: in 1791 on the production of various useful materials, including a wood preservative, by distillation of pyrites (this work was one of the tributaries that fed into his later work on coal-gas production and gaslighting) and in 1799, when he patented the D-slide valve, a device that came into wide use in locomotives in the nineteenth century.[21]

In 1798 Murdoch was brought back to Soho as Boulton & Watt interests in Cornwall declined and the nature of the business changed. He had earlier been involved in the establishment of the Soho Foundry, in which Watt Jr. and Robinson Boulton played the leading part. Murdoch, together with John Southern and Abraham Storey (the foundry manager), designed much of the machinery to be deployed there. When he returned to Soho permanently, Murdoch worked on various projects including a boring mill and the design of a smaller steam engine, the "bell-crank" engine, before his work on gaslighting began in earnest in 1802. The terms of Murdoch's employment changed, reflecting the value placed upon him. From 1800 his salary was set at £300 per year, and he also received 1 percent of the value of all orders for Soho Foundry. As the gaslighting business developed, he was also given 1.5 percent of the value of all the gaslighting apparatus made by Boulton, Watt & Co. He was offered a partnership in 1810 but took a salary of £1,000 per year instead of a share of profits, an arrangement that lasted until 1830, when Watt Jr. finally abandoned it in view of Murdoch's advanced age and fading contribution.

What, then, of the relationship of the work of Watt and Murdoch? Watt spent much time in Cornwall, and he and Murdoch overlapped there for long periods between 1779 and 1782. There was thus much opportunity for cooperative work to occur. So far as the sun-and-planet gear system is concerned, it is clear that there were contributions from various quarters. The sun-and-planet arrangement for producing a "continued rotative motion" was the subject of patent 1306 in Watt's name, which was specified on 25 October 1781 and enrolled on 23 February 1782. As we saw, the hunt for a means of converting reciprocating motion into a continuous, smooth rotative motion had been on for some time. Watt was much frustrated in this, first by Wasborough's rotative engine and then by Pickard's opportunistic patenting of the crank for the purpose. According to Richard Hills's convincing account, Murdoch does deserve credit for suggesting the sun-and-planet method, which became the fifth means of achieving rotation specified in the 1781 patent in addition to the four methods that Watt had contrived. Writing to Boulton, however, Watt described this method as one of *his* "old plans" that had been "revived and executed" by Murdoch.[22] The idea itself was only worth something if it could be effectively executed, and if Murdoch had done so, we would traditionally award credit to him. But there remained a problem with the arrangement as it emerged from the hands of both Watt and Murdoch—how to ensure that the wheels running on each other remained in contact at all times. Some method of confinement was needed, and it appears to have been Matthew Boulton who came up with a solution on Sunday 13 January 1782 when he should have been in church.[23] So the patent that was enrolled in Watt's name described a method that was a joint effort between him, Murdoch, and Boulton. It would be a mistake to attribute the invention to any of them singly, but in the event Watt gets the credit because his name is on the patent.

John Southern

Another recruit who became not only a faithful servant but also (like Murdoch) a close collaborator to a degree not fully appreciated was John Southern (1758?–1815). Southern came from Derbyshire, the son of Thomas Southern; he was apprenticed to his brother as a surgeon when he came to Boulton's notice in 1781 as a possible draftsman and assistant to Watt.[24] Watt observed: "If you have a notion that young Southern would be sufficiently sedate, would come to us for a reasonable sum annually, and would engage for a sufficient time, I should be very glad to engage him for a drawer, provided he gives bond to give up music, otherwise I am sure he will do no good, it being the source of idleness."[25] Perhaps this is another example of Watt's Presbyterian slip showing, since suspicion of musical recreation was common in the culture of Presbyterianism.

Southern did well, whether or not he gave up his musical interests. He worked primarily as a draftsman but was good with figures generally and acquired considerable scientific knowledge. He did more personal work for Watt. When Watt and Annie were extracting money from the business in 1787, and weighing up investments, Watt left his wife to do as she thought proper "with Mr S's advice of the values of the Houses & annuities." Southern was apparently involved in negotiations for them regarding house purchases.[26] His involvement in domestic affairs, and closeness to much of Watt's work, was no doubt encouraged by the fact that until 1790 Watt operated primarily from home in Regent's Place, where his first assistant, William Playfair, and then Southern also worked. Long periods of toil in close proximity and a consequent often-shared table must have cemented their relationship. When it came to engine design, Southern had important insights that Watt trusted, an example being the decision to abandon use of a hot-water pump in the rotative engines, which was based on Southern's argument that the pump did nothing to improve the vacuum.[27] Southern's expertise with engines and mathematics became such that he was a frontline defender in much of the litigation of the 1790s that Boulton & Watt conducted against those they regarded as "pirates" in violation of Watt's patent. Thus, in 1792 Watt wrote to him: "as they [Hornblower and Maberley] have brought Mr. Giddy, the High Sheriff of Cornwall, an Oxford boy to prove by *fluxions* the superiority of their engine, perhaps we shall be obliged to call upon you to come up by Thursday to face his fluxions by common sense."[28] In that case Southern was not, in the end, needed, but Watt recognized in him an ingenious outsider in his own image.

Southern, like Murdoch, was involved in authoring a device or technique usually associated only with Watt. In Southern's case this was the recording steam indicator, a device that could record the variation in the pressure in the cylinder of a steam engine (see figure 5.2). From that record the performance of the engine could be determined, giving clues to adjustments needed to improve it. While Watt had experimented from the mid 1780s with devices that could give insight into what was happening inside the cylinder of a steam engine, the device he had developed involved the real-time monitoring of a pointer attached to the piston. This instrument was used in 1793 to try to measure the performance of a Boulton & Watt steam engine at the Salford Cotton Mill of George Lee. Southern and James Lawson also worked on this engine, and in 1796 we find Southern writing to his colleague: "Tell Mr. Lee I have contrived an instrument that shall tell *accurately* what power any engine exerts"; shortly thereafter he included a hand sketch of a closed indicator diagram in a letter to Lawson, as he instructed the latter about how the device's platen and card to receive the inscription should be configured.[29]

Historians have made a great deal of the importance of the steam indicator and the indicator diagram. Mid-nineteenth-century physicists used it to conceptualize some of the foundations of thermodynamics, and Watt, being credited with the device and the production of the diagram, was often consequently seen as a progenitor of energy physics and thermodynamics. I have argued that this is quite mistaken because Watt conceived of what the steam indicator was measuring in chemical terms rather than

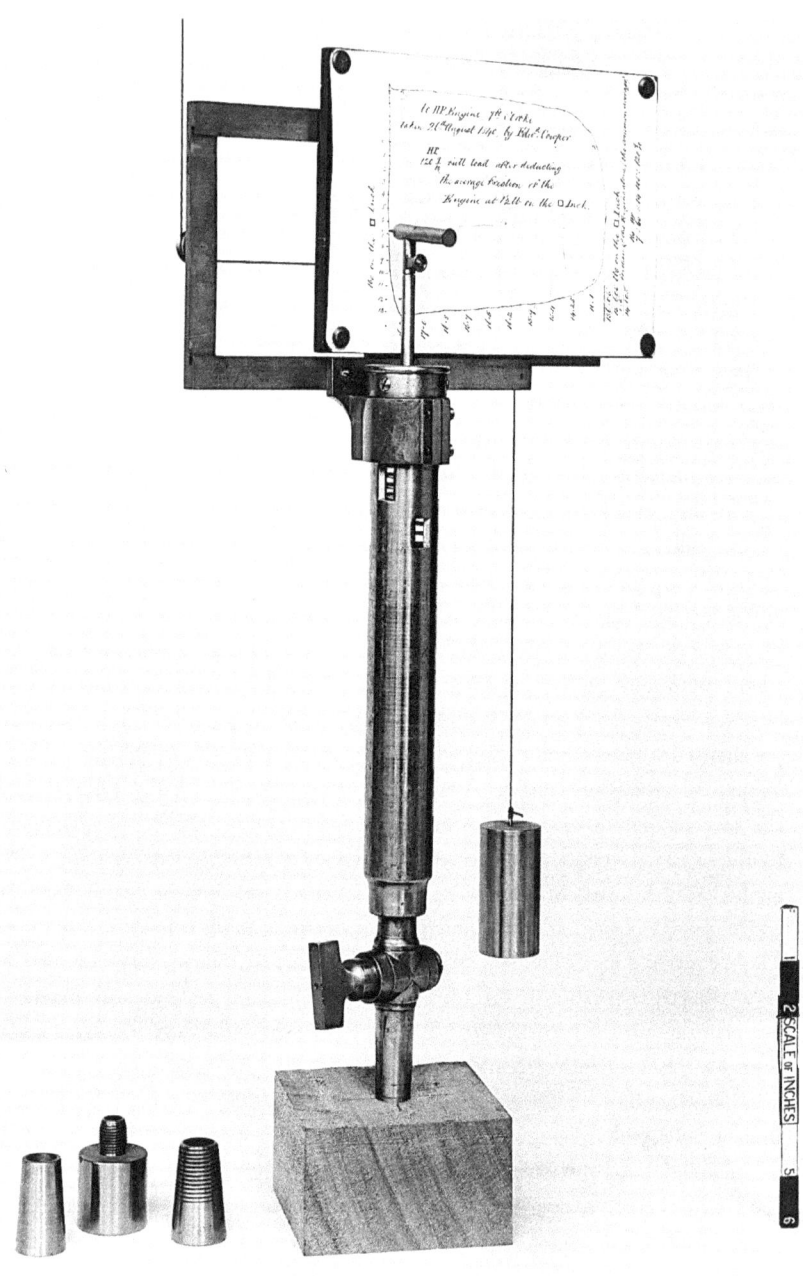

FIGURE 5.2: A recording steam indicator, circa early nineteenth century, of the type devised by John Southern and Watt. © Science & Society Picture Library.

in terms of heat energy.[30] But in any case it appears that Southern, rather than Watt, was responsible for the crucial innovation that created the self-recording device that generated the closed indicator diagram. At the very least the steam indicator should be regarded as the product of their combined efforts.

Southern was also a natural philosophical assistant to Watt. In particular, he helped after Watt's retirement with further investigations into steam and with the repetition of Watt's earlier experiments of that sort. These experiments were prompted at least in part by inquiries from important figures in the scientific world (including Laplace on behalf of the French Institute) for accounts of those experiments, which had been much talked about but not published. Watt was, in effect, being asked to justify a natural philosophical reputation that was now bringing him recognition such as election as one of the corresponding members of the French Institute. But the further steam experiments were also prompted by the difficulties involved in providing historical accounts of Watt's work on heat at a time when important new developments were taking place in that field. Southern was instrumental in helping Watt to do this. In that sense he was not only an important member of Team Watt as a collaborator (and one evidently willing to sublimate his own achievements) but also as a retrospective builder and defender of the Great Steamer's reputation.[31] Watt was less than generous in acknowledging Southern's contribution. When Southern died in 1815, still in harness to Boulton, Watt & Co., Watt was preoccupied with the important business of engravings of his portrait and the completion of his bust by Chantrey, and he offered the following observation on his departed assistant: "Mr Southern's death is undoubtedly a very distressing circumstance, but I hope James will get some help soon perhaps more efficacious than that of Southern which was principally that of calculation & the mechanical knowledge attained from experience more than skill in the management of business."[32] This undoubtedly rational and accurate assessment of Southern's role in the firm by that time left much unsaid about his earlier contribution to achievements that were subsequently attributed to Watt.

We see in the story of Team Watt repeated examples of a theme that has emerged concerning Watt's character. While Watt certainly exhibited a fecund imagination, powerful intellect, mechanical skill, considerable organizational sense, and even financial acumen, his relations with others were often problematic and required constant management by a number of people who were close to him.

Watt relied heavily on his wives in connection with business. Though we have only sparse evidence concerning his first wife, Peggy, in this regard, we have seen that she helped a great deal with the shop as well as domestic arrangements during her husband's frequent absences from home. His second wife, Annie, organized his life in many ways from the time, before they were married, when she helped Betty Millar select his wardrobe for the first trip to Birmingham. Annie tried, disastrously as we will see shortly, to manage Watt's relations with the children of his first marriage and certainly to manipulate his relationship with them. More benignly, she intervened with Boulton to say things about her husband's emotional state that Watt perhaps could not bring himself to say to his business partner. We have seen, for example, that she explained to Boulton how the weight of debt hung heavily on Watt and contributed to his bouts of anxiety and depression. When what Annie saw as Boulton's profligate ways continued, she persuaded Watt to separate their finances to some degree by taking money from the business and squirreling it away in Scotland with Gilbert Hamilton. At one point, when the rotative engines were first being developed, Annie even encouraged Watt to reconsider the terms of the partnership with Boulton: "I think in Justice you ought to have at least half the profits that arise from them [the rotative engines], the[y] certainly were no part of your original agreement with Mr B."[33] In 1786, when Boulton and Watt's trip to France seemed to promise much engine business abroad, Annie was similarly encouraging Watt to have Boulton send orders to Pearson (the keeper of the books at Soho) "to enter all Engine business done out of Britain as you having an equal share with him. I think you can now do it better than when any Engagements for erecting Engines are made."[34]

Annie was always active on Watt's behalf and certainly saw herself as protecting him. Two months earlier (September 1786), in a heated exchange about Annie's troubled relations with her stepchildren, James and Peggy, she reproached Watt for having "withdrawn all friendship & confidence" from her despite her earnest wish "to release you of the Burden of business that seems to hang so heavy upon you."[35] The close knowledge of, and attention to, matters of business that is evident in Annie's letters to Watt during his frequent absences from Birmingham indicate significant involvement in, and care for, business that also perhaps pertained even when Watt was at home and at his desk. Annie occasionally acted independently in business matters. For example, again in 1786, having received a letter requesting an abatement of the premium from George Walker of Chester, whose mill, and therefore the Boulton & Watt engine operating it, were lying idle for want of work, Annie told Watt: "I wrote them that Mr B & you had set out for the continent and would not return for a month or six weeks but was afraid the abatement cou[l]d not be complied with as it wou[l]d be so very bad an example."[36]

Boulton sometimes had to manage Watt's relationship with members of their growing industrial organization. Because of his insecurities in dealing with men, Watt often overcompensated by taking a very stern, if not uncompromising and unreasonable, attitude toward those in their employ, including men like Murdoch and Southern who were close collaborators. While Watt tended to fight, or flee from, the inventive efforts of others, Boulton both controlled and cajoled their leading hands. He ensured that they were rewarded well and recognized internally even as they were persuaded, for the most part, to sublimate their own inventive ambitions and achievements to the commercial importance of depicting Watt as the philosophically driven inventive genius whose reputation ensured the preeminence of Boulton & Watt engines. As we will see shortly, in the patent trials of the 1790s, former employees of Boulton & Watt who had decided to pursue their own engines were challenged in part by the testimony of those leading hands who had stayed with the firm and were willing, unlike their former colleagues, to put to one side the complexities of their own involvement in the incremental inventive development of the steam engine and to join in the chorus of praise for the philosophically informed inventions of Mr. Watt.

As Watt's active involvement in the business was reduced and the key patents in his name expired, the role of collaborators was more explicitly recognized. Thus, the new terms of engagement after 1800, already noted for Murdoch, also applied to James Lawson and to John Southern. Their contracts contained a clause stating that the firm would have "full use and property in any invention which may be made by him, tending to the improvement of the construction of steam engines."[37] These gentlemen were not rewarded directly for their inventive contributions but were given generous salary and profit-sharing arrangements in return for ceding their rights to those contributions. This, I suggest, formalized an understanding that had long existed.[38]

For the most part, these arrangements enabled the creative energies of employees to be drawn upon and their resentments contained. We see this in the case of the Creighton brothers, engine erectors for Boulton, Watt & Co. who, by their own lights at least, frequently came up with improvements to engines. Their letters give a direct, if late, insight into the frequent sublimation of tensions by humor. Thus, remarking on an improvement regarding steamboat engines, Henry Creighton joked to his brother that if the suggestion was adopted, he "should get a patent for discovery and be dubbed FRS or LLD."[39] This jibe reflected the elements of both resentment and good-humored resignation at the contrast between the rewards of a junior member of the group and those received by the man himself.[40]

James Watt Jr.

It is well known that numerous tensions and difficulties also attended Watt's relations with the boy and man who was to be the most important member of Team Watt, his eldest son, James Watt Jr. The activities of Watt Jr. provide the most startling example of management of the great man's social relations and reputation, as well as show us Watt's failures as a father to deal effectively with a rebellious and strong-minded son. In a strange twist, those failures helped to originate a chain of events in Watt Jr.'s life that eventually transformed him into the most devoted son imaginable—devoted to the point of an almost deranged filial piety. He became the pivot around which the team turned from the mid-1790s until his own death in 1848. To understand how this came to be, we need to take an extended excursion through the early life of Watt's eldest son.

We have seen that when Watt Jr.'s mother died in September 1773 and his father began the process of moving to Birmingham, the boy was left, along with his sister Peggy, initially with grandfather Watt in Greenock but subsequently with an aunt on his mother's side, Betty Millar, in Glasgow. Betty's correspondence with Watt after his departure painted a picture of the children as well behaved and devoted to their learning, if very anxious to be reunited with their father. After Watt's visit to Scotland in 1775, during which his courtship of Annie McGrigor probably began, he took his son back to Birmingham with him but left Peggy with her aunt. Watt was very busy with his engines and in the anxious early stages of establishing himself at Soho; it was not an ideal time to devote attention to his son, so recently bereaved and now deprived of the constant and caring company of his Aunt and sister, who had stepped into the gap left by his mother. Young James began to show signs of behavior that shocked his family. It appears that shortly after his arrival in Birmingham he was caught stealing—perhaps not unusual behavior, in our eyes, in a troubled boy of six or seven years of age, but to Watt and Betty a devastating occurrence. Responding to a letter from Watt conveying the news, Betty replied: "Words cannot express what Greif I felt My Dear Jame at reading your accounts of poor little Jamy & how in the name of wonder could it come in his head to covet the 2 pence for I never so much as knew him take any thing but what he thought was his own not so much as a bit of Bread."[41] It appears that Watt's response was severe and included an invocation of hell that "would make him tremble." Betty agreed with Watt's approach: "I hope God will Bless your instrouction [sic] and make him ever fear to do the like again." There is a suggestion in Betty's immediate promise to take care of Peggy and to "make

FIGURE 5.3: *James Watt Jr.*, pastel portrait by L. de Longastre (circa 1800). The radical is here turned businessman at the time of his father's retirement. © Science & Society Picture Library.

sure she has no correspondence with the Calton children" that Watt may have con-sidered James's behavior to be a result of contact with the children in the Glasgow neighborhood where Betty lived.[42] But we might wonder, too, whether the disrup-tions in the young boy's life were more to blame.

There is no surviving evidence that young James continued to be troublesome in the years immediately after this incident; indeed, Watt reported subsequently to Bet-ty that he was doing well in Birmingham. But in the following year his new mother arrived unannounced by Watt, along with his sister, perhaps another shock for the young boy. Initially, it seems that Annie willingly took on the care of her stepchildren and in following years had much to do, given Watt's frequent and long absences on engine business. James Jr. attended the Reverend Henry Pickering's school in Bir-mingham but had perhaps moved elsewhere when, in 1780, his father requested that he be disciplined for reported "insolence, sauciness and disobedience."[43] Once her own children had arrived (Gregory was born in 1777 and Janet—known as Jessy— in 1779), Annie's attitude toward her stepchildren changed. She often made unfa-vorable comparisons between them and the younger members of the family. When she was in Glasgow in 1779–1780 during one of Watt's absences in Cornwall, her frequent letters to him scarcely fail to paint a flattering picture of young Gregory's progress and his attachment to his father. James Jr. and Peggy wrote to her in dutiful fashion, and she dutifully reported to Watt that they were well. But she was always quick to find fault even at that stage. For example, as she and the young children were leaving Birmingham for Glasgow, she reported a parting with the older children that was hardly satisfactory. While Gregory was upset, wanting his father to come home again ("you never saw such weeping and wailing"), Annie was surprised that James Jr. took leave of Gregory "without a tear," but, she continued, "at Peggy doing so I was not the least surprized she did not even bid any one in the house good-by."[44] Betty Millar, on the other hand, missed Peggy and saw a chance to see her again courtesy of an impending visit of the Marrs to Glasgow. Annie put a spoke in the wheel of any such arrangement, advising her husband that Betty wanted Mrs. Marr to bring Peggy to Glasgow but would not do anything without Watt's permission. In forestalling any such move, Annie suggested that "Peggy is not so much improved as one would wish to have her when she visits Glasgow."[45] As Watt Jr. entered adolescence, tensions with his stepmother grew and, partly as a result, he began to argue with his father also. Again, this is not atypical behavior from a teenage boy, but there are signs that Watt, torn between loyalties to his wife and to his son, managed to satisfy neither.

By the mid-1780s the difficulties of Annie's relationship with her stepchildren had come to a head. James and Peggy no doubt appealed to their father about Annie's dealings with them, and as a result Watt would have offered discipline and indulgence in equal measure—the latter, becoming known to Annie, driving a wedge between Watt and his wife. From Annie's point of view, she had been left to manage Watt's children while he was away. As he made educational and financial provision for his older children, Annie became anxious that her own children's futures be provided for. Having expressed herself glad that Watt had written James a "trimming Letter" to curb his "Arrogance conceit & Insolence," she continued:

> While he is under control nobody does behave better but the moment the reins are let go his good conduct is forgot forgive me my Dear Jamie for what I have said & may say I have only to entreat that you will make such an allotment of your affairs that Gregory & Jessy may be intirely independent of him when they shall have the Misfortune to lose us their Parents. I trimble to think what they might suffer from so over-bearing a temper and if it is not corrected one that has not the least regard to the property of another excuse me my Dear Jamie for saying so much. As for myself as I have no wish to out live you I trust I have nothing to fear from him.[46]

The tensions had become too much when, in September 1786, Annie wrote to Watt that she was essentially washing her hands of her stepchildren:

> In justice to myself I have strove to correct and inform both James & Margaret that my regard for you made me bear with many things that no other power on earth would have done that I have carefully concealed faults that must have been very painful for you to know they are now older & I hope for your sake will be wiser & better than they have hitherto been But from henceforth I am resolved to give up all management of them they are your children do with them what seems best to you nor will I put you to the disagreeable task of concealing from me what you do for them for believe me I am determined to be pleased and satisfied with whatever it is nor shall I ever complain you have a right to dispose of your own. The lecture that you gave me about I acting unjustly I now see was in regard to J & M but I was not unjust to them I complained from facts but no more of that As they are now at a distance from me they will be happier than it was possible for us to be together & may she be more useful & agreeable to her Aunts than ever she was to me.[47]

The "distance" to which she refers that had been established between Watt's older children and Annie involved, as we will see, James Jr. being sent to the Continent for his education and then later to Manchester to learn the ropes of business. Keeping Annie and her stepchildren apart seems to have become Watt's ambition as well as Annie's. Peggy, whom Annie by now always called Margaret, avoiding the affectionate diminutive, had been left behind at Regent's Place when Annie and the two younger children visited Scotland in 1785. Then in August 1786, Peggy left Birmingham to live with her aunts Betty Millar and Agnes Marr, the widow of Watt's old friend Major John Marr, in Scotland. She never returned to her father's household. From the copy that Watt kept of a remarkable letter he wrote to Peggy, we learn that the proximate cause of the break with Annie was a letter Peggy wrote to her when Watt was recently in London. It appears that, in a disrespectful letter to Annie, Peggy had questioned her stepmother's right to superintend her conduct and, in particular, to demand "an account of your expenses."[48] Once again the regular keeping of records of one's personal expenses was heavily loaded with moral weight. Watt advised that "any want of respect or of due deference" toward Annie was "a want of respect to me." Watt told Peggy that Annie's conduct toward her was consequent upon this disrespectful demeanour:

It is not always possible for us to acquire the love of others but it certainly is in our power to force them to esteem us. In almost all cases I have found it better to suppress my feelings of what I thought injuries than by resenting them to provoke or increase the resentment of others. . . . I earnestly entreat you to use every means to acquire your Mamma's friendship. . . . If your own interest cannot prevail with you to do this I hope that you will do it to prevent the evening of my life, which has been a life of trouble and anxiety, from being rendered unhappy by the differences of those whose study it ought to be to render it pleasant.[49]

Watt's self-pity is unattractive to say the least, but he protested that nobody had her happiness more at heart than he did. Annie, who left for Glasgow with Jessy and Gregory the day after the letter was written, delivered it to Peggy. The young woman must have felt the powerful affirmation of her stepmother's and father's solidarity. Peggy's reply, dated almost a month later, was contrite but professed surprise that she had been judged disrespectful. She praised her aunts for their "Goodness, Kindness and Freindship" and thanked Watt for placing her with them. This perhaps confirmed Watt's view that the kind of deference required from his daughter in order for her to remain peaceably at Heathfield was unlikely to be forthcoming. By December, when he was in Glasgow, he wrote to Agnes Marr concerning Peggy's permanent placement with them and offered the sum of £50 per annum for her "board & Clothes," providing that such an arrangement was agreeable to them. There had clearly been a prior tense meeting, at which some remark must have been made about Watt's wealth and implying favoritism in how he chose to distribute it, for he wrote to Agnes: "I hope you will upon cool reflection do me the justice to believe that I am not capable of making an unfair distribution of the fortune which providence has or may bestow on me, whatever may be its magnitude, and that I do not fail in true affection to any of my children, however little I may be disposed to shelter their faults or failings."[50] Peggy remained at distant daggers drawn with her stepmother, with the result that Watt rarely communicated with her. She married James Miller in June 1791 in Edinburgh. Although Watt had been consulted and gave his permission for the union, he did not attend the wedding and was angry with Watt Jr. for doing so. The marriage settlement of £500, though a substantial sum, was not the height of conceivable generosity on Watt's part. Over the next few years, Peggy gave birth to Watt's only grandchildren, then died of fever at the age of 29 after the birth of the last of them in June 1796, in an eerie echo of her mother's fate. Watt supported his grandchildren in various ways, and they were to be provided for in his will.

For his part, Watt Jr. spent time at John Wilkinson's ironworks at Bersham, near Wrexham, before being dispatched to Geneva by his father some time before this exchange about his and his sister's conduct. Watt Jr. crossed the channel in November 1784 and traveled through Paris to Geneva. He was only fifteen years of age at the time. There he received tutoring that had been organized through Watt's friend and chemical correspondent, De Luc. The boy was inundated by instructions from his father to concentrate upon his studies and avoid "frivolous" activities and the expenditure that they involved. Watt Jr. had an appetite for fun, but his father had, in Peter Jones's words, "a horror of frivolity and extravagance, be it emotional or material."[51] The son had broken with the father's dour Presbyterian habits. Watt decided that Geneva was not the best environment for Watt Jr., and so he was moved in 1785 to Stadtfeld in the Upper Saxony region of Germany, where he was tutored by a clergyman, the Reverend M. Reinhard, and showed signs of much greater attention to his studies in a place of fewer external distractions. Even as his father detected these improvements, Annie decided, perhaps without consulting Watt and with an enthusiasm powered by her own frustrations, to maintain the chorus of disapproval directed at the young man. In subsequent correspondence James Jr. alternated between attack and defense, Annie commenting that she was "sorry he makes such strides towards Lord Chesterfield's system of fawning & flattery which in my opinion is still worse than the insolent strain he formerly wrote in."[52] In a letter to young James of March 1786 she once again reproached him for his extravagance and his arrogance.[53] It seems likely that it was exchanges of this sort that eventually led Watt to side, however briefly, with his son and to castigate his wife about acting unjustly, thus producing the ultimatum that we saw Annie deliver in September 1786.

James then left Reinhard's tutelage and moved to the School of Mines in Freiberg, where he learned a great deal about mineralogy before returning to Birmingham via London; his father met him there in late October 1787 and reported that he was "much improved" and had changed a good deal: "I did not know him when he came to me last night at the Albion Mill."[54] Annie Watt, who was in Glasgow at the time, argued that Watt Jr. should be sent to work in London at the Albion Mill, but Watt considered this too harsh, allowing his son to return to the family home, Regent's Place, in Birmingham.

I know not how to answer you he has behaved with propriety & obedience since he came home & attends to what he is set about. His expenses abroad seem to have been occasioned more by his thoughtlessness than from any vicious habits. I shall do my endeavour to inspire him with habits of reflection on the consequences of his actions, and as he has good sense, there is reason to hope for success. If he does not apply properly to business, or persists in anything he should not do, I shall certainly place him somewhere else, but while he gives no cause of offence why should I deprive myself of any benefit his future assistance may be of to me.[55]

Watt's sense that his son, once he had acquired proper habits of accountability, would in the future be of great assistance to him was to prove prophetic, but for the moment matters did not improve, and by late 1788 Watt had decided to apprentice James to a Manchester calico firm rather than employ him directly in the family business—though Manchester was clearly a strategic location for the introduction of steam engines in mills and factories. Watt Jr. did act as a representative of Boulton & Watt in that city, and correspondence from his father in 1790 was coaching him in the niceties of the engine business.[56]

Watt Jr.'s employers in Manchester at the firm of Taylor & Maxwell were closely involved in the burgeoning city's intellectual life, including with the Manchester

Literary and Philosophical Society, founded in 1781, of which Watt Jr. became co-secretary. He published in the society's *Memoirs* and developed a network of contacts and correspondents in Britain and beyond. But at the "Lit. & Phil." he also befriended Thomas Cooper and Thomas Walker, ardent admirers of the early stages of the French Revolution. In 1791 Cooper, Walker, Watt Jr., and other members sought an official statement of support from the Society for Priestley, whose house and laboratory had just been destroyed in the Birmingham riots. They resigned when the society refused to endorse their proposal. Watt Jr. was restless to return to France, and he found the perfect excuse as a commercial traveler for the firm of T. & R. Walker, whose proprietors were among his political associates in Manchester.

On 4 March 1792 Watt Jr .landed in France with Thomas Cooper and headed straight for Paris, where he met various scientific figures, including Lavoisier. Most of them were "mad with politics" and on the side of the revolutionaries, though not, of course, the ill-fated Lavoisier, who went to the guillotine in May 1794.[57] Watt Jr. and Cooper delivered an address of fraternal support from the Manchester Constitutional Society to the Jacobin Club in Paris. News of this spread, and Edmund Burke attacked Cooper, Walker, and "James Watt" in the House of Commons on 30 April in his famous speech on parliamentary reform. As reported, Burke's speech spoke of agents from England, "men of some consideration in this country," sent to enter into a federation with the Jacobin Club of Paris: "the names he alluded to were Thomas Cooper and James Watt" and he read to the House the address presented by those gentlemen to the Jacobin Club on 16 April, which, as he saw it, supported the Jacobins in their regicides, approved their conduct, and acted in concert with them. Moreover, and outrageously, in doing so they claimed to represent "all England."[58]

Neither Burke's speech nor Thomas Cooper's reply made a distinction between Watt and Watt Jr., and as news of the speech spread Watt was beside himself. In Watt's case the news did not have to spread far because he was then in London testifying to a House of Commons' committee in opposition to Jonathan Hornblower's application for an extension on his steam engine patent. The situation needed little fanning from Annie, but she nevertheless wrote to her husband that if she had not known that he was in the English capital it would have been easy to wonder whether it was him to whom Burke referred.[59] Despite his father's pointing out the damage that he could do to his reputation and hence to the cause of Boulton & Watt in the struggle with Hornblower and more broadly, Watt Jr. continued to make public statements in support of the Jacobins for quite some time. However, by October 1792 business, which could not be totally ignored, took Watt Jr. outside Paris and its political maelstrom. Jones argues that Watt Jr.'s renunciation of his former allies was neither so quick nor so decisive as later family biographies painted it.[60]

The story that emerged in those biographies, with partial corroboration elsewhere, was that when in Paris Watt Jr. emerged as a moderating force against the growing violence. First, he reputedly prevented a duel between Danton and Robespierre by acting as second to one of them and persuading him that the cause of liberty would be the loser if either died.[61] Then, when blood was spilled in the Tuileries and the Paris streets, Watt Jr. resolved that he must try to keep revolutionary violence under control. In response to Watt Jr.'s moderating efforts, Robespierre implied in an address at the Jacobin Club that he and Cooper were agents of the British government. Watt Jr. responded "in a brief but impassioned harangue, delivered in French," rebutted the charge, and carried the audience with him. It was from this point that Watt Jr. realized that his life was in danger if he remained in Paris, and so he headed south and eventually arrived in Italy.[62]

This may have been the story that Watt Jr. subsequently told his father. But at the time no such dramatic change of heart was evident to the residents of Heathfield. In 1793 Watt came close to breaking with his son entirely, as Watt Jr. seemed to persist in views that even his close allies were abandoning in the face of the French Terror. Watt may have been almost persuaded that his son was one of the "monsters" that populated the correspondence of Watt and Annie whenever they discussed the French revolutionaries. But Watt reached out to young James. He advised strongly against returning to England at that point. But he also advised strongly against going to America. (Watt Jr.'s "comrade in arms" Thomas Cooper had fled to the United States before he could be arrested in Britain on charges of sedition.[63]) Watt suggested that if James could manage to be a "silent spectator" of Spanish politics and religion then he might spend time there and superintend Boulton & Watt business from Cadiz. Watt also gave his son world-weary, pragmatic advice about the need to take people as they are, not expect too much, and "combat" his own passions and prejudices. Noting that millions, including himself, lived unmolested in England, safe in their lives and property with "reasonable freedom of speech," Watt considered that "whatever my opinions may be, I am only *one* & have no right to dispute the will of the majority, or to disturb the peace of the country by propagating them. If I dislike another man's opinions I keep out of his company & seek quiet in my Study."[64]

Watt Jr., for his part, seems to have seized this opportunity and seen in his father's olive branch an occasion to mend relations, as his political views moderated from Jacobinism to mere republicanism and he began to see his future tied, after all, to the family business. He confided to his friends and employers the Walkers that his father's recent letters had touched him: "They breathe a warmth of affection I never before experienced from him and prove to me that with time and moderation on my side we shall live in that harmony I so much desire."[65] Hostilities on both sides may also have been softened by Jessy's final decline from the ravages of tuberculosis.

Boulton probably played a role in the rapprochement. He certainly helped Watt Jr. behind the scenes at this time, using his connections with Pitt to ensure that Watt Jr. could return without prosecution.[66] Both he and Watt were considering what would happen to the business as they sought to wind down their activity. The idea could have early suggested itself that Robinson Boulton and Watt Jr., whose young lives had long paralleled each other in many respects, might pool their strengths to take the business over.

Watt Jr. returned quietly to England in February 1794, met his father in London, and slipped easily back into the business of Boulton & Watt. On 22 February 1794 he was writing to Thomas Wilson in Cornwall on their behalf about the purchase of copper cake and the aftermath of the Edward Bull patent lawsuit.[67] A month or so later he returned to Birmingham, lodging for a period with Boulton. As early as April he told the Watts' Dutch friend Jan van Liender that he felt "extremely comfortable & happy at home after all my rambles and probably shall never quit this country any more, particularly as I find my friends, Dr Priestley excepted, determined to remain here. Indeed, bad as this country may be, it is the best I know."[68] That could have been his father talking. Indeed, two days later he wrote to his exiled friend Thomas Cooper, "You can scarcely credit upon what excellent terms I am with my father. I do not understand it at all myself but so it is."[69]

Watt Jr. never lived at Heathfield House—the reconciliation only went so far where Annie was concerned. But Watt Jr. probably spent some time at Heathfield. Jessy died on 6 June, and the intense bout of pneumatic medicine experiments with the breathing of airs began in the laboratory at Heathfield some time later, lasting until October. It seems likely that Watt Jr. joined his father and Gregory in taking refuge in experimental natural philosophy to deal with their grief. Watt Jr., as we have seen, certainly joined in his father's disgust at the reception of the pneumatic medicine project among the metropolitan elite. His radical reformism and his father's "purity of principle" led them to the same stance.

During the same period, discussions must have been held about the future of the engine business, for it was at the end of 1794 that the new engine partnership called Boulton, Watt & Co. was established with a view to manufacturing complete engines.

Watt funded the purchase of the land that became the site of the Soho Foundry but did not enter the partnership as such, which involved the Boultons, father and son; Watt Jr.; and Gregory, which would have pleased Annie. Watt Jr. took the leading role very enthusiastically. He also took on much of the graft involved in Boulton & Watt's legal pursuit of the engine pirates, which had begun seriously in 1792 but was not successfully concluded until the end of the century. He was given charge of the copying machine business, improving its prospects by devising a portable machine that proved popular. The amorphous task of building and protecting his father's reputation, to be examined in a later chapter, also began in this way. Watt Jr. came to regard his custody and enhancement of the businesses begun by his father as one part of that task. The management of publicity concerning Watt's inventive and scientific claims to fame was another part of it. As he put it to his father's French eulogist, François Arago, many years later, "your appeal to my filial duties is not made in vain. It has been the study and endeavour of my life to cause justice to be done to my father's merits."[70] Thanks to this member of Team Watt (and with the cooperation of other members) the great man has been given sole credit for much that was done in collaboration with others, including with his prodigal eldest son, the only child to survive him.

The first, and economically crucial, reputational fight of this sort in which Watt Jr. played an important role was the litigation in defense of Watt's patent in the 1790s. This did not begin with Watt Jr., of course. The possibility of litigation against the pirates had been discussed through much of the 1780s before finally being launched in the 1790s. It was not until 1799 that the final victory was achieved. It seems unlikely that Watt alone would have completed this stressful legal marathon, but as we will see, Team Watt carried him across the line.

Patent Trials and Pirates

Boulton & Watt was far from the only game in town when it came to steam engineering in the later decades of the eighteenth century. An indication of this is that Watt had only four of the fifty-nine steam engine patents taken out between 1769 and 1800. A wide variety of consulting engineers and iron founders developed their own improvements on the Newcomen engine. Some of these, though far from all, Watt regarded as infringing on his patents.[71] Although they were constantly alert from the beginning to such infringements, eventually Boulton and Watt resolved to assert their patent rights and bring infringers to court. Watt and his partner had long been keen students and observers of the patenting scene and had earlier intervened to assist others, notably Richard Arkwright, whose patents were threatened by pirates. Watt even drew up a revision of the patent law that he thought would serve inventors better.[72] So, Watt was well prepared as a "bush lawyer" by the time of the 1790s patent trials. But those trials were in fact a team effort. Boulton and Watt, and somewhat later Watt Jr. and Robinson Boulton, worked at the strategic level with their lawyers, especially Ambrose Weston. But in presenting their case through witnesses in court Boulton & Watt drew on a roll call of members of Team Watt who testified to the nature of Watt's genius; how, when, and where it had expressed itself, the sufficiency of his specification; and the public utility derived from his steam innovations. Given the state of patent law at the time, it was still possible to sway the verdict by such means. The army of collaborators, both scientific and artisanal, that Watt had accumulated over the years was of great value and importance in winning these cases.

The first public skirmish of Boulton & Watt with engine "pirates" in the 1790s was not a legal trial but a parliamentary contest with the Hornblowers. Jonathan Hornblower Jr. had taken a patent out in 1781 for an improved steam engine, a double cylinder expansion machine.[73] Boulton & Watt had maintained that this infringed Watt's patent by incorporating, in effect, a separate condenser and by employing expansive working. They had long threatened legal action against any who might use it. For his part, Hornblower cast aspersions on the Watt patent. In 1788 he published a pamphlet in which he maintained that the fourth principle of Watt's patent specification (the use of the expansive force of steam to press on the pistons) was at least fifty years old and so that patent was invalid, with the result that the premiums being charged by Boulton & Watt in Cornwall had no legal basis. He further charged that Boulton & Watt had failed so far to prosecute "because they find it more their interest to threaten, than really to prosecute; for the former [i.e., threats] seems to have deterred you from adopting our Engine, and they know the latter will not stand the Test of a Court of Justice."[74] Pirating engines was one thing; such public challenges to the legality of their patent and Boulton & Watt's veracity were quite another. This got under Watt's skin, and Annie encouraged him in his feelings of outrage. But outrage was not a good basis for making judicious decisions about how to deal with such challengers.

Hornblower's efforts to build a machine were not very successful. Only in 1791 did he erect a viable engine at Tin Croft Mine in Cornwall. Watt kept an eye on this development through the simple expedient, at Thomas Wilson's suggestion, of buying a small share in the mine in Wilson's name so that he could attend meetings and learn what was afoot.[75] Though Boulton and Watt maintained that this machine was unsuccessful in that it was inferior to Watt's and infringed his patent, the Hornblowers were encouraged into thinking that, given more time, their engines might enjoy success. With Cornwall looking on, the competing sides conducted trials of engines, each claiming superiority over the other. Wilson wrote the only published account of such engine trials. He spent little of this twenty-five page pamphlet discussing Hornblower engines, preferring to laud the superiority of Boulton & Watt engines over the Newcomen machines that had previously operated in the Cornish mines and to enumerate what we would call the "economic multiplier" effects of Watt's engines in Cornwall. In a fine example of how such arguments from within Team Watt typically simplified the attribution of credit for these developments, Wilson stated that these economic effects "cannot but convince every candid mind, that it is an improvement causing more beneficial consequences then [sic] were ever effected by the genius of one Man, since the commencement of Mechanic arts."[76] An objective comparison between the engines was not to be had as each side sought in various ways to weight the trials to their advantage. While the incentive remained to set up, and to interpret the results of, these trials in divergent ways, they could be so interpreted quite reasonably.[77] A circuit breaker was required. For a while the contest shifted to the parliamentary arena.

Hornblower Jr. sought in 1792, with the patent's fourteen-year term approaching its end, to extend that term by Act of Parliament. His petition came before the House of Commons for its first reading on 24 February 1792. In presenting a case to the committee to which the bill was referred, Hornblower asserted the superiority of his engine over Watt's, noted the mischief that Boulton & Watt's threats had wrought on his project, and sought the extension to cover the time thus lost. For their part, Boulton & Watt presented a petition to the committee in opposition, challenging Hornblower's claimed invention as not new and his engines as inferior plagiarisms of Watt's invention.[78] They lobbied hard, knowing that in the end they would have to win a vote in the House, and no doubt Boulton was in his element through all this, cultivating his wide political circle. Watt's friend De Luc, whose position in the Royal Court as reader to Queen Charlotte gave him considerable contact with influential politicians, was also recruited to lobby for the cause.[79] The suggestion was made in Parliament that Boulton & Watt, rather than blocking Hornblower's bill, should negotiate with him or prosecute him. There was a vote on whether the objections of Boulton & Watt should be heard, which went strongly in the company's favor. This meant that witnesses were called—first Watt himself, and then John Rennie, yet another member of Team Watt.[80] Both made simple but effective comparisons between the engines, demonstrating, as they saw it, the inferiority of Hornblower's. The Hornblower camp, knowing that another vote would likely go against them even had the evidential contest been less clearly unfavorable to them, decided to withdraw the bill. Watt cut to the central political point when he told Gilbert Hamilton at the end of May 1792 that "it was no small satisfaction to find we could make such a majority against the whole interest of Cornwall."[81]

Watt's correspondence with Annie throughout this affair shows that he was made ill by the tension and uncertainty of it all. She thought him "sinking under a weight that is too heavy for you" and suggested that he free himself from the business that was tormenting him so much by selling his share in the engine patent.[82] Even as she ferried Wilson's letters concerning the Hornblowers to Watt in London, and argued with him about Watt Jr.'s antics in France, she had to offer reinforcement continually, and the reassurance that they would be fine even if all were lost against their Cornish enemies. He should not be uneasy, she said, but be mindful of his own integrity and "rise above the malice of worthless men as far above their vile insinuations as you surpass them in every thing else."[83] When John Southern arrived with the good news that the Hornblowers had given up their cause, she wished Watt "joy of your victory" observing "may vanity perjury and ingratitude ever meet with the same fate."[84] In the Watt household these were always the faults of others.

To win such a victory without going to law was pleasing. As Watt put it as the next fight was brewing, "it is truly said that law is a bottomless pit and we would fain see some good firm ladder to get out before we venture to go into it." While he was, naturally, confident of the justice of their cause, "the labour, the uncertainty, & the expence of a suit are not light things." Nevertheless, if they had to go to law to deal with the "gross injustice" they were subject to, then they would do so.[85] And so they did, but not, initially, against the Hornblowers. For the Hornblower engines had failed to attract customers and were judged to be less of a threat than those of another engine builder, Edward Bull.

Bull is a rather fugitive character. We know that in 1779 he was an engineman at Bedworth colliery with a prime responsibility as a stoker of the boiler. It was as a stoker that Boulton & Watt referred to him in the legal struggle, but he had clearly acquired considerable knowledge of engines, because he worked for Boulton & Watt as an engine erector in the early 1780s. Thereafter, he probably worked as an engineman again in Cornwall. He appears in the late 1780s at Wheal Virgin mine, where he may well have engaged in unauthorized tinkering with the engine, for Watt told Wilson that he should warn Bull that he owed his position to Boulton and Watt, who could just as easily remove him.[86] These tinkerings were the subject of talk in the Watt camp of "Bull's Scheme." Bull was evidently competing with the Hornblowers, who were themselves contemplating action against him. Watt, who reported himself as being "much better in my spirits than I have been for a 12 month," told Wilson in September 1789: "In relation to Bull's scheme, the H's [Hornblowers] should first wait its success before they attack it & then they will not be very wise to meddle with it as it will appear on the trial that their whole invention is a robbery, & I apprehend they have not weight of metal to withstand such an attack, but Wheal Virgins gaining the victory will I think rather confirm our patent than annul it, as they will probably hear the law laid down in a different manner than they think of."[87] At this stage Watt was glad to see two sets of troublemakers at each other's throats: one, as he saw it, with neither the merit nor the money to achieve a successful legal outcome, and the other (Bull) a minor player whose tinkering offered little challenge. But then, in 1791 Bull emerged as an engine builder who offered an engine with an inverted cylinder, which enjoyed success. Even more importantly, it appeared that by this time he and Hornblower were part of a concerted campaign by the Cornish adventurers to force Boulton & Watt to law in hopes of successfully challenging Watt's patent, which they saw as vulnerable so far as the specification was concerned. The Cornwall combina-tion did exert considerable pressure: as rival engine builders became more credible and their purchasers ceased payments of premiums to Boulton & Watt, it became pro-gressively easier for other concerns still employing Boulton & Watt engines to refuse payment also. The vicious economic downward spiral had to be stopped, and law was the only way to do it.[88] Importantly, it was clear from the beginning that a successful prosecution by Boulton and Watt for infringement of their patent would immediately raise the issue of their own specification.

So, finally, on 22 and 23 June 1793, after significant delays, the action against Bull came to trial in the Court of Common Pleas located in Westminster Hall. Watt would have made his way there on those cloudy days from no. 6 Green Lettice Lane, the house of his bankers William and Charlotte Matthews, with whom he routinely stayed when in London on business.[89] Serjeant Adair opened the case for Boulton & Watt as plaintiff. He first outlined the nature of Watt's invention and the great saving in fuel that it made possible. Adair then emphasized the fact that Boulton and Watt had devised and operated a very fair system for working their patent, the taking of a third of the fuel savings. Despite this fair way of working, Adair charged, Bull had sought to work around their patent. First he had sought permission to erect en-gines to their design and pay them the savings, knowing full well that Boulton & Watt could not cede such control. When that was denied Bull adopted the expedient of simply turning Watt's engine upside down and calling it his own. Boulton & Watt had made working models of both engines for the trial. The model of Bull's engine had been prepared from drawings made by William Murdoch. The provision of models was itself an indication of the well-organized corporate initiative that lay behind the court action. The jury were invited to inspect both these models in the yard outside Westminster Hall and see with their own eyes, it was claimed, the essential identity of the engines.

Adair called a long list of eminent witnesses who were examined and cross-examined on the originality of Watt's engine and the savings that it made possible. These included several prominent Fellows of the Royal Society—Watt's friend and collaborator De Luc, the great astronomer and instrument maker William Herschel, Dr. James Lind, the engineer Robert Mylne, Secretary of the Society of Arts Samuel More, and the instrument maker Jesse Ramsden. Also among the list of witnesses were Alexander Cumming—the longtime model maker to the Duke of Argyll who had testified at the hearings in Parliament on the extension of Watt's patent back in 1775—and finally John Southern. In case the evidence presented to the jury's own eyes was not enough on the question of Bull's infringement, Adair called William Murdoch, Richard Mitchell, and Peter Godfrey as witnesses to that, and to the crucial fact that Bull had actually erected such an engine.[90]

Peter Godfrey had worked for Edward Bull. As such, he was a valuable witness, providing his credibility was not compromised in any way. When his affidavit was to be used in a later phase of the proceedings against Bull, Ambrose Weston advised Thomas Wilson that Godfrey might be detached from Bull, but only "if this can be done Without Bribery, or any unfair means."[91] That the Watt camp had to be advised to avoid such means on this occasion suggests that they were not above tactics of that sort.

The use of witnesses working in the engine business directly was often a delicate matter. Weston and Boulton & Watt were keen to obtain affidavits from "indifferent witnesses," that is from people who were not transparently part of Team Watt and even, as in the case of Godfrey, people employed by the opposition. Their statements on basic matters of fact concerning the activities of infringers could be very useful. But such people could be exposed to significant risks to their employment, or worse, if they lent their evidence to the cause. On the other hand, such exposure could itself be useful to Boulton & Watt, especially if it led to threats, abuse, or even violence that could itself become the subject of affidavits. Watt himself, observing this rough and tumble courageously but increasingly from the safe, congenial sidelines, put it this way to Wilson:

> We think that Murdock and all our friends should by no means avoid Bull & Trevithick's Company. They *dare* do them no personal injury, and as to their insults, threats and abuse, they *may* become the subject of future affidavits. . . . Are the threats & ill-language that has been used to yourself, Murdock, Rogers, Mitchell and our other friends such, as if made a matter for one or more affidavits, would go to prove the existence of the combination against us? Do they amount to putting your persons in danger and are means employed to render you obnoxious to the County for having told the truth? If so give it to them in Chancery; that at least will stop their mouths in future.[92]

In any case, the witnesses paraded by Team Watt against Bull in 1793 proved effective, at least so far as the value of Watt's invention and the fact of the infringement by Bull was concerned.

Serjeant Le Blanc's defense of Bull relied upon a group of witnesses without the scientific credibility of Boulton & Watt's supporters in terms of the infringement. But Le Blanc began to turn the case in another direction. An infringement was only material if the patent that was supposedly infringed was itself valid. We have seen that Watt and his supporters and collaborators had long recognized that the specification of the patent might be a weak point. In one sense it was deliberately so: recall that Watt had framed the specification in quite abstract terms as a statement of the principles of his improved engine. Following this tack Le Blanc was able to use some of the evidence amassed in cross-examining the opposition's scientifically eminent witnesses. He had obtained from some of them the concession that they would need to perform additional experiments in order to build an engine from Watt's specification alone. If that was so, he contended, how could the specification be sufficient for a humble mechanic to do so, as the law required?

Le Blanc's line of argument was cut short, as was Adair's response, when the jury foreman intervened to say that the jury was perfectly satisfied in finding for the plaintiffs. In this sense, Boulton & Watt won a victory in 1793, but the judge, Justice Eyre, took the shine off it by declaring a point of law that should go to a special case. This, as Boulton and Watt feared, and as their Cornish opponents had long hoped, ensured that the matter of the adequacy of Watt's specification would have to be settled in a separate proceeding.

This legal prolongation of the struggle had important practical consequences. It gave greater confidence to the users of engines in defying Boulton & Watt by purchasing pirate engines and denying them premium payments. Boulton & Watt was forced to take further legal action and ultimately to face a trial of the specification if it was to avoid this downward spiral of business. In the interim, an injunction was taken out against Bull pending resolution of the special case. On 22 March 1794, the Lord Chancellor issued an injunction preventing Bull from erecting any new engines of Watt's design or completing those under construction. Existing Bull engines were not covered by the injunction, but Boulton & Watt sought to negotiate terms with those employing them that would see arrears paid on the old Watt engines in exchange for a license from Boulton & Watt to continue with Bull's engines.

Matters moved slowly. The special case resulting from Justice Eyre's referral concerning the adequacy of the Watt specification was argued in the Court of Common Pleas in June 1794 without result; it was resumed in late January and early February 1795, and again in May 1795. The arguments concerned whether Watt had been allowed, illegitimately, to patent a principle, or whether he had patented an application of a principle. In either case the adequacy of his description to enable a person "versed in the art" of steam engine building to realize Watt's improvements was at issue. Considerable research was done by both sides, including by Watt himself and Watt Jr., on the specifications of other steam engine patents and of other patents in trying to make the case for and against the reasonableness and validity of Watt's specification. The four justices before whom the case was being argued divided equally in their judgments in May 1795. Watt described the situation to Joseph Black: "The Court has sate in Judgement in our cause, the Ld Chief & Mr Justice Rooke made very able arguments in our favour. Justice Buller and Justice Heath against us, (Cornwall & Devonshire own these gentlemen) The votes being equal the court cannot decide we are therefore hung up until some Law lie or other expedient can be devised to loosen the knot."[93] Pleased with the effectiveness of the injunction process in stymieing further piracies in Cornwall, Boulton and Watt proceeded to widen their net to other parts of the country where piracies of rotative engines, now the most dynamic part of their engine business, were becoming a problem. This was a task that Watt Jr. and Robinson Boulton took up with an energy that Watt ascribed to "the eagerness incident to well disposed young men."[94] Watt Jr. toured northern England serving injunctions, especially in Leeds and Newcastle. For the most part these northern pirates, not so well organized as the Cornish variety, submitted to the injunctions. The exception was Jabez Hornblower and his employer J. A. Maberley, who had erected engines in London and in Durham that were widely regarded as piracies of Watt's engine.[95] A lawsuit against them was instituted, and the jury trial took place in mid-December 1796 before Chief Justice Eyre.

As before in the case against Edward Bull, in a jury trial Boulton & Watt was able to marshal a stellar cast of witnesses to testify to the original character of Watt's invention, and to the sufficiency of the specification. These included, again, De Luc, Ramsden, Cumming, and William Herschel. It was also decided to seek support from Joseph Black, John Robison, and the son of Watt's first partner in steam, John Roebuck. Both Black and Robison were ill. In the event, Black did not attend the trial but provided his "History of Mr. Watt's Invention of His Improvements on the Steam Engine," a document that has colored much subsequent history of their relationship.[96] Both Robison and Roebuck Jr. did attend the trial. Boulton & Watt decided not to call Robison, partly because of his ill health but also because the arguments over priority of invention and infringement had gone so well. But Hornblower and Maberley's counsel called Robison, seeing, as they thought, an opportunity to exploit a weak witness. In the event, Robison's evidence was a rhetorical triumph.

When asked whether a person versed in the art could realize Watt's improvements on the basis of his specification, Robison brandished a paper bearing a sketch drawn by a Mr. Model at St. Petersburg, which he claimed had proved enough for a Watt engine to be constructed there. He also delivered a panegyric to Watt's scientific approach and capacities, the resulting originality of his invention, and the tremendous economic value to the country of his and Boulton's engines. Robison was correspondingly hard on the engine pirates as deceiving and discouraging those with capital to invest in economic development.[97] In so clothing Watt and his engines in scientific garb and Boulton & Watt's business in that of public economic benefit, and apparently demonstrating so strikingly the sufficiency of the patent specification, Robison made considerable impact on the jury, which immediately found in Boulton & Watt's favor.

When Watt wrote to thank him for his exertions and their effectiveness, Robison reported that on his return to his class of students at Edinburgh University he was received with a plaudit. When he explained to his audience what had happened and "the simple turn which the Cause took at last, which made it not more the Cause of Watt versus Hornblower than of Science against Ignorance . . . I got another plaudit that Mrs. Siddons would have relished."[98] Robison was not above egging the pudding, but the image of the youth of scientific Scotland relishing Watt's success is plausible and striking. Robison the showman had enjoyed himself, in court and in class.

Still, the legal machinations continued. Hornblower and Maberley applied for a Writ of Error to be heard before the King's Bench. This required argument first about whether such a writ should be granted, an argument that Boulton & Watt lost on a ruling by Eyre. It was not until January 1799 that the business in King's Bench was taken up again, and Boulton & Watt gained a declaration in their favor by all four judges. That, finally, was it, and Watt rejoiced to his partner Boulton: "We have WON THE CAUSE hollow."[99] The task of trying to recover arrears of premiums was now all that was left.

Through all this, Watt was in a state of almost perpetual anxiety. By 1796, as matters moved to a crescendo, relations with his wife were strained to the breaking point because, despite the availability of the "young Gentlemen"—Annie's term for Watt Jr. and Robinson Boulton—Watt refused to join her in Scotland, where she kept a long vigil over her dying father, James McGrigor.[100] Despite an awkward start when he was courting Annie, Watt had enjoyed a good relationship with his father-in-law. But he insisted on giving priority to legal research and being available in case legal proceedings resumed. Watt was also dealing, in his own strange way, with bereavement. Only two years after losing his youngest daughter, Jessy, a loss that he and Annie still felt keenly, Watt lost his estranged daughter, Peggy, who left her husband James Miller with three young children. Watt sent a letter of condolence to Peggy's aunt Agnes, which was delivered by Watt Jr. "who comes to offer any consolation & assistance he can to you & Mr Miller on the present afflicting occasion." By way of comfort to Agnes, Watt offered the following: "All that can be said is that the misfortune has befallen upon us, that our part is now to submit to the will of providence without murmuring & to turn our affections & cares to the dear infants. . . . James will concert with Mr Miller & you what can be done for their welfare & comfort." That said, Watt reported that his wife was reasonably well, though troubled with rheumatism, and that he was in his ordinary state of health, "feeling much the advances of age," which vexations of various kinds (the patent trials) accelerated. Gregory and James were in good health and spirits. Watt noted that by James's "exertions in business my life is rendered much very much more comfortable. . . . May he long continue a comfort and ornament to the family & enjoy more happiness than has fallen to my lot!"[101] If we were not already convinced of it, this thoughtless segue would lead us to conclude that emotional sensitivity was something that Watt simply did not understand and was incapable of feigning. His own troubles and anxieties were real, debilitating, and oft remarked on, but those of others were to be borne without a murmur!

Watt Jr.'s significant role in the ongoing legal activities had begun in earnest in mid-1794, some six months after his prodigal return. An indicator of this is the extent to which he is represented in the correspondence with Thomas Wilson in Cornwall, who was involved in most of the legal discussions. By early 1795 he is the major point of contact with Wilson on legal and many other matters. Boulton advised Wilson in January 1795 that Watt Jr. was in London "& you may write to him & Weston on Law points."[102] Watt was crushed by the death of Jessy and subsequently engaged in the legal struggles only episodically, with Watt Jr. being the source of constant attention to them. Certainly Watt Jr.'s correspondence shows a strong command of Boulton & Watt's tactics and strategy as well as his own increasingly significant role in shaping them.[103] We have noted that he took to the road to serve injunctions upon infringers in the north of England in 1795 as attention shifted to the piracy of rotative engines. He engaged in a great deal of legal research, consulting with Ambrose Weston in particular. When the trial of Hornblower and Maberley came forward, he recruited and coached witnesses.[104] He also put forward a new attitude in the business. While his father oscillated between a very cautious optimism and the depths of despair about legal matters, Watt Jr. announced that "We are all in good health & spirits, neither too much elated with hope, nor depressed by anxiety, but prepared to meet every event with stoical indifference."[105] He tried to persuade his father that they had been focused too much on the patent monopoly as the basis of their competitive advantage over rivals: even if the legal proceedings went against them they had such a technical lead over their competitors with the new machinery and efficient organization of Soho Foundry (which of course Watt Jr. was also instrumental in establishing with the acknowledged help of his father's reputation and capital) they would easily see off competition in the engine business.[106]

We have seen that Watt's inventions, the business activities that effectively exploited them, and the attempts to secure rewards from them were far from a singular venture. Watt's success was a result of the work of numerous collaborators. The collaboration with Boulton has, of course, long been recognized; indeed, it became legendary. But the extent of Watt's indebtedness to the support and work of family, friends, employees, and hired legal help has not been fully recognized or appropriately emphasized. Not the least of the roles of these collaborators was to cope with, and work around, the anxieties and vexations that plagued the principal lead in Team Watt.

Boulton & Watt had spent a fortune on the legal battles of the 1790s, a sum variously estimated at between £8,000 and £10,000, without accounting for the time and energy they had diverted from business to fight them.[107] By this time, however, Boulton & Watt could readily afford it.

Six

The Fruits of Success

Material success was very important to James Watt. This has rarely been emphasised, since historians have tended to follow the heroic, rather oth-erworldly tropes concerning his character that were set out during his lifetime by Team Watt and by nineteenth-century biographers. Depicting Watt as a philosopher and emphasising the differences between him and Boulton (seen as the entrepreneur, financier, and "networker") has encouraged this tendency. Yet there are many clues in the correspondence of his intimates that Watt was materially ambitious: recall, for example, Betty Millar, his sister-in-law who took on the children immediately after Peggy's death, pleading with him not to put the pursuit of fortune above all else. There was surely more than a grain of truth behind Joseph Black's jocular sugges-tion in 1779, reported to Watt by Annie when she was visiting Glasgow, that "you would get free of your headachs when you get rich."[1] Annie's favorite theme in their intimate correspondence—especially when they were separated by his absence in Cornwall, where he was frantically but miserably drumming up business—was that "riches won't give happiness," so he should not work so hard to the exclusion of en-joyment. They should, she said, "enjoy life in the most pleasant form we can dress it in, let us even adorn it with agreeable triffles if it will make it more engaging."[2] The same themes followed him in their correspondence during his numerous absences in London. Watt placed hard work and making money above almost everything else.

Part of him disliked debt, though he had no compunction about incurring his fair share of indebtedness early on to friends and relatives in order to pursue his goal of material independence. The great variety of ventures and styles of work in which Watt engaged early in his life, in the days that he paid himself £35 a year from the shop in Trongate, Glasgow,[3] was itself a symptom of the restless urgency of Watt's material ambition, as was his momentous decision to move to Birmingham.

A degree of material success was not unknown to Watt's childhood. His father had been moderately successful if only in the frame of Greenock, although it does appear that at some stage he suffered a severe reversal of fortune, perhaps as a result of the loss of a ship in which he was heavily invested. His mother's side of the family, the Muirheads, were of the solid Glasgow merchant class with landed estate. Such was certainly the case with Uncle John Muirhead, at whose Croy Leckie estate Watt spent some time as a youth. There he would also have gained insight into the material base that enabled a level of comfort and gentility that his proud mother apparently strove to imitate, in a paler fashion, in the Watt household in Greenock. But even the Muirheads were vulnerable. Uncle John was hit hard by financial troubles, and at one time it was touch-and-go whether the estate at Croy Leckie could be retained. It was held on to and passed down to cousin Robert. So Watt had before him both examples of success and a sense of its precariousness; his partner Roebuck's experiences were another case in point. This sense, I think, gave him ambition to succeed financially on a grand scale that would place him and his family beyond worry.

The urge to make money, however, was not always all consuming: Watt denied it sometimes in darker moments and also once the success that ensured security, and fewer headaches, had been achieved. At a low point after the loss of the Albion Mill to fire in 1791, Watt told Watt Jr., "The ambition both of acquiring fame or money are dead in me, all I wish for is tranquility."[4] At other times Watt was determined to soldier on. A year later when he was again in low spirits amid the early legal tussles with the Hornblowers, Annie had to entreat him, "We have already enough to make us live comfortable . . . do not therefor torment yourself but at once free yourself from a business that is to you an endless source of vexation let us enjoy the remainder of our days in peace."[5] The return of Watt Jr. and his prodigal enthusiasm to take over much of the family business gave Watt the courage to take his wife's advice—sometimes. In September 1794 Watt Jr. wrote that his father was "upon the point of retiring from business and committing the active management to me."[6] But Watt did not let go so quickly, and there were times in the mid-1790s when, from Annie's point of view at least, the process was too gradual. She encouraged her husband (without punctuation) to let the "young Gentlemen" take on the work and the worry of it: "I am sorry you make yourself uneasy about Maberley Albion Mills or any thing else of the kind throw all such plagues upon the young men's shoulders let them fight as you have done they are both able & seem willing and they have much more in their favour than you had when you begaun [*sic*] the contest with me therefore your excuses go for nothing a willing mind is all that is wanted."[7]

Watt's more decisive retirement, in 1800 at the age of sixty-four, when the partnership with Boulton concluded put the stamp on a process that had been going on, albeit haltingly, for some years. From around the turn of the century, though no doubt still accumulating from his investments, Watt was focused on investing money, including on a country estate and other properties in the Welsh borders. How successful was he? What were the fruits of his success? We might begin at the end, by considering the will of "Dr. James Watt," an obviously important document in the dispersal of his fortune, though far from the whole story of that process.

Watt's Will

In July 1819, during what turned out to be his final illness, representatives from Watt's London solicitors visited Heathfield Hall to assist in drawing up the final version of his will.[8] Reading this convoluted document, we might be forgiven for thinking that it was the last will and testament of a landed gentleman. Watt disposed of Heathfield and his other estates; his money in government funds, shares, and mortgages; his cows, pigs, and horses; and his gardening and farming implements. He made provision for the cutting of the timbers of his estates. We gain a glimpse, certainly, of a man of learning as he disposes of personal items: his books on travel and history go to his wife, Annie, and the rest to his son James along with the contents of his laboratory and study. James also inherits his father's books of account, his letters (other than those that Watt exchanged over the years with Annie, which go to her), and his other papers. Beyond this the will ensures Annie's propertied and financial comfort for the rest of her life. Annie was given the use of Heathfield for her lifetime and an annuity of £1,400[9]; this was in addition to the independent income that Watt had established for her by purchase of cottages in the 1780s. There was probably also a substantial income from a trust deed of settlement established in 1808. Though Annie was very well provided for, her will and that of James Jr. confirm that most of Watt's accumulated wealth, however it may have been detained in the meantime, ultimately passed to his only surviving son.[10]

Watt also made substantial bequests, however, to his surviving granddaughters, and smaller ones to his cousins and their families as well as his friends, colleagues, and servants, with the residue of his estate going to James Jr. There is little sign here of Watt the engineer and entrepreneur, the progenitor of the industrial era. There are many reasons for this: Watt had already transferred much of his surviving business interests to his son James, and so they bypassed the formal will. What we would today call Watt's intellectual property (notably his lucrative patents) had of course long ago lapsed. Watt had also long before established a trust to help provide for his wife. Watt, then, had converted the fortune he made from his business ventures into investments and landed property. The legacy of those ventures could be traced in his papers and correspondence, in the continuing businesses overseen by Watt Jr., and in the skilled persons who had worked for, and been trained by, them. It could be seen more broadly in the hundreds of machines that littered the British industrial landscape. But it is undetectable in his will. The fruits of success were there reduced to property and money. So how much money had Watt made?

Watt's Earnings Estimated

It is impossible to reconstruct definitively the exact extent of Watt's fortune, but we can gain a sense of its scale and its likely lower and upper limits. The will itself was proved in October 1819 for upward of £60,000.[11] This is the lower limit of Watt's fortune since, as noted, there had already been major transfers of wealth.

Beginning with inputs to wealth rather than dispersals of it, we can try to estimate Watt's income during the crucial years of his partnership with Boulton, that is from 1775 to 1800. Jennifer Tann has estimated the money earned by Boulton & Watt from the Cornwall engines alone via the premiums that they were paid as amounting to a total of £219,990 during that period.[12] The large reciprocating engines used in Cornwall were, when they were operating, a major source of overall engine revenue. There were forty-nine engines of this sort supplied to Cornwall between 1777 and 1801. When we consider the total orders for Boulton & Watt reciprocating engines over that period, Tann records 183. Had all engines yielded at the same rate as the Cornish engines, which they clearly did not because of the very large fuel savings in Cornwall, then calculating pro rata they would have earned Boulton & Watt £821,595.[13] If non-Cornish reciprocating engines yielded at one-quarter of the rate, then the total premium payments from reciprocating engines would be £150,401 plus £219,990, that is £370,391. This last figure is probably realistic. We have figures for earnings on non-Cornish reciprocating engines in the year 1788 that allow calculation of a figure comparing them with Cornish engines that indicate that the former returned 23 percent of the return of the latter.[14] Obviously, whatever these earnings were they were not all profit since, even though the partners' early modus operandi sought to leave to the customer the expenses of producing the engine, they were not always successful in doing so, and there were in any case substantial expenses involved in Boulton and Watt overseeing the design, construction, and early operation of the engines themselves or by the efforts of the growing band of engine erectors whom they employed. If we put those expenses as eating up, as a guess, 20 percent of earnings, then on the latter earnings figure the partners would have pocketed something around £296,000. This takes no account of the profits that were made from the manufacture and supply of parts in which Boulton & Watt engaged to an increasing extent as the years passed. By the 1790s annual retail sales often matched or exceeded premium income. By that time the sales of rotative engines were becoming substantial. The reciprocating engines earned much more money per engine than the smaller rotative engines that were supplied by Boulton & Watt from the mid 1780s, but the latter were much more numerous: Tann counts 266 ordered between 1783 and 1800.

The total of annual retail sales for all Boulton & Watt engines from 1775 to 1800 was, again according to Tann's calculation, £689,061. If their markup on such sales was 50 percent,[15] then that would be a profit of the order of £230,000. Another consideration is that, thanks to Boulton's activities, the engine business and Boulton's other businesses were not always kept as separate as they might have been.[16] But it is hard to allow for that even by guesswork. We do know, however, that Watt with Annie's encouragement began to draw funds from the engine business and sequester them in Scotland (with Gilbert Hamilton) precisely to avoid being drawn into investment by default in Boulton's perpetual schemes. So, adding our educated guesses of profit from activities generating premium income (£296,000) to profit from sales income (£345,000) we reach a very tentative figure for total profit to the partners from reciprocating and rotative engines of £526,000 between 1775 and 1800. If accurate, then these numbers would mean that Watt's one-third share of profits would have amounted to something like £175,000.[17] There would have been other income flows from other partnerships and projects, such as the copying machine business and from the Delftfield Pottery, for example, but none of these were significant compared to the scale of the engine business. So something short of £200,000 might be a reasonable guess at an upper limit for the income that Watt received during the period of his partnership with Boulton.

That figure, though, is very much an upper limit since, as Richard Hills indicates, much of the money due from Cornwall was not collected. Indeed an accounting made on 22 February 1799 by Thomas Wilson, Boulton & Watt's representative and agent in Cornwall, suggests that the shortfall was very large indeed. Wilson calculated that the total fuel saved in Cornwall over the whole period of operations had been valued at £803,869 19s. 8d. One-third of that had been due to Boulton & Watt as premium, amounting to £267,956 13s. 0d. Wilson calculated that the amount actually received in payments was only £105,904 9s. 5d., leaving a shortfall of over £162,000. Watt Jr. and Robinson Boulton were very enthusiastic in their efforts to recover this short-fall once the legal verdict came down in the partnership's favor, but they only managed to regain a fraction of it. Although damages of £6,000 were awarded to them, little of this was obtained, and Boulton & Watt's legal expenses were of the order of £10,000.[18] So, it does seem that the fortune Watt received from Cornwall could have been as little as £40,000. Combining this with the earlier figures calculated for profits from reciprocating engines, Watt's total "take" might have been closer to £130,000 than to a quarter of a million pounds.

Watt's living expenses, for what became a large household, and his property purchases would have come out of this figure. The provisions for his wife and children would have as well, especially the transfer of business capital to Watt Jr., which must have been considerable. Given that all this was prior to the £60,000+ leavings of his will, it appears that an estimate of Watt's earnings up to 1800 as somewhere around £150,000 might be a reasonable one.

There is another way to estimate earnings. This starts with a set of figures that we have for earnings in the year 1788, from which average premium income of different engines can be calculated and then multiplied by the numbers of engines and their likely working lives.[19] This produces a figure for total premium income for 1775–1800 of approximately £370,000. When we add that to income from retail sales of engines, we arrive at a similar figure to that previously estimated, leaving us again with an estimate for Watt's share of £150,000 to £200,000. It is unclear whether this order of magnitude is what Watt meant when he described himself to Sir Joseph Banks as having retired from business with "a very moderate fortune."[20]

Another clue is provided by Watt's statement of his income tax liability. The income tax was a new phenomenon, having been introduced in 1799 by William Pitt as a temporary measure to cover the cost of the Napoleonic Wars.[21] Initially the tax was voluntary, and Watt, ever the patriot, volunteered. In a return for 1799–1800 Watt declared various sources of income:

I James Watt of Heathfield . . . do declare that I am willing to pay the sum of Three hundred & Forty pounds for my contribution from the 5th day of April 1799 to the 5th day of April 1800 . . . and that I am willing to pay the sum of Two hundred and twenty eight pounds as the joint contributions of myself and my son Gregory Watt who is concerned with me in the business of manufacturing steam engines and also that I am willing to pay nine pounds fifteen shillings as my wife Anne Watts contribution for her separate income and I do declare that the said sums are not less than one tenth of my income.[22]

This declared income tax contribution of £578 15s. od., being "not less than" 10 percent of Watt's income for that year, would point to that income amounting to close to £6,000. The year 1799 was perhaps a bonanza year for Watt, but if his income had averaged that level over the period of the partnership (admittedly an unlikely situation) then that would have represented a total inflow of £150,000.

This income tax data also perhaps gives us some clue as to the investments made up to that time on behalf of Annie. Annie Watt's separate income of approximately £100, if it had been realized from investments in government funds at, say, 3 percent, indicates that the sum invested would have been over £3,000, though it is likely that this income came from the rental of cottages that Watt had bought in her name beginning in the mid-1780s at the time of his wife's announced insecurity about provision for herself and Jessy and Gregory if Watt were to die and Watt Jr. was left in charge of the family fortune.

In 1803 we find further records of tax affairs in the form of Watt's declaration to the Commissioners of the Property Tax acting for the City of London of his property in the government funds.[23] That declaration detailed £19,000 in his name in 3 percent consols and £15,633 16s. 4d. in 3 percent reduced stocks, being a total of £34,633 16s. 4d. on which the dividends were just over a thousand pounds and the duty payable about fifty pounds. In a further declaration at about the same time to local Commissioners acting for the Hundred of Offlow in the County of Stafford, Watt noted the prior declaration regarding money in the funds and added that the assessment of income from his shares in the Birmingham Canal Company would be part of the company's declaration, as would his gains from the Delftfield Company.[24] Watt advised Gilbert Hamilton that there was no legal requirement to declare interest on monies lent through mortgages and bonds. So, Watt clearly had further wealth in the form of such loans, the extent of which is unclear. In February 1804 Watt thanked Hamilton for his account of the "balance in my favour" that Hamilton held in Scotland, that balance being about £6,400.[25] Taken together then, Watt's liquid assets in the early years of the nineteenth century may have been of the order of £40,000 plus his mortgages, bonds, canal shares, and stake in the Delftfield Company. He had long expressed a preference for investment in land rather than stocks and had already invested a substantial amount of what would be several tens of thousands of pounds in property in Wales. This, together with the wealth transferred to Watt Jr. via the engine business, could well have added up to a figure of the same order of magnitude (certainly £100,000 plus) as we estimated above from earnings.

In 1808, when relations between Annie and Watt Jr. had improved, Watt was contemplating establishing a trust involving the transfer of £32,500 in 3 percent stocks, the trustees being Annie, Watt Jr., and Robert Hamilton. This was perhaps in connection with avoiding legacy tax. The dividends were intended to go to Watt during his lifetime; then dividends on £17,000 would go to Annie for her lifetime, and on £15,500 to Watt's grandchildren during Annie's life. Watt subsequently decided to provide for his grandchildren through his will, perhaps because legacy tax was dealt with differently in Scotland, where the grandchildren lived.[26] This reveals the availability of substantial sums and Watt's inclination to use trust arrangements to provide for the dispersal of his wealth before his will took effect. If this money in stocks is the *same* money as that on which Watt declared his earnings in 1803, then our estimate of Watt's fortune based on information from that year remains.

We can also look at Watt's large capital expenditures, such as the sums paid for Heathfield House and for the Doldowlod and other estates in Wales. Watt's decision in 1789 to build a new house, to be called Heathfield Hall, was at the juncture of a number of developments. The situation of his existing residence, Regent's Place, on Harper's Hill was changing. When Watt and Annie first occupied that house it was a genteel distance from the expanding jewellery quarter of the town. But by the mid-1780s the town was encroaching upon it. Negotiations began in early 1787 with Mr. Birch for land that eventually became part of the Heathfield estate, which would put the Watts in a more rural setting but still quite close to Soho. Annie evidently wanted more of a country existence and also to be nearer her friends, and Birch's land would fit the bill. However, at that point there were problems regarding both the price Birch was asking and issues of access.[27] Failing to agree with Birch, later that year Watt considered a more distant existing house, Ladywood, in what is now Edgbaston but was then a rural area, certainly meeting one of Annie's criteria. Dr. William Withering had informed Watt that the house was available and "presses me much to take it." Watt regarded it as "too far from Soho" but asked Annie her opinion of it. She considered the situation very good and thought that a ready-built house would be better than building from scratch. But she agreed that its distance from Soho was a problem.[28] She left the decision to Watt, and he must have declined Lady-wood since they remained for the time being at Regent's Place.

In 1789 Watt paid a Mr J. Blythe £900 for the lease on a site on Handsworth Heath adjacent to Birch's land with an existing house upon it. In May of that year Watt wrote to Samuel Wyatt (who had designed Soho House for Boulton) explaining the situation: "The present building is small but new & neat, on the other page you have a drawing of the additions I propose to make to it, which will make it perfectly convenient for my family, I have only 6 acres of ground in all."[29] The extensive additions that resulted in Heathfield House were executed by Charles Glover for a cost of about £1,400 between October 1789 and August 1790, the final balance being paid to Glover in December that year. The family moved into the house during September 1790. The difficulties of Annie's relationship with her stepchildren, James Jr. and Peggy, had, as we have seen, come to a head some time before. So, the family that moved into Heathfield in 1790 was Watt's second family only. They had plenty of room.

In 1793 Watt bought Birch's adjacent land and cottages for £1,700, a sum much reduced from that initially proposed. Watt drove a hard bargain! Parliamentary Acts of 1791 and 1793 saw the enclosure of the commons of Handsworth Heath as part of the rage for enclosure of common lands throughout the country at that time, and Watt had purchased his lands as part of that process. These purchases produced an estate of some forty-three acres. The total cost of acquiring land and building the house was just over £4,000, to which the cost of employing Wyatt would have to be added. During ensuing years Watt also spent a considerable sum on planting trees and landscaping the grounds. Larch trees that were started there were soon to be used in some of the planting of the estates that he acquired on the Welsh borders.

Watt's total outlay on buildings and grounds must have been of the order of £5,000 before he and Annie began to furnish and equip Heathfield. In the early 1790s the tedium of Watt's patent trial business in London was relieved by the tedium of numerous commissions from Annie connected with furnishing the house. In 1820 Charles Pye's guide to modern Birmingham described the house as an "elegant mansion" and observed that only a few years back, "the adjacent ground was a wild and dreary waste, but it now exhibits all the beauty and luxuriance that art assisted by taste can give it. Woods and groves appear to have started up by command, and it

FIGURE 6.1: Heathfield House, Watt's home from 1790 to his death, pictured in the late nineteenth century. It shows the "gay trellis" veranda upon which Maria Edgeworth remarked during a visit. Reproduced by kind permission of Handsworth Historical Society.

may now vie with any seat in the neighbourhood, for rural elegance and picturesque beauty."[30]

Heathfield was a large, neoclassical-style building with a stucco finish. The additions had included dining and drawing rooms, a library, a hall, and a dressing room. The house incorporated kitchen, larder, brewhouse, and stables. Watt's famous workshop was relegated, tradition has it by Annie, to the distant heights of the house in its southwest garret. But there was also a large room described as the "Elaboratory" next to the brewhouse on the ground floor of the house where experiments would have been conducted, notably those on the breathing of airs that Watt performed with Gregory in a fit of experimentation, including self-experimentation, after Jessy's death in 1794.[31]

Some years before his patent came to an end, bringing with it the conclusion of his partnership with Boulton, Watt was preparing for retirement. Indeed, even as Heathfield was being completed and the grounds planted, Watt was contemplating

life after business. He had not participated in the Soho Foundry partnership established in 1794, Boulton, Watt & Co. That arrangement was between Matthew Boulton, Watt Jr., Robinson Boulton, and Gregory Watt only. Watt's role was as a landlord
for the new company. In mid-1795 he bought the land on which Soho Foundry was to
be built for the sum of £1,340, leasing it to the new partnership for £65 per year until they paid him back.[32] This indicates Watt's preference by this time for investment
in property rather than in business. At this time of Watt's and Boulton's lives, the
contrast between their financial strategies to deal with the wealth they had accumulated is notable. The initial plan had been for Boulton to participate in the acquisition
of land on the Welsh borders. But he withdrew and, in a characteristic move, decided
upon yet another business venture, which, perhaps less characteristically, was to be
a success. This was the minting of coinage and the design and supply of coining machinery.[33]

Watt's decision to invest substantial sums in land was in one way simply the usual
strategy of those who made fortunes in business, banking, or law. However, his decision was also affected by the troubled times—by the ongoing French wars and the
industrial and political unrest that flared up regularly in Britain during that period.
As Larry Stewart has discovered, Watt was an active counterrevolutionary, even to
the extent of secretly informing to the authorities.[34] As a man of fortune Watt decided that land was the most secure investment in these circumstances of unrest. Watt
appears to have anticipated serious social and economic breakdown in Birmingham
and acquired his Welsh estates partly as an insurance against this. His concern with
timber was probably encouraged by the high demand for it from the British navy. So
Watt was not just sheltering his money but making an expedient, and perhaps as he
saw it a patriotic, investment.

Watt's correspondence in the late 1790s onward, especially with his banker, Mrs.
Charlotte Matthews; his lawyer, Ambrose Weston; and his Scottish friend and agent
Gilbert Hamilton is peppered with reflections on the best ways to invest money in
those troubled political times. Those reflections are interesting for the light they
throw on the man.

The gold crisis of 1797 worried Watt, as did the unlimited issue of notes, because
it made it hard for people to flee from danger "if democracy & other works of satan should prevail."[35] The danger of democracy and invasion both inclined him to
investment in land, but "terra firma" had its problems too, which Watt, as we might
expect, had studied seriously. He wrote to Ambrose Weston:

Lands near towns objectionable as depending upon the prosperity or increase of the town—cow feeders bad payers in general—. . . Farmers the most cunning of men. . . . My opinion on land would be to buy that in where the major part is grazing land & of small value per acre, but lying near more cultivated countries—where luxury has not yet made so much progress—the more moderate a farmers living, the more rent he can pay. Estates distant from great towns are of more fixed value than near them, & subject to less heavy taxes, or depradations of mobs or invaders. Lands in Wales I am told pay better than lands in England & are managed at less expense.[36]

This is a sample of how Watt reasoned himself into investment in land in Wales.

Watt's first purchase in Radnorshire in the Welsh borders, in 1797, was an estate belonging to Thomas Harley M.P. known as Stone House.[37] This consisted of the Stone House itself, just southwest of the village of Gladestry; the Stone House Water Mill; and the farms of Llanakiddy, Yr Abbir, and Pantglas, the latter two farms being detached from the main acreage. In all, this estate consisted of 277 acres with timber valued at over £500 and rents from tenants of £185 per year. After long and hard negotiations, Watt paid £4,800 for it. In these early days of his acquisitions Watt told Francis Garbett, who was advising him on them, that he had an immediate £10,000 for speculation in land and within a year could have £4,000 more if all went well.[38] Watt's best-known later purchase was the Doldowlod estate on the River Wye consisting of 557 acres, having a timber worth over £700, and yielding rents of £296 per year.

According to the ledger book of James Crummer, who acted as Watt's agent in all his Welsh affairs, by 1819 (the year of his death) Watt owned twenty-seven properties and estates in Breconshire and Radnorshire with a total rental roll of £1,368 per annum.[39] If the rent received per acre over all this accumulated acreage was the same as that for the Doldowlod estate, then we could calculate that Watt's total rental roll of 1819 would be the product of holdings of 2,574 acres. Further, if we assume that the price paid per acre for all this acreage was the same as that for the Stone House estate (that is £4,800/277 or £17 per acre), then Watt's total expenditure in purchasing his Welsh estates would have been 2574 x £17, or approximately £43,700. This is perhaps a reasonable estimate of the scale of his investment in the Welsh borders, at least in terms of the purchase of the assets.

Although Watt's estates were one of the fruits of his success, he did not treat them as a passive source of income. The way that he reasoned himself to the purchase of the land in the first place, as well as his general love of puzzle solving, made such passivity unlikely. Thus while he could, like many other landlords, have allowed an agent to run the show and sat back to enjoy the returns, it would have been out of character for Watt to do this. Instead he became an active improving landlord, deeply involved in all decisions concerning not only the acquisition but also the use of his estates. This rational and businesslike approach to landed status would also have contrasted in Watt's mind with the lazy approach to such matters by the worst of the landed classes. Just as he had earned his landed property, he would work it in a carefully researched and rational fashion. It is clear that Watt transferred substantial resources into building up these Welsh estates. Whether he eventually invested most of his liquid assets in this way, or whether he retained substantial sums elsewhere, is unclear.

Overall, the financial fruits of Watt's success may never be accurately known. It does appear, however, that the £60,000+ valuation of his will at probate was a significant underestimate of the fortune that he had amassed during the years of his earnings from business. What is certain is that he had achieved his goal of putting his own and his family's finances beyond reach of the sense of impending disaster that had dogged his early years. He had left his eldest son in charge of a very valuable and profitable business; acquired an elegant mansion, extensive estates in the Welsh borders, and numerous other properties in England and Scotland; and had other investments and significant balances in the bank. The son of Greenock had done very well indeed, though his health and his closest relationships had been taxed by his single-minded pursuit of riches.

With the acquisition of these estates in Wales, Watt began a process of gentrification within his family. He and Watt Jr. became heavily involved in improving their farms and in the pursuit of silviculture and horticulture, though Watt Jr. remained devoted also to the engine business. As we will see, Watt resisted nomination to the office of high sheriff, but Watt Jr. succumbed in 1829–1830, when he became high sheriff of Warwickshire. Unlike his father, Watt Jr. lived in a lavish style, mainly at Aston Hall, which he occupied a few months before his father's death.[40] Perhaps ironically, Watt Jr. fought the passage of railways across his land. Watt's grandchildren, great-grandchildren ,and down the generations (the Gibson-Watts) became leaders in county affairs in Radnorshire, given to military service.

When Watt was working his hardest in the 1780s, Annie took frequent refuge back at Clober House, her father's residence in Milngavie near Glasgow, constantly encouraging her husband to forsake the cares of business for a while and join her there. In 1787 she reported feeling much better than she had for a long time, "perhaps laughing ten times more than I have done for years back for we have had such a group of visitors at Clober since I have been here that the weeping philosopher must have relaxed a little. A thousand & a thousand times have I wished you with us, for chearfulness is the balm of life."[41] In retirement, Watt was able to achieve a better balance in his life and take his wife's advice.

Watt in Retirement

Among the major fruits of Watt's success must be counted a long, and in many respects happy, retirement. Watt's life underwent a remarkable change in his last twenty years, and the "weeping philosopher" seems to have disappeared or at least gone into hiding. This was undoubtedly the result of his escape from business. The continual plague of commercial negotiation and of attempts by all and sundry, as he saw it, to steal the fruits of his inventive genius were now largely behind him. The conclusion of the legal cases of the 1790s drew a line under the saga that had begun with the taking out of his patent of 1769. Also over was the partnership with Boulton, whom he admired in many ways and relied upon in more, but who was of a very different temperament to him and with whom Watt and Annie clearly felt they were struggling, almost as much as with any outside threat, to realize and preserve the fortune that their steam engines had made possible. Watt had been persuaded that he left the business in the capable hands of his once-errant son James, of whose efforts he became very proud, and also, to a much lesser extent and for only a brief while before Gregory's death in 1804, in the hands of his youngest son.

In his mid-sixties when he retired, Watt still possessed considerable creative energy. But his financial security and burgeoning reputation enabled him to control how he spent his time. He could finally lead the life of ingenious indolence that he had coveted since his early days in Glasgow, but he and Annie traveled a great deal and the indolence was less ingenious than it might have been. They developed an annual routine, spending the winters and spring at Heathfield, going to London in the early summer, often taking a holiday in Cheltenham or Tenby into the early autumn, and then heading to Scotland to spend time with friends and family there. Given the usually futile efforts she had expended over many long years to get Watt back to their native land on a regular basis, Annie must have been particularly pleased with the latter aspect of the itinerary.

Cheltenham was a spa town that had developed from the 1740s as a place to take the waters. Its popularity with the wealthy and emulative (the Watts qualified on at least one count) was much increased after George III and his family visited there in

1788. Watt and Annie were both valetudinarians—she suffered much from rheumatism—and they were thus medical tourists to a degree. Their other favored resort
town, Tenby on the south Pembrokeshire coast in south Wales, was a small port that
underwent a rapid revival in the early nineteenth century thanks to investment by
Sir William Paxton.[42] Against a backdrop where the French wars had long prevented
grand tourists from visiting the spa towns of Europe, Paxton built sea-bathing baths
at Tenby that opened in July 1806 and quickly became fashionable with high society.
So here, like Cheltenham, was another destination for the creaking couple, involving a manageable journey from Heathfield that would allow a visit to Watt's Welsh
properties en route. Watt and Annie were partaking here of well-established circuits
of internal tourism connecting Birmingham, London, Cheltenham, and various des-
tinations in Wales. There were published guides to the best inns and coach services
on those routes, such as the *Cambrian Directory*.[43] Watt would have read these, no
doubt; anything with which he engaged he made a serious study of, and tourism was
no exception. His correspondence in his retirement years shows him ever ready to advise friends and relations on routes, modes of travel, places to stay, and sights to see.

There were visits to other places too. Even before his retirement Watt and Annie made an extensive tour in 1797 of the south coast, initially via Bath and Bristol
(where Watt saw Beddoes). Watt liked Lymington especially for its salt-water baths,
obliging people, and pleasant roads. The Isle of Wight (Cowes and Newport) he found
expensive, and he disliked using bathing machines on the open beach. The couple also
went to Plymouth and spent a week at Southampton, which Watt considered a "dull
place," before heading home via Winchester and Oxford.[44] After a three-week trip
to Brighton, Watt observed that "the hot baths were of service to Mrs W's rheuma-
tisms & I thought were of use to some of my complaints." But he did not much like
the place, telling Robert Muirhead one of his Frenchman jokes: "A French man who
was asked his opinion of [Brighton] gave a very just one 'I no like land without tree,
nor sea without ship.'"[45] It was Annie who liked the spas. Despite their occasional
medical value, Watt thought there was "nothing to me more insipid than a watering
place." He liked London better because there was always something to do and "the
mere going about the streets is entertainment." But even in the metropolis there were
disadvantages: "there is no avoiding invitations to dinner & consequent late hours."[46]
Watt's celebrity, as perhaps also his growing reputation for amusing conversation,
would gain him many invitations. But his reputed need for nine or ten hours sleep a
night rendered the celebrity circuit a problem too!

Friends and relatives often accompanied Watt and Annie on their travels. Annie's unmarried sister, Janet, was constant company since she seems to have lived at Heath-field much of the time, presumably as a companion to her sister. Watt's cousin Rob-ert Muirhead and his wife visited Heathfield regularly and were fellow travelers on jaunts to Wales, Bristol, Bath, and the south coast, as well as to London.

More extensive travel had been severely restricted by the French Revolution and the French wars. Like many others, Watt and Annie, despite their growing infirmi-ties, were eager to take advantage of the brief peace brought by the Treaty of Amiens to visit the Continent in 1802–1803. They were in Paris for five weeks, and their visit intersected with that of the Edgeworths, with whom they spent considerable time. Watt reported to Robison that they had been well received by old friends in-cluding Berthollet, Monge, and Laplace, as well as by de Prony and Hassenfratz. Any tensions with de Prony about his lack of attention to Watt in his work on the steam engine seem to have diffused. De Prony "appeared to be sorry that he had not taken more notice of me in his book on the steam-engine, and has offered to publish in a suc-ceeding volume, anything I please to furnish him with on the subject."[47] The peace certainly saw an intense revitalization of the networks connecting French and Brit-ish purveyors of enlightenment, but it was all too brief, and the resumption of war prevented further travel and reintroduced communication difficulties for Watt with his friends across the channel.

Death of Gregory Watt

We are accustomed to the regular intrusion of family tragedy into the various stag-es of Watt's life, and the early years of his retirement were no exception. The death of their youngest son, Gregory, on 16 October 1804 struck Watt and Annie very hard. Like his sister Jessy ten years before, Gregory died from consumption.

We have seen that during his studies at Glasgow College in the early 1790s, Greg-ory had begun to distinguish himself as a versatile and talented student, and at the college's Commemoration Day in May 1796 he carried off a number of prizes. Fran-cis Jeffrey, who was present, gave a sense of Watt's character as well as that of his youngest son when he described Gregory as having "all the genius of his father, with a great deal of animation and ardour which is all his own."[48] Gregory certainly had strong natural philosophical talents and interests, which Watt was keen to encour-age. But in 1793, when Gregory was sixteen, Watt seemed resolved that his youngest son would enter the family business. At the time Watt was estranged from Watt Jr., and he told Gilbert Hamilton that his plans for Gregory involved giving him "all the instruction in my power so that the knowledge I have acquired may not be totally lost to my family."[49] To this end Gregory was to spend three half-days a week in the Counting House at Soho and "the other 3 days to be in the drawing apartment where he will learn practical drawing of machines & mechanisms."[50] Spare half-days were to be devoted to experiments in mechanics or chemistry. All this, together with necessary exercise, would completely fill up the boy's time. Watt was determined to prevent idleness, as he had been with Watt Jr.

It is not clear whether Gregory's education at Glasgow College, which quickly intervened, was part of the plan for his ultimate future in the firm or a diversion from it made feasible by Watt Jr.'s return to the fold. In fact, not long after he finished his studies in early 1797, Gregory himself still seemed determined to work for the family business as a partner in the Soho Foundry.[51] But almost immediately came the first serious concerns for his health, with an attack of influenza and other ailments leading to pursuit of a better climate than Birmingham for a while. Gregory traveled with the family group (Watt, Annie, and her sister Janet) to Clifton, Bath, Leamington, Cowes, and Southampton, getting plenty of exercise and also pursuing at various opportunities his growing mineralogical interests. Although he was much improved as a result, it was decided that he should winter in Cornwall, though Watt issued further instructions for his business education to continue by visits to mines in the company of Thomas Wilson and William Murdoch. Gregory was to report on the Boulton & Watt engines working there.[52] Watt was particularly keen for Gregory to shadow Murdoch as much as possible to learn how business was done, an acknowledgment on Watt's part that Murdoch possessed skills for on-the-ground negotiation that he lacked. Apart from studying the performance, operation, and repair of engines, Gregory could also pursue his mineralogy.

So, in December 1797 Gregory went to Penzance, where he lodged with a widow, Mrs. Davy, and became acquainted with her son Humphry. It was through the Watt connection that the future president of the Royal Society was introduced to Thomas Beddoes as an assistant, beginning his rapid rise in the scientific world.[53] Gregory became part of a circle of young men, including Coleridge, Southey, and Tom Wedgwood as well as Davy, who engaged together in all manner of experimentation, including with their own bodies (most famously with nitrous oxide), in a way that was probably not beneficial to Gregory's consumptive frame.[54] When he returned to Birmingham in June 1798, Gregory once again resumed direct involvement with the family business for a period in the staid confines of Soho.

In the summer of 1801 Gregory began touring the Continent, aiming to acquire knowledge and health. The Watt family, like many others at the time, regarded travel in itself as salutary. Gregory traveled through Germany and Switzerland until November, when he established himself in Paris to study the sciences. Suffering ill health once again, in February 1802, with his father's encouragement, he traveled through the South of France and then into Italy seeking a better climate. William Maclure, an American whom Gregory had met in Paris, accompanied him on this journey. As we have seen, in August 1802, partly because of fears for Gregory's health but also seizing an opportunity that the Treaty of Amiens offered, Watt and Annie decided to travel on the Continent rather than head to Scotland as had been their recent habit. However, Gregory's anxious mother and father failed to meet up with him. He traveled back by Paris to Leipzig and Amsterdam, arriving home in England in mid-October 1802.

Gregory made yet another attempt in early 1803 to resume work at Soho, but in June 1804 he resigned from the firm, citing his ill health as by now constitutional and his unwillingness to remain "a dead weight upon the concern."[55] By that time he was in Bath with his mother and father, where they remained through most of the summer; they then traveled to Sidmouth in September. Gregory's increasingly hopeless state induced Watt to try to take him back to Clifton or to Bath, where he might find better medical attention. They stopped at Exeter on 8 October because Gregory was too ill to continue. Eight days later he died. His distraught parents buried him on 22 October in St. Peter's Cathedral, Exeter, then struggled home to Birmingham.

When he resumed his correspondence, Watt described himself as "assailed by the many afflicting recollections" of his poor son and haunted by the experience of watching "a fine young man wasting away without having the power to prevent it."[56] To his old friend Gilbert Hamilton he declared, "My loss is no common one, for I believe few have such a son to lose & I should probably not have borne up under it so well as I do had not the misfortune been long anticipated. I cannot describe the anxiety & other feelings which have been my lot for these last twelve months."[57] Writing to his cousin and close friend Robert Muirhead, Watt tried to maintain control, praising his dead son's genius, ability to adapt to any science, and activity and industry despite long illness. He continued with the stoic façade: "I cannot weep, but I must ever lament his early fate. We must however console ourselves as well as we can & remember that we still have duties to fulfil in the world & that we still have a son affectionately attached to us, whose abilities do not fall short of his Brothers though differently directed."[58] But this resignation to Providence then gave way as Watt confessed to Muirhead the "tryal" he had faced and the fortitude he had had to summon in Exeter, consoling Mrs Watt while "among strangers without a friend, except those which humanity brought to our assistance, obliged to look out a grave for him who in the course of nature should have seen my head laid in it." One senses Watt's horror at his loss of emotional control in writing this: "But I must stop, I did not mean to have said so much & what I have said is only for your own reading."[59] Watt did not want a break in his habitual stoicism to be known even in his immediate family circle. In ensuing months his lament continued, and even as he repeatedly invoked the necessity to submit to the will of Providence and be thankful for remaining blessings, he confessed again, this time to Hamilton, that he had lost "all desire for my customary amusements & . . . feel that I have little more to do in the world."[60]

Nevertheless, Watt and his remaining son busied themselves with necessary tasks such as obtaining and distributing mourning rings and, for the older ladies, lockets containing Gregory's hair.[61] Gregory left no formal will, only a letter of intent, but legal recognition of this was obtained. Watt Jr. took the lead in this even to the extent of himself paying the tax on all the legacies. Gregory had requested his brother to settle £3,000 upon his late half-sister Peggy's children and £100 per year on "Miss McGrigor" (probably his Aunt Janet) during her life. Exactly how to achieve this was left to Watt Jr., as was the task of redistributing Gregory's interest in the business, which went to Watt Jr. himself.

For all his distress Watt was alert as ever on financial matters. In sorting out Gregory's private debts, Watt listed those owed to him: "a years board & the extra expence [sic] of his living at Bath for both which I could not put down less than £200 . . . money paid by me to his Doctors over what was charged to his account above £100. His funeral charges at Exeter, £60 exclusive of mournings & servants to which add his monument." Finally, Watt added that "Mrs W says she lent him £10 or guineas at Bath." This precise accounting seems strange in the circumstances, but it illustrates, once again, the degree to which Watt saw punctiliousness about personal accounts as both a duty and high among the moral virtues. It was necessary to Gregory's honor that these matters be sorted out fully. It was also important to make sure that the taxman did not gain unduly—"the charges I have mentioned are fair claims on his estate I do not mean to make them against you, but my free gift should not be taxed."[62]

Amid all of this, of paramount concern to Watt and Watt Jr. was Gregory's literary remains. They were proud that Gregory's paper on basalt was published in the *Philosophical Transactions of the Royal Society of London* not long after his death.[63] This took the form of a letter to C. F. Greville, vice president of the Royal Society, a leading figure in the mineralogical and geological community and a good friend of the Watt family. Gregory dated that letter 10 April 1804, meaning that he finished it just before his final sojourn in Bath. The investigations it reported focused on Gregory's use of an iron foundry reverberating furnace to conduct large-scale experiments on the production of basalt and its characteristics, an interesting example of manufactory natural philosophy. But the paper also drew upon observations during his extensive travels. Watt thanked Greville for his lament of the talented young man, and he advised that "James and myself have been engaged in examining his papers but perhaps from some jealousy of his fame we find little except an unfinished memoir on vitrification, which with some things which have appeared in print though not under his name, his correspondence & his mineralogical map of Scotland & whatever else we can pick up, we have some thoughts of printing for the use of his friends."[64] They were scouring Gregory's Commonplace Book for original material and wrote to his friends asking for his letters, or copies of sections of them, dealing with science or arts. The printing project was not proceeded with. It is not clear whether Watt had any intimation of the challenging scatological and homoerotic correspondence that his golden boy exchanged with William Creighton, whom we have met as an employee of Boulton & Watt, and who accompanied Gregory on some of his geological excursions. Gregory enjoyed similar friendships when a very young man with William Wilkinson and at Glasgow College with Thomas Campbell, the poet, and his tutor Thomas Jackson. He also had a spirited time in Cornwall with Davy, Tom Wedgwood, Robert Southey, and S. T. Coleridge.[65] Had Gregory's transgressive writings, and possibly behavior, become known to his father, an open rift such as that which had earlier occurred with Watt Jr. would surely have been inevitable. As it was, the gathering together of his son's literary remains did nothing to cloud the fond memories preserved in the trunk of Gregory's possessions that sat in Watt's garret workshop at Heathfield until his death, and long beyond. It was a testament to his devotion to a son whose loss haunted Watt's years of retirement.[66]

Yet another loss of a young life to consumption was that of Watt's only grandson, James Miller Jr. Despite his strained relations with Peggy before her death and his less than flattering opinion of her husband, Watt had kept a benevolent eye on their children, as had Watt Jr. and Gregory. Watt was clearly very fond of the boy and in 1815 remarked of him, by now a young man of twenty-one: "Although James does not show much genius yet he possesses good sense and is remarkably affectionate, docile and obedient . . ."[67] Watt might have added that he was a refreshing change from his late mother, and others of Watt's own children!

Learning that his grandson was in likely consumptive decline, Watt, and Watt Jr., moved with imperious generosity, as so often in family affairs. They sent the invalid to Plymouth and ensured that the young man made his will. Most extravagantly, they decided to send him on a voyage to Madeira or the Cape of Good Hope. Watt asked Sir Joseph Banks to inquire which might be the best for James's health. Watt Jr. went to Plymouth to see James off to the selected destination, Madeira. A young apothecary had been hired to travel with the invalid. Despite this mobilization of the wealth and contacts of the Watt family, James Miller Jr. died not long after his arrival in Madeira.[68]

Less untimely deaths were, inevitably, a frequent occurrence in Watt's years of retirement, and he found himself offering condolences to relatives and friends and routinely remarking on their need to accept the work of Providence. Two of his close associates during the Glasgow years who had remained lifelong friends were Joseph Black, who died quietly in his chair in December 1799, and John Robison, who died after a long painful illness a few months after Gregory Watt. When Watt consoled Robison's son that his father had "lived long enough to have by his writings raised monuments to himself that will long outlive the date of frail mortality," the implied contrast with Gregory's lot was clear.[69] When Robison's own lectures and other works were very belatedly also collected together and published, Watt was, as we will see in the final chapters, to have a hand in correcting some of his friend's "misunderstandings," especially about the origins of the steam engine improvements. But Robison had been a great friend, particularly in attending the patent trials when in poor health and making a decisive contribution. He had thus played an important part in securing Watt's reputation and fortune.

Gilbert Hamilton—Watt's faithful friend, relative by marriage, and manager of affairs in Scotland—and Thomas Beddoes both died in December 1808. Then on 17 August 1809, Matthew Boulton, too, passed away at age eighty. In Scotland at the time, Watt missed the large funeral that conveyed Boulton's remains through the streets from Soho House to his grave at St. Mary's Church in Handsworth with most of his Soho workforce in attendance and thousands of spectators straining to see.[70] According to Watt Jr.'s report to his father, the affair was "well-conducted" and was an "awful and impressive" ceremony.[71] The man, after all, had created considerable employment and prosperity in Birmingham.

Watt played a part from a distance. He wrote a moving tribute to his late partner in a letter to Robinson Boulton and agreed to the son's request to write a memoir of his father. Now, Watt and Annie had certainly had their differences with Boulton, often anxious that his perpetual risk taking threatened to bring the whole enterprise down. They were critical of Boulton's ostentatious tendencies and his cultivation of the rich and powerful. Since his own retirement Watt had reported to numerous friends on Boulton's stubborn insistence on working even when debilitated by illness and pain. But for the purposes of the memoir, all this was forgotten. Watt acknowledged his partner's great qualities as a manufacturer, engineer, businessman, and promoter. Watt attributed their joint success to Boulton's optimism and enthusiasm, which counterbalanced his own "despondency and diffidence." In all, it was a judicious and generous tribute.[72] As such it perhaps inevitably made poor history and gave Watt's imprimatur to a view of their relationship that was to mislead a great deal thereafter.[73]

Watt was to survive his partner by a decade, and he crammed a great deal into those years even as his circle shrank along with his correspondence. Watt complained increasingly of "rheumaticks," including in his hands, which caused him pain when writing. His letters became fewer and generally shorter. He found comfort in his relationship with Watt Jr. and a small circle of friends. He still traveled, though not as extensively.

Relations between the Watt and Boulton families seemed for a while as if they might be cemented in a new generation, not only by business partnership but also by marriage. This was the prospect, which in 1810 seemed very real, of the union of Watt Jr. and Boulton's daughter Anne.[74] By the time this episode began, Watt Jr. was forty-one years old. The issue of his bachelorhood must have been a concern not only to him but to his father and his stepmother. With Gregory's death in 1804, Watt Jr.'s eligible bachelorhood became perhaps even more of an issue if a direct male line was to be continued.

Watt Jr. and Anne Boulton (1768–1829) were close contemporaries. They had spent much time together as children, and perhaps there were positive echoes in this of Watt's relationship with his first wife, Peggy. Anne was her father's favorite. She had a limp because one leg was shorter than the other, but she was characteristically determined to minimize the impact of that problem upon her life. Anne felt the loss of her father greatly, and she was watched over carefully by her brother Robinson, her particular friend Annie Watt, and Watt Jr. But once the initial grief subsided, Anne had to settle into a new role within the family, and she found herself facing a future as housekeeper to her brother, a common fate of unmarried women.

Anne had not been without proposals of marriage. She had been engaged in 1803 to Dr. John Carmichael, but the affair had fallen through. In 1810 she rejected a proposal from the naval man Sir Isaac Coffin, perhaps because she and Watt Jr. were already discussing marriage. In June 1810, Watt Jr. and his stepmother were in London waiting for Anne to join them there. The plan was clearly afoot, with Annie Watt playing a role. She told Watt Jr. that she had written to his "fair friend urging her very much to return no more to Soho a single lady, you may add your forces to mine and try if we cannot prevail."[75] When Anne did arrive in London, Watt himself seems to have discussed the issue with her, and he then wrote to his son: "A certain person has informed Mrs W that her brother seems to think that the affair is entirely broken off between her and her lover & that she cannot assume courage enough to speak to him on it, it seems therefore incumbent on the gentleman to come to an eclairisse-ment with the Brother without further delay."[76] Watt Jr. and Anne clearly wished to marry, but Robinson Boulton objected, or at least required some persuasion. It is also clear from Watt Jr.'s consequent letter to Robinson Boulton that the latter's reluctance was matched by Anne's unwillingness to discuss the issue further with her brother. Watt Jr. asked Robinson directly to raise the subject with Anne.[77] In the same letter there is a long, convoluted passage that relates to one of the possible objections to the marriage that had been discussed at some stage between Watt Jr. and Boulton Jr. This related to Anne's health, probably a gynaecological or obstetric prob-lem or at least concern. In addition, there were issues about the financial terms of any marriage under the complex provisions of Boulton's will. Watt Jr. wrote at length to his father about these questions.[78]

For whatever reason, by the end of 1810 Watt Jr. had fallen silent on the question of marriage, a neglect for which his stepmother reproached him in February 1811.[79] Annie Watt explained that Anne Boulton was upset, fearing that James had changed his mind, but herself was as ardent as ever. Anne was putting up with her brother's ill treatment in the hope and expectation that Watt Jr. would come to the rescue. Annie Watt clearly thought Robinson Boulton the great culprit and obstacle—he had from the beginning, she considered, found every objection he could to the match in order to keep Anne in his household.

There the affair ended. Watt Jr. seems to have abandoned the idea of the marriage in the face of Robinson's opposition, this despite his own and Anne's feelings and the support of his father, his stepmother, and a few close friends who were party to the secret. Whether for financial reasons, because of fears about Anne's ability to become pregnant (or the risks involved in such an event), or, more prosaically, because Rob-inson wanted to retain his housekeeper, the houses of Watt and Boulton were not conjoined in marriage. Somehow, Anne and Watt Jr. remained friends, but the episode probably soured the relationship between Watt Jr. and Boulton Jr. for the rest of their lives. Anne Boulton had the independent means, thanks to her father's will, to set up her own household, which she promptly did.

Although it is clear that Watt had favored the marriage, he made no visible fuss about the decision to abandon it. Once again, it appears, Annie had taken the lead in emotional matters while Watt hovered in the background. It is surprising that the future of the male line of the family did not exercise him more than it apparently did.

Projects in Retirement

At Heathfield, Watt spent much time in his workshop on projects there, especially during the winter months, when the difficulty of travel and the threat to health kept the Watts at home. Perhaps the best known of Watt's retirement projects was the development of his sculpture machines, that is machines to make faithful copies, either at the same or reduced size, of sculpted materials. These are well known because they were such a prominent feature of Watt's Garret Workshop at Heathfield, as subsequently preserved at the Science Museum in London.[80] At first blush this might seem a rather strange departure for Watt, but it had continuities with his other inventive work. Mechanically, these machines were a development of other linkage inventions that Watt had come up with, beginning with his perspective machine during his Glasgow days and including various aspects of his steam engine linkages and regulators.[81] At another level they were a different manifestation of an interest in mechanical copying that Watt, and Boulton, had exhibited in other ways. Watt's machine for producing press copies of letters and other writings or drawings was the most obvious, and successful, example. But Boulton, with Francis Eginton, had also sought to produce mechanical pictures, a venture that was less successful, though it may have given Watt crucial ideas about his letterpress copier.[82] Of course, the copying of designs onto ceramics was another aspect of the art of reproduction that interested them. Watt had a long-standing interest in the production of portraits in the form of plaster casts from moulds of the sort commercialized by his friend Wedgwood in his portrait medallions. He also was closely aware of the more general business of applying moulded relief work to objects of various kinds. Finally, his interest in clays and plaster dated back to his early days with the Delftfield Company.

As in the case of Wedgwood, this interest in the reproduction of relief images and three-dimensional representations was not a matter of idle curiosity but of real market value as the consumption of such objects expanded significantly in the late eighteenth and early nineteenth centuries. Watt's major retirement project, then, had strong commercial possibilities underpinning it. It appears that he prepared drawings with a view to patenting his machines, but that plan was overtaken by growing frailty and remained unrealized at his death.

The sculpture machine project may have begun in Watt's mind during his trip to France with Annie in 1802. While in Paris he visited an exhibition in which he saw a machine for copying medals and conceived the idea of doing that in full three dimensions.[83] Two years later he designed a machine for copying sculpture in a reduced mode, which came to be called an "eidograph." The pointer tracing the rotating item to be copied was linked to a tool for working the rotating blank, which became the copy, both being mounted on a pivoted rod and the whole apparatus powered by a treadle. Watt had various collaborators: his nephew Robert Hamilton, himself a potter, assisted, as did William Murdoch with the metal parts of the machinery, while Peter Ewart in Manchester helped with tool bits. Artists also became involved. Peter Turnerelli, a highly reputed sculptor, supplied busts of Watt himself, and also of Sappho, Socrates, and the Princess of Wales, which may have been used as test pieces.

Watt also designed and built a machine for making same-size copies, which was much more complex than the eidograph, and he managed to produce a machine that could make multiple copies simultaneously. He reported to Ambrose Weston in 1809 on his considerable progress with the "carving machine" and playfully tried out possible Greek names for it: "What do you think of the following? Iconspaia Iconurga Iconoglypta Agalmatopoia Glypta Polyglypt Glyptick Machine . . ." He asked Weston to observe secrecy on all this.[84] By 1810 the machines were well developed, but finetuning continued for some time as well as practice in their use.

In July or early August 1814 Watt sat for his bust for Francis Chantrey. Ever alert to improvement, and learning that Chantrey was unhappy with the plaster he was using, Watt advised Robert Hamilton to send some of his product to Chantrey. By November Chantrey was preparing to send the head of the bust to Watt, partly so that it could be tried on the copying machine. Watt told Hamilton that Watt Jr. had seen the bust and, though he felt it might still be improved, considered it "by much the best likeness which had been made of me either in sculpture or painting."[85] The marble version was exhibited at the Royal Academy in 1815. Watt spent many hours in his garret workshop making copies of this bust of himself, a victim more of obsessive concern for improvement than of narcissism.

The sculpture machine project, though technically successful in that the machines worked reasonably well, was never completed, or at least never reached the stage where Watt took out the patent that he had been working on for some time. The fact that Watt conceived the project in commercial terms, as did the artists with whom he collaborated, means that we should not see it merely as a hobby of his retirement years.

Interestingly, Watt probably spent less time in the laboratory at Heathfield after his retirement than before it. The laboratory on the ground floor of the house most likely found other uses, perhaps filled with produce from the gardens. Significantly, in February 1810 Watt wrote to his London supplier of French publications, Joseph Charles De Boffe, canceling his subscription to *Annales de Chimie*, "as the greater part of these Annales are occupied with subjects that have ceased to interest me."[86] He showed little interest in philosophical experimentation during these years, except in connection with literary pursuits related to the issue of his reputation as a natural philosopher, notably the work for Brewster on Robison's *Britannica* essays on "Steam" and "Steam-Engine." This reinforces our impression that Watt was not a closeted natural philosopher but someone who was driven to experimentation, at least initially, by practical objectives and as an escape from the cares of business. During his retirement, without these imperatives, Watt the natural philosopher went into abeyance. Also, perhaps he felt overtaken by rapidly moving events in the fields to which he had contributed. In 1815, in the wake of Watt's election as a foreign member of the French Institute (he had been a corresponding member since 1808), Jean-Baptiste Biot inquired through Sir Joseph Banks about Watt's experiments on heat and where he might learn of them with a view to referring to them in some of his own work. Watt responded very cagily, directing Biot to the commentary that he had written for David Brewster on Robison's memoir on the steam engine: "I put therein some account of all I have done on the subject of the chaleur specifique or in plain English of the heats at which water boils under various pressures, on its Latent heat &c." But the publication of this was likely to be long delayed because of Brewster's many commitments: "Mean while nothing relating to the subject is in my possession and if it were, I should not do either Mr. Biot or myself justice by sending much less than the whole I have written, which would be to me, *infandum renovare dolorem*, and without it were accompanied with Dr. Robisons text would not be always intelligible."[87] This, especially the Latin quotation from book two of the Aeneid about reliving unutterable grief, certainly reads as if Watt was at the very least tired of this material. I suspect that he was probably no longer in command of it, and even somewhat embarrassed by the fact. Then there was his health. He wrote to

M. Lévêque in 1810, "My health does not permit me to make chemical experiments, but I still do a little in mechanics . . .".[88] The latter was, of course, a reference to the sculpture machines.

As Watt's interests were shifting, his retirement was not an idle one. But his work regime was kept under control in various ways. First there was travel, to Wales and the southwest, to London, and to Scotland, which took him away for a good part of the year. Even when he was at Heathfield, conditions in Watt's garret workshop were not always conducive to spending long hours there because the remote upper reaches of the house could become stifling hot in summer and very cold in winter. Watt's rheumatism was not so severe as Annie's, but it did affect him badly sometimes. The pain in his hands that sometimes created problems in writing must have interfered also with operation of his sculpture machines. Watt spent much time on less taxing pursuits such as reading. His reading material was of all sorts, but horticultural catalogues and light fiction, particularly of a romantic Scottish variety, figured significantly in it. There was also time spent with friends and visitors in conversation, a pastime that Watt enjoyed and at which he excelled.

As early as 1796 Watt found himself in Annie's bad books for exercising his conversational skills. She was in Scotland attending to her dying father and learned that while telling her about his hard work at Heathfield on legal matters (which he said prevented him from joining her), Watt was conducting informal gatherings there with various young people, including Ann (or Anna) Campbell, the daughter of his cousin Marion, who was visiting. Annie was angry that Watt had turned congenial "School master after supper" and that Ms. Campbell flattered herself on "engaging the attention of such a man as Mr Watt . . . who lets slip no opportunity to *amuse* and entertain her."[89] This suggests a character of Watt that we have not seen before—the avuncular sage. Perhaps here, in 1796, with Watt reasonably secure despite his ongoing worrying, the valetudinarian mask with which he greeted the world (especially Annie) began on occasion to slip, such as when he had the opportunity to play the sage. But it was not just young ladies that Watt entranced. Two years later, as the end of legal proceedings in the patent trials approached, Watt's lawyer Ambrose Weston remarked upon the way that those trials had distracted the sage from his central function. Weston told Watt how much he enjoyed his company and receiving the benefit of "those stores of wisdom & knowledge which you are constantly & without effort, pouring out for the entertainment & instruction of your friends."[90]

Other reports from the period of his retirement seem to reveal such a character. In 1805 he visited Scotland and spent time in Edinburgh, where Henry Brougham recalled him as a regular attender at the "Friday Club" that included Francis Jeffrey, Professor John Playfair, Leonard Horner, and others. Watt was the life of this gathering of the "Northern lights." Walter Scott recalled another occasion, probably in 1814, when Watt held his own among the northern literati whatever the topic of conversation: "The alert, kind, benevolent old man had his attention alive to every one's question, his information at everyone's command. His talents and fancy overflowed on every subject."[91]

This Watt, however, was not entirely new. Such qualities were responsible presumably for attracting the circle of young men that Watt gathered around him in his workshop at Glasgow College forty years before, and they must also have found expression in the Lunar Society gatherings of his maturity. But during his retirement this Watt was seen more often and had a lifetime of readily recallable information at his fingertips with which to instruct and entertain. As his fame as engineer and natural philosopher grew and was consolidated, Watt's capacity for everyday conversation and easy communication of his stores of wisdom became remarkable and remarked upon. It appears that many were flattered and delighted to engage "the attention of such a man as Mr Watt." Such stories about the great and the good contribute to the establishment of their legendary status.

Seven

The Living Legend

*B*ack in the autumn of *1792*, Gregory Watt had traveled to Glasgow with
his mother, leaving his father behind at Heathfield to worry about the
threat by the Cornish interest to his patent and his profits. Gregory was about to
turn fifteen and attend classes at Glasgow College. Annie subsequently reported that
the boy was well received by Watt's college friends "as being your son" and beyond
that he had "been introduced to some of his fellow students by the proffessors as
son to the Great Mr Watt."[1] Later, advising that Gregory was doing very well in his
studies, Annie noted that he had "become not a little proud of his father, the many
compliments that he hears paid you seem to give him great delight. So you see if you
have not got profit you have got fame in your own country."[2] Annie's emphasis on
Gregory's pride in his father might well have been intended to contrast with the attitude of Watt Jr., whose behavior at about this time was causing Watt no end of grief.
But whatever it might indicate about family dynamics, this episode was certainly a
sign of Watt's growing celebrity by the 1790s. In the last twenty years of his life, he
became a living legend not just in his own country but further afield.

Things happened to Watt that happen to, and help to create, legends: he was repeatedly written about, frequently honored, and popularly portrayed.[3] He also began
to play the part, relishing, as we have seen, a role as avuncular sage in social gatherings and engaging in public acts of philanthropy, which were certainly genuine but at

the same time intended to mark his achievements and perpetuate his memory. While money was long Watt's main measure of success, he was always concerned, if often behind a cloak of diffidence, with reputation and fame. As his finances were thoroughly secured, reputation became more important. Celebrity, as always, was constituted by an amalgam of bestowals by others and cultivation by self. Not everyone, however, celebrated Watt. His enemies in the steam business and disgruntled employees propagated a different image of Watt as a ruthless, unprincipled monopolist who stifled the inventions of others and rode to glory on their backs. These countercurrents were in themselves a sign of the fame—or, for some, the notoriety—he had achieved.

Watt in Print

Watt found himself and his works represented in various literary forms, including poetry.[4] The stanzas of Erasmus Darwin's famous poem *The Botanic Garden* of 1791 featured the engines:

NYMPHS! You erewhile on simmering cauldrons play'd,
And call'd delighted SAVERY to your aid;
Bade round the youth explosive STEAM aspire
In gathering clouds, and wing'd the wave with fire;
Bade with cold streams the quick expansion stop.
And sunk the immense vapour to a drop.—
Press'd by the ponderous air the Piston falls
Resistless, sliding through its iron walls; Quick moves the balanced beam, of
    giant-birth,
Wields his large limbs, and nodding shakes the earth.[5]

But in his philosophical notes to the poem (note XI), Darwin presented the achievements of Boulton and Watt—the product of their "united ingenuity"—as the culmination of the long history of the steam engine and the basis for its remarkable future as he anticipated it. Darwin compiled this history from information supplied at his request by Boulton and Watt. While the former had confidently supplied the facts, Watt, in low spirits, presented a self-effacing account and asked that Darwin bestow no personal praise upon him: "In my own conscience," he said, "I do not think I merit much praise. I am in fact a mere compilator. I have tacked and sewed together the seraffo & remnants of many a man's ideas, altered & fitted them to purposes he never thought of, but I have invented or discovered little."[6] Darwin, in response, gently chided Watt for his false modesty: "Have not you more mechanical invention, accuracy, and execution than any other person alive?—besides an inexhaustible fund of wit, when you please to call for it? So Misers talk of their poverty that their companions may contradict them." He proceeded to prescribe remedies for Watt's lack of cheerfulness, and in what he wrote promptly ignored his friend's diffidence about his originality.[7]

Watt's diffidence, however, inevitably became a feature of the legend. Another piece of poetry, less exalted than Darwin's, in which Watt featured was penned by a Birmingham barrister in 1794 and appeared in William Hutton's history of that town. It exhorts Watt, in the same manner as Darwin, to abandon his self-deprecating ways:

> Rise, lowly Watt, retiring merit, rise!
> Erect thy honest front, assert thy claim,
> Illustrious partner of thy Boulton's fame
> With perseverance keen your eye explores
> Nature's mysterious law, and latent pow'rs.
> At length the secret dawns—behold at length
> The engine heaving with augmented strength.
> United glories round your heads shall stream,
> While bubbling boilers generate their steam.[8]

A steady stream of local literature began to feature Watt from thereon. But he also began to appear as subject to a wider public.

The *Encyclopaedia Britannica* in its early days did not include biographical entries, and when it did introduce them it did not at first cover contemporary figures, but not surprisingly Watt appeared nevertheless in its pages. He did so first in the third edition of 1797 in the articles "Steam" and "Steam-Engine" written by his old friend Professor John Robison. He appeared in the entry "Water," which gave an extended history of the discovery of the composition of water. This was almost certainly written by Robison too. Watt also made cameo appearances in the articles "Soho," which waxed lyrical about the public and national benefit of Boulton's manufactory and Watt's engines, and "Writing," where the "ingenious Mr Watt's" copying machine was described. Watt is presented as a transformative engineer and as a profound philosopher. Stories that had hitherto been told only to the juries and judges in the steam patent trials were now retailed. His steam engine improvements are attributed to the careful philosophical approach that Watt supposedly took, and they take up a third of the thirty pages devoted to the steam engine. The many and varied engines that Watt had erected were products of his "fertile genius," and the numberless attempts to improve on Watt's engines had largely failed because "our engineers by profession are in general miserably deficient in that accurate knowledge of mechanics and of chemistry which is necessary for understanding this machine." Robison justified the detail into which he entered as enabling the reader to judge engines better and "not be deceived by the puffs of an ignorant or dishonest engineer."[9]

Robison, certainly by his own lights, was intent here upon recording Watt's achievements. But there were aspects of his account with which Watt and his collaborators were not happy. We have already noticed that Robison's description of him as a "pupil" of Black was something Watt took particular exception to. There was also another problem: Robison had given a positive assessment of a Hornblower engine in the article; indeed, it was the only engine besides Watt's to be given a plate to illustrate it. Robison made an exception here to his blanket condemnation of other engines and engineers, detecting "ingenuity and real skill" in the Hornblower construction. As to that machine's indebtedness or otherwise to the Watt engine, Robison left open the possibility that, depending on precisely how and where the condensation had been conducted, Hornblower had "erected engines clear of Mr. Watt's patent which are considerably superior to Newcomen's."[10]

In the midst of initiating plans for Robison to give testimony in the patent case against Hornblower & Maberley, Watt expressed his view about Robison's concession in the *Encyclopaedia Britannica* article:

I would thank you for the very flattering things you say of me in the Encya but I cannot find phrases. In respects to the Hrs [Hornblowers] they are mere thieves which you could not know. The steam collar was mine and in use before their patent, one of them saw it, also the expansive mode of using steam, their engine would not work till they also stole the air pump and lastly the condenser also. After all they were immensely under par when compared with ours, and by dint of fair trials and computations we have made them be laid aside. They have however the impudence and malice of devil.[11]

Robison must have been persuaded by Watt that he had made errors concerning the Hornblowers, for in a supplement issued in 1801 to the third edition of the *Britannica*, Robison informed his readers that expansive working had been introduced by Watt in 1775–1776, which would place it conveniently in priority to Hornblower's claim to have produced a model of an engine working in that way in 1776. The article also contained the remark that Hornblower had been working for Boulton & Watt before a certain key date, which put in doubt his claims to independent invention: "We think it fully more probable that he has in this respect profited by the instruction of such intelligent employers."[12] While these matters were thus "corrected" in the 1801 supplement to the *Britannica*, Robison's original articles lived on in the 1810 edition of the encyclopedia.

In 1813–1814 Watt took an opportunity to revise and annotate those original articles with a little urging from David Brewster, who had undertaken the publication of Robison's scientific works.[13] Watt represented this as an unwelcome chore, telling Sir Joseph Banks, for example, that the task had been undertaken "at the instigation of my friends, & to my own great annoyance," having taken up all his working hours in the winter of 1813 and spring of 1814.[14] But this exercise was, as Hugh Torrens showed us, the occasion for some radical surgery to Robison's articles and also for some creatively misleading justification for that surgery. The article omitted the account that Robison had given of Hornblower's engine, providing the following reason as a note from the editor: "Dr Robison's account of Hornblower's engine has been omitted by the Editor, as unsuitable to the present work. Notwithstanding the great ingenuity evinced by Mr Hornblower in the construction of his engine, yet it was found by a court of law to be a plagiarism of Mr Watt's invention, and on this account Mr Watt could not have made any commentary upon it, even if it had appeared deserving of insertion."[15] Watt publicly presented himself as doing a favor to posterity—in his annotations to the article "Steam," he did present the first published account of some of his experiments on steam—but even as he did so he was carefully shaping the historical record in a rather devious fashion. Not only did he hide behind the editor, David Brewster, in excising Robison's positive account of Hornblower's engine from the article "Steam-Engine," but the excision was made on quite spurious grounds. He stated that Jonathan Hornblower's engine had been found to be a piracy of Watt's when in fact that decision had been against Jabez Carter Hornblower's quite different machine.[16] The making of the living legend was a matter for considerable intervention by the legend himself. The young man of almost fifty years before who had seemingly fretted about the patent oath and truth-telling was not in evidence here.

The original *Britannica* articles in which Watt appeared would have brought him to the attention of some more erudite readers, but his wider reputation was made in other ways. In 1803 an entry on Watt appeared in the fifth volume of the publication *Public Characters.*[17] Boulton had appeared in volume three a couple of years earlier, and some attention was given to Watt as his partner there, but much of the credit for their steam engine exploits was given to Boulton. This would not have gone down well in the Watt household, especially with Annie, nor would Watt Jr. have been pleased.

Perhaps they had something to do with the fact that Watt soon appeared in his own right, described as of great celebrity among mechanicians and men of science but as a person whose "character and talents . . . are not known so extensively as they merit." Moreover, he was presented as an exemplar of the kind of individual that *Public Characters* sought to bring attention to, aiming "to hold up the benefactors of mankind to the view of that society which they have benefited" and "to endeavour to appreciate their talents, and thereby to excite a sense of gratitude."[18] The account of Watt's life was rather garbled in some respects, especially regarding his supposed apprenticeship as an instrument maker in Glasgow. Watt would not have liked being referred to once again as a "pupil" of Black. But the article paid Watt homage as "a real philosopher," a benefactor to society whose steam engine was sufficient to "immortalize" his "genius and superior talents."[19] His penchant for reading novels was also mentioned. The article finished on a jingoistic note with a complaint about the French. Prony's recently published work had described Watt's engine in detail but never mentioned him once, and it had given the impression that the Périer brothers, who had bought an engine from Boulton & Watt, had actually invented it.

As a symbol of inventiveness, Watt became a cultural resource used by many. One was Sir Walter Scott, who gave a long character of Watt in the prefatory matter to *The Monastery*, published in 1820. That character, which, apart from indulging in the usual hyperbole about the significance of Watt's steam engines, highlighted Watt's varied information and his pleasure in conveying it in intellectual and everyday conversation, was much used subsequently by Watt's biographers. But Scott was not only praising Watt; he was also using him. Watt was invoked in Scott's fictionalized response to the fictional Captain Clutterbuck. Clutterbuck is an antiquarian who despises fiction and its purveyors, such matter being, so he thinks, of no interest or value to serious minds. What better answer to him than to invoke Watt the great inventor and man of such considerable knowledge and information, of whom it was also true that "no novel of the least celebrity escaped his perusal. . . . The gifted man of science was as much addicted to the productions of your native country, in other words, as shameless and obstinate a peruser of novels, as if he had been a very milliner's apprentice of eighteen."[20] We have seen that in his retirement Watt did gain a reputation for an interest in light fiction, one that is borne out by some of the contents of his library as well as by reminiscences of him. This interest seems to have been a long-standing habit.[21] Walter Scott is having fun here, but he is also using Watt to sell the acceptability of his novels as part of a well-rounded intellectual culture, the perusal of which was not inconsistent with great intellectual achievement. Indeed, according to some ideas promoted within Watt's circle, such diversions might even promote that achievement.

A strong thread of interest in juvenile education ran among the Lunar Society and other of Watt's associates. Notable in this regard were Joseph Priestley, Thomas Day, Richard Lovell Edgeworth, and Thomas Beddoes. In the 1790s Beddoes pursued a program for educational reform within which "rational toys" were of central importance. He helped to inspire the work of his sister-in-law, Maria Edgeworth, who, building on the earlier efforts and writing of her father and her stepmother, Honora Sneyd, wrote *Practical Education,* which had had a long gestation from the late 1770s but was finally published in 1798.[22] In 1825 Maria published *Harry and Lucy Concluded,* in which she made the case not only for rational toys but also for "nonsense," or childish play, in which trivial and domestic circumstances could exercise and develop a child's inventive and experimental capacities.[23] Watt was referred to in *Practical Education,* in which the protagonist children are encouraged not just to read about Watt's invention of the separate condenser but to invent it for themselves beforehand.[24] While the general romantic tendency was to treat Watt's accomplishments as the product of individual genius, the suggestion here was that the capacity to invent and discover could be inculcated in children through education of the right sort.[25] Mary Schimmelpennick in her reminiscences of Watt suggested that he was also a central character in *Harry and Lucy Concluded:* "When I recollect Mr. Watt's philosophic mind, and calm truth and loving-kindness, I have often thought that Miss Edgeworth, in her story of Harry and Lucy, had, in the character of Harry, depicted what she conceived the childhood of Mr. Watt might have been."[26] Maria Edgeworth was certainly a great admirer of Watt. She visited him during the last months of his life, finding him "in perfect possession of eyes, ears, and all his comprehensive understanding and warm heart." Mrs. Watt was crippled with rheumatism, but both were glad to see her. Maria discussed with Watt the issue of forgeries of banknotes and whether a method could be invented of avoiding them. She considered Watt "at this moment himself the best encyclopaedia extant" and admired the Chantrey bust of him. The Watts pressed their visitors to come again, "Mr Watt almost with tears in his eyes; and I was ashamed to see that venerable man standing bareheaded at his door to do us the last honor, till the carriage drove away."[27] Watt died a few months later.

When we recall that Marion Campbell's daughter Jane wrote down her mother's anecdotes of Watt's childhood in the year that *Practical Education* was published, and three years before the first of the Harry and Lucy books appeared, we can see that the business might have gone full circle. Watt as remembered may well have been cast by Marion in the character of Harry, only for Harry to later remind Maria and Mary of Watt! Whether or not this was the case, it was certainly true that by the early nineteenth century Watt's legendary status was such that, for many, he had become the model of the inventive mind, however that mind might be formed. His family and others were anxious to preserve records of Watt's formative years or to create them through the filter of what they considered they must have been like. The temptation to tell stories that conformed to emerging stereotypes would have been strong, as would the tendency to gild the lily—hence the need for the caution we displayed in using these memoirs as historical documents in recounting Watt's early life.

A Legendary Authority

As a symbol of inventiveness, Watt also embodied considerable authority, which many others making claim to invention were keen to make use of. As a result he found himself approached regularly for his help or endorsement. A perhaps typical example of an importunate inventor involved a Mr. Hendricks and his waterproof cloth. Hendricks did not approach Watt directly but through Watt's old acquaintance Sir John Dalrymple, a prominent legal figure, member of the Scottish literati, and projector of many things, including the making of soap from herring.[28] On first being asked for his advice on Hendricks's invention, Watt replied that he was happy, health permitting, to advise on patents, and that if samples were sent he would look at them. Samples arrived at Heathfield, and Watt conducted various tests on the cloth, which he reported to Dalrymple along with the advice that Hendricks conduct further tests and do more work on the project.[29] Pressed for yet more advice and faced with Hendricks's apparent unwillingness to reveal his "secret," Watt lost patience: "I am very glad Mr Hendricks does not desire to communicate his secret as I am by no means inclined to learn it from him, conceiving, pardon the presumption, that I already know several ways of doing the same did I think it worth while."[30]

Even though he was seemingly offended by the implied questioning of his honesty (always a very sensitive point with Watt), he still offered some advice but asked pointedly that his name not be used in any public discussions of the invention. Finally, he regretted his inability to help further. His health and other troubles left him "necessitated to suffer dozens of schemes of my own to be neglected" and in the circumstance "you ought not to expect I should make exertions in those of other men." In case he was still not understood, Watt said that he would be happy to see Dalrymple "when you come this way, but you must not consider me as a correspondent, unless upon *very urgent* occasions."[31]

By contrast, Watt would go to considerable lengths to assist those who had what he considered a just cause, especially if he was in debt to them in some way. Such a case was that of Captain Joseph Huddart and his patent on rope manufacture in 1803. In a memoir of Huddart we find the following:

> With the late Mr Watt of Soho, eminent as a philosopher, and the greatest contributor to the advancement of practical science that this country or even Europe has ever known, Captain Huddart had the happiness of being well acquainted; and although their individual pursuits had not in their lives led them to much personal intercourse, so great was Mr Watt's estimation of him, and so anxious was he to evince it, that at a very advanced period of life, and without solicitation on the part of Captain Huddart, he presented himself in London, purposely to attend as an evidence in support of a patent which had been infringed, and which was at the time under litigation.[32]

The purpose here was to give Huddart's inventive activities the same gloss as that given to the recently departed Watt. The friendship of Watt and Huddart (and John Rennie) "united them, added lustre to their individual acquirements, and was at once a pledge of the utility of their undertakings, and proof of the liberality of their sentiments."[33]

Watt's "unsolicited" help for Huddart was actually requested by Ambrose Weston, who was acting for the captain in the case concerning his rope manufacturing patent. This was *Huddart v. Grimshaw* (1803), and the case relied on the decision in *Boulton & Watt v. Bull* that a method was patentable.[34] Watt was usually happy to help his old friend Weston in a patent case, but he found himself in a difficulty. He was reluctant to visit London in the winter because he always caught a bad cold, which tended these days to linger. Having had "no intimation of being needed," he had already arranged for his cousin Robert Muirhead and his wife to visit Heathfield on their way to Bath. Watt also described himself as "nearly quite ignorant of the point on which the question turns." He feared that this and his reduced facility of comprehension might mean that as a witness he might do more harm than good. Nevertheless, he wanted very much to help Huddart, who he thought "very hardly used, by a similar kind of knave to those with whom we were plagued."[35]

Shortly afterward Watt wrote to Weston again, "enclosing a hasty opinion upon paper" and still doubting that he could attend, thinking it very unlikely. But in the end he did go to London, with Annie, for the patent trial, missing cousin Muirhead as a result. To add to the frustrations, the trial was compromised and the witnesses were not required. At least Huddart won the case, and he was very grateful for the trouble to which Watt had gone.[36] Weston, who had worked tirelessly for Boulton & Watt in its patent trials, parlayed that experience into a lucrative professional trade in patent litigation. One of the things that he could offer to his clients was the possibility of technical advice, and even testimony, from the great Watt himself.

On only one occasion that we know of did Watt respond positively to a request for not only technical and patent advice but also financial assistance. This was in the case of a fish oil project. At the end of 1806 Ambrose Weston approached Watt asking for his assistance for William Speer (also rendered as "Spier" or "Spear"), who was developing a process for purifying fish oil. Speer issued a specification, dated 13 December 1806, for his patent for "a new Art, Method, or Process of purifying, refining, and otherwise improving Fish Oils and other Oils, and converting and applying to Use the unrefined Parts thereof."[37] His method involved treating the oil with an infusion of tannin (from natural sources such as oak or other barks, tormentil root, gall-nuts, catechu, or artificial tannin as described by Charles Hatchett in the *Philosophical Transactions of the Royal Society*). Heating and agitating the mixture produced a separation of impurities, and then the oil could be drawn off the top. The impurities themselves could be used in preparing various commodities including cement, paint, varnish, putty, and blacking for leather. Although the process could be used with any oil, Speer found that the impurities drawn off fish oils were particularly useful for these applications. Speer took out another patent (No. 3325) in April 1810 for increasing the inflammability of oils, especially those produced by his refining process. The oils with which he dealt were intended primarily for burning for illumination.

When first approached by Weston, Watt carried out some experiments that proved to him the value of the process. He also gave advice about the patent. In all probability Watt would have known of Speer, who was, or at least had been, supervisor and assayer of spirits at the Port of Dublin. He was a person of some technical credibility, having earlier taken out a patent on an improved construction of a hydrometer. His advice had been used by Nicholas Vansittart, joint secretary to the treasury, in dealing with the inaccuracies of excise determinations that became of considerable consequence after the Act of Union between Britain and Ireland in 1801.[38]

When he approached Weston and Watt, Speer was trying to raise substantial capital of £15,000 for his oil venture. By 1808 he was refining oil and sent Watt five gallons of it in 1810. Because he thanked Watt for his help with the refinery, it looks as if Watt probably did invest in it.[39]

While Watt was cagey about approaches for assistance from individual inventors, he was on occasion more receptive to requests of a more public, or institutional, nature. A number of these requests came from Scotland, and in acceding to some of them Watt was in one sense returning in his mature fame to the role of jobbing civil engineer of his pre-Birmingham days. Having made his fortune, and having trumpeted much the public benefit that had flowed from his activities, Watt perhaps felt some pressure to lend his expertise and authority to good public causes.

One of these projects was the plan for an improved water supply for Glasgow. Watt had prior experience in this area, having been centrally involved in Greenock's water improvements at the time he left for Birmingham. Thirty years later, Watt's close friend and agent Gilbert Hamilton was a member of the committee set up to oversee Glasgow's new waterworks, and it was probably through him that Watt was drawn in.[40] Thomas Telford and John Rennie were the engineers consulted, but Watt was asked, initially in 1805, to suggest the best type of pipes to use in the main streets. He recommended that either wood or cast iron pipes be used but noted that if the earthenware ones were employed then they should be bedded in clay. Through Hamilton he gave informal advice on various aspects of Telford's plan.

The first version of the new waterworks experienced problems with its filter beds, and the idea was floated that a natural gravel and sand feature on the other side of the Clyde might be used that would not experience the clogging that plagued the original filter beds.[41] The disadvantage of that plan was that it meant that the water would have to be piped across the Clyde so that it could be distributed by the pumping engines, and the riverbed was a difficult place to lay a pipeline because of its rapidly shifting sand and mud. In 1808 Watt was consulted again in hopes that he could suggest how to deal with this problem. His solution was an ingenious one that modeled the pipeline on a lobster's tail. It was a flexible, articulated structure, a thousand feet long, that could shift with any changing profile of the riverbed. The structure was manufactured by Boulton, Watt & Co. and succeeded admirably in its purpose.[42] Watt did not charge for his services, no doubt regarding his contribution as recognition from his current prosperous vantage point of the support that Glasgow had given to a young struggling merchant and engineer. He did receive, however, the gift of a large silver soup tureen and stand "Presented to James Watt L.L.D. by the Company of Proprietors of the Glasgow Waterworks." In later years part of the local remembrance of Watt was as an ingenious benefactor of the city in various ways, the development of its water supply being one of them.[43]

Another project of Watt's retirement years that revived the consulting engineer of his early life was a scheme for the heating of the Hunterian Museum at the University of Glasgow, an institution that was subsequently to house a Chantrey statue of Watt, the supposed model Newcomen engine that featured in the apocryphal stories of his steam engine inventions, and other Watt relics. The museum's original hot air heating system had a number of problems, seeming insufficient to purpose and also polluting the building with smoke. Watt was consulted, initially in 1808, to see what improvements might be made. The request came from James Mylne, a professor of moral philosophy at Glasgow who was involved with the museum.[44] One obvious possibility was to devise a system for heating with steam, but the danger of accidents, unwanted humidity, and anticipated expense worried Mylne. So, although Boulton, Watt, and their associates had considerable experience of using steam to heat their own houses, the Soho Manufactory, and a number of mills, Watt ended up recommending a hot air system of the sort that had been developed by Matthew Murray and William Strutt, and his suggestion was accepted. Although Watt was asked to implement it, the scheme seems to have been prosecuted by Boulton, Watt & Co. with particular assistance from William Murdoch. Watt did, however, attend trials of the system in September 1809 during the same visit to Glasgow when he learned of Boulton's death.

Watt's involvement in this became known to Lord Frederick Campbell (1729–1816), a son of the 4th Duke of Argyll who was Lord Clerk-Register of Scotland and in the market for a heating system for Register House in Edinburgh. Watt was recruited to consult on the issue and submitted a report in 1810. The architecture of Register House made for severe difficulties in heating it. Watt's report rehearsed the pros and cons of various options but was unable to come to any definite recommendation, suggesting that someone of great experience in the area be consulted. The Glasgow waterworks, Hunterian Museum, and Register House consultations drew on Watt's expertise and bore varied fruit, but they certainly reflected Watt's reputation, and his willingness to engage with them constituted, perhaps, a kind of prodigal return.

On other occasions it was Watt's authority rather than his expertise as such that was at a premium in trying to settle disputes. This was the case with his involvement in 1812 in giving an opinion on civil engineering works at Sheerness Dockyard, located on the Isle of Sheppey at the mouth of the River Thames.[45]

The naval dockyards in general had become a matter of great concern to the Admiralty and the government in the early nineteenth century, with the threat of invasion hanging over the country. In 1806 Watt's old employee and friend John Rennie was asked by the Board of Naval Revision to inspect and report on all the royal dock-yards and the improvements that ought to be made in them. Reporting on 14 May 1807, Rennie found that most of the dockyards were in a poor state and many of them inappropriately laid out, badly equipped, and generally unfit for purpose. Rennie proposed the construction of a new great naval harbor with large wet and dry docks and modern, well-equipped workshops for the building and repair of ships. The Admiralty asked him to report on what he considered the best site for this great naval harbor and arsenal, and in particular to consider a suggested site at Northfleet on the Thames. Rennie inspected the site accompanied by Watt, John Southern, and Mr. Whidbey, who was acting master-attendant at Woolwich Dockyard. Presumably Watt was there, assisted by Southern, at Rennie's request and perhaps at this stage to add his gravitas to Rennie's views. Rennie concluded that Northfleet was the ideal location in many respects, and he submitted a detailed plan for what might be built there. There was a long period of inaction because of serious disagreements about the Northfleet option as opposed to a large investment in Sheerness.

In 1812 Watt was formally requested to attend a meeting of the navy board about the matter, along with Captain Joseph Huddart and Josiah Jessop. It seems that those on the other side of the argument were now trying to recruit him. Watt tried to back out, but Rear-Admiral Sir Joseph Yorke, lately appointed one of the Lords Commissioners of the Admiralty, wrote to persuade him to see it through. Yorke probably recruited Sir Joseph Banks, as a known close friend of Watt, to add his persuasive powers. Watt was needed, Banks told him, to "help subdue the dragon of the Dock Yards." Banks suggested that Yorke "wants only the succour of Science in the venerable form in which you be[a]r it about you." It was a patriotic cause, Banks suggested, in which Watt's scientific authority could serve an important purpose: "As you are in all matters Science actuely [sic] invulnerable in Public opinion and as the dragon is a most vulnerable animal I want to urge you on to be a sharer in the glory of a victory which with your aid will be in my mind a certainty."[46] The purpose appears to have been to promote the case for Sheerness against the Northfleet option further up the Thames. We must assume that the "dragon" is either Rennie, as a continuing strong advocate of Northfleet, or possibly the dilemma of indecision in which the question had been mired for so long. Watt gave in and inspected the works at Sheerness in early July. With Captain Huddart he drew up a report that seems to have contributed to the desired objective. With the weight of Watt's authority on their side, the Admiralty decided that Sheerness would be the option. Rennie was engaged to oversee the construction of his less-favored plan, and Watt received the thanks of the Admiralty for his contribution. This role in the apparent triumph of politics over technical advice seems to have been Watt's final public engagement as a consulting engineer. It is a clear indication that Watt's reputation, wanted by all parties as authoritative support, was of legendary proportions, and his pronouncements, as the president of the Royal Society put it, "invulnerable in Public opinion."

Honoring the Great Steamer

Among the greatest honors received by Watt during his retirement was election to the National Institute of France (successor during the Napoleonic period to the Académie des Sciences), first as a corresponding member in 1808, and then in 1814 as one of only eight "Associés étrangers" that the Institute had at any given time. These awards were made even though Britain and France were at war through most of the revolutionary and Napoleonic period from 1793 to 1815. Then, as subsequently, many might have maintained that "the sciences were never at war," so that such mutual exchange of honors was possible. But that rhetoric of scientific internationalism and universalism disguised a complex and changing set of relations between science and the state that was reflected in many aspects of scientific activity and exchange.[47]

Watt had had a long and complex association with France and its scientific and commercial elites from before his first visit there in the mid-1780s. He had enjoyed friendly face-to-face relations and correspondence with many French savants. Some of these characters fell foul of the revolution; others survived and formed close relations with the Napoleonic state, like his friend Berthollet, for example, as also Laplace, Monge, Prony, and Delambre. But there were also tensions in these cross-channel relations, including the ever-present concern about industrial theft and espionage. Boulton and Watt had learned that foreigners bearing gifts, or cloaking themselves in the mantle of the Republic of Letters, had to be dealt with very carefully.[48] Thus while Watt was no doubt glad of the recognition involved here, he was to a degree wary of its motivation and possible consequences. After Watt's death François Arago was to use his life story as told in an eloge for the Académie to try to promote the steam industry in France, and this, as well as the desire to honor Watt, was likely in the minds of his French electors when they recruited him to their standard. The French government had tried to recruit Watt's expertise in 1786–1787 during his and Boulton's visit to Paris. That attempt had failed. Now in Watt's declining years, they might at least harness his fame.

Whatever pleasure Watt derived from the honor was clouded by confusion because of the long-standing poor state of communication between the British and the French as networks collapsed.[49] Watt learned of his election from a letter of nomination and a certificate of election sent to him by J. B. J. Delambre, the astronomer, then permanent secretary for mathematical sciences of the institute.[50] But these were delivered out of the blue one day not to Watt himself, but to Watt Jr. by an unknown American who presented himself in Birmingham while Watt was on the Welsh borders attending to his estates there.[51]

Watt was left with a puzzle: How to communicate his acceptance? The American had been charged to take Watt's reply back to Paris, but this was impossible because of Watt's absence from home. Watt in the end entrusted his reply to Sir Joseph Banks, hoping that the great intelligencer would find a chance to transmit it. Watt's response asked Delambre "to communicate to the Class [of Mathematical & Physical Sciences of the Institute], my acceptance, the grateful sense I entertain of the honour they have done me, assuring them at the same time, that whenever I shall have any thing to communicate which shall appear worthy of their attention I shall feel much pleasure in laying it before a Society who have so eminently contributed to the extension of science."[52]

Watt remained uncertain for a long time whether or not the institute had received his reply. In March 1810 he took the opportunity of a letter to the academician Pierre Lévêque to convey thanks for his election and advise that he hoped his original acceptance had been received.[53] That letter to Delambre had been carefully worded. In the state of international relations, such an honor was best not too loudly treasured, as Sir Joseph Banks had found out when his effusive letter of thanks for his election to the institute in 1801 became the subject of public controversy, challenging among other things his fitness to hold the office of president of the Royal Society.[54] Though the challenge came to nothing, Watt, aware of the trouble his friend experienced, would have been anxious to avoid anything of the sort in his own case. Both Watt and Watt Jr. seem to have refrained from remarking on the honor, and Watt did not mention it at the time even to close friends.

In 1814 Watt received the even greater honor of election as one of only eight foreign members of the institute, though it was March 1815 before he learned of this via a letter from his old friend Comte Berthollet: "Nous avons eu le plaisir à l'Institut de vous donner un témoignage de la haute considération que nous avons pour vous, en vous choisissant pour l'un des huit Associés Étrangers."[55] Watt broke this news to his grandson, James Miller Jr., on 19 March.[56] But otherwise he seems again to have made little fuss, judging by his correspondence. This is understandable given the state of play between Britain and France at the time. Though Napoleon had been exiled to Elba in May 1814, he returned to France on 1 March 1815. So, the institute's decision to elect Watt was made, and Berthollet's letter was written, under a restored monarchy in France, but by the time it arrived in Birmingham Louis XVIII had fled the country in the face of Napoleon's return. It was not an auspicious time to celebrate the honor, no matter how much scientific affairs might have been regarded as immune to the state of political and military conflict. Watt Jr. was probably delighted at this ultimate recognition of his father's philosophical credentials since it would add considerable heft in his ongoing promotion and defense of his reputation, especially once the decisive victory over the French forces at Waterloo on 18 June 1815 finally brought peace between the two nations.

Another early honor during Watt's retirement was a more parochial one—the award in 1806 of an honorary doctor of laws from Glasgow College, a title he used thereafter, especially in legal documents such as his will. The award as described in correspondence from the college was made in recognition of his "character" and because of the value of his "important discoveries & improvements in the useful Arts."[57] The college, understandably, was keen to claim Watt as its own despite his relatively short-lived, and in some ways tenuous, relationship with the institution. Throughout the nineteenth century the inaugural addresses of rectors of the university were to routinely invoke Watt's name; thanks to the honorary degree, they could claim him as an alumnus, not just as a former employee.[58] John Robison had early emphasised the link, much to Watt's displeasure, when he described Watt as a pupil of Black in his introduction to the volume of Black's lectures that he edited. We saw that Watt denied that his great invention of the separate condenser was made at the college, and remarked wryly on the institution's tendency to claim him, when the Hart brothers encountered him in Glasgow in 1813–1814.[59]

Despite his annoyance about such matters, Watt was well prepared to go along with the university's celebration of his achievements. In founding the Watt Prize in 1808 by donating £300 that would be invested to provide an annuity of £10, Watt wrote to the principal of the university: "Entertaining a due sense of the many favours conferred upon me by the University of Glasgow, I wish to leave them some memorial of my gratitude, and, at the same time, to excite a spirit of inquiry & exertion among the students of natural philosophy and chemistry attending the College, which appears to me the more useful, as the very existence of Britain as a nation appears to me in great measure to depend upon her exertions in Science & in the arts."[60] The prize was to be for a student essay on a rotating list of topics. In the negotiations about the detailed specification and conduct of the prize, Watt had his own ideas about how it should be set up. As always, Watt thought that he saw the best arrangement. But in this case he did allow his perhaps over-elaborate plan to be simplified, though the agreement required that Watt himself, or on his death his nominee, be sent a copy of the winning essay each year. Whether it was the education of his children, the conduct of his farms, or the direction of his philanthropy, proper supervision was required.

There were also honors that Watt was very reluctant to receive, especially when they involved work. An interesting case in point is the position as sheriff of Stafford-shire. Such positions were in one sense coveted as a definite step in the hierarchy that joined sheriffs, Lord Lieutenants, the Privy Council, and the monarchy. They had status. But during the period of the Napoleonic Wars, they also, at least potentially, involved work and action in the maintenance of public order. Watt was very keen on public order, had a horror of the mob, and—so far as we can tell—no compunction about its violent suppression. He was very enthusiastic about the raising of volunteers in Birmingham in 1803 when invasion was feared. He reported to Hamilton that Robinson Boulton was to serve on horseback and Watt Jr. on foot. They had all attended meetings in Birmingham about raising volunteers. Watt also supplied swords to his cousin Robert Muirhead, who was arming volunteers in Glasgow.[61] But Watt was distressed when, a few months later, he found out that his name was on a list of three nominees due to be put forward to the Privy Council for the position of sheriff of Staffordshire. Perhaps he should not have been so surprised, given his recently demonstrated enthusiasm.

He took advice from his lawyer Ambrose Weston about how he might be excused nomination or avoid service if chosen. Watt sought Sir Joseph Banks's assistance, and Banks wrote to the Duke of Portland, a member of the Privy Council, which would make the ultimate decision. Watt wrote also to other members of the Privy Council including the Right Honourable Chief Baron. His plea is an interesting one. Watt declared himself "extremely unfit" for the position as he was nearly seventy years old, ill, and confined to the house much of the year. He had of property in the county only his house and forty acres. He was also "little acquainted with the affairs of the County & never possessed that firmness decision & activity, which the present juncture of affairs seems to require in a Sheriff, who should be something else than an infirm old man." Watt continued: "Perhaps I may be allowed also to plead my long continued exertions in the line of my profession, which I flatter myself have been of some service to the nation & that as a reward I may be permitted in my old age to enjoy that domestick quiet to attain which I have undergone so much anxiety & care during the best part of my life."[62] Watt's declaration of his legendary status and the exemption it ought to gain him seems to have worked, though Weston advised him close to the time of decision that he might possibly employ another excuse—the fact that he was a Presbyterian, a member of the Church of Scotland and therefore a dissenter from the Church of England, and so disqualified from such office. Weston told him that doing so might involve the need to specify "of what particular Congregation of Protestant Dissenter you belong to and that you have been accustomed to receive the sacrament among them."[63] Watt's response is an interesting one, especially since he seems, through his time in Birmingham, to have supported the Church of England, and he was to be buried in the established church at St. Mary's, Handsworth[64]: "I am not a member of any dissenting congregation here [in Birmingham]. These Gentlemens politics & mine differed so widely that when I came to England I thought it most prudent not to enlist under their banner, yet in 1791 my house was in the list to be burnt for my being a Presbyterian. I nevertheless remain a member of the Church of Scotland and should not for any office of *profit* proffess myself any other–If therefore I am *compelled* to communicate with the Church of England the sin must lie on the head of the compellers."[65] Watt hurriedly communicated his claim to dissenting status to Joseph Banks, Charles Greville, and others, indeed to everyone who would listen, as he became increasingly anxious that he might be appointed. In a letter to Weston the next day, Watt added that he had "never conformed to the church of England, & cannot conscientiously take the sacrament in that church."[66]

It seems likely that a fundamental disqualification of Watt on religious grounds was not registered in the system because there was another effort to recruit him to the shrievalty, this time for Radnorshire, in 1816. Watt's reasons for declining had not changed, except he considered them even more compelling given that he was by then eighty-one years old. In explaining to Robert Muirhead, Watt again emphasized his unfitness because of age and infirmity but also the extent of his public service that had earned him rest: "I have spent a long life in improving the arts and manufactures of the nation; my inventions at present, or lately, giving employment to best part of a million people, and having added many millions to the national riches." It sounds as if he had done the sums! Watt was well rehearsed in the reasons for his legendary status. Once again, Sir Joseph Banks and other friends were mobilized to successfully repel the approach.[67]

Watt also, at first sight rather mysteriously, refused another honor that was reputedly offered to him perhaps at this time. This was a baronetcy. His biographer

J. P. Muirhead states that an offer was communicated to Watt via Banks that "if he chose to express a wish to that effect" then a baronetcy was his. Watt discussed the matter with his son, and "it occurred to both, that there were circumstances and considerations that rendered it ineligible."[68] It was left unclear just what those circumstances were. Into the vacuum much speculation has moved over the years, including the view that Watt had refused it because he was "essentially a democrat."[69] As we have seen, nothing was further from the truth. Watt thought democracy the work of Satan. Watt Jr., who would have inherited the baronetcy, had certainly been a loud democrat and quietly retained such sympathies even as practically he became more conservative. Perhaps most importantly, Watt disliked aristocracy as an institution precisely because of its hereditary character, which in his view led to exercise of unmerited power.[70] This seems the most likely reason for Watt's refusal of this hereditary honor.

Countercurrents

The relatively rare negative press that Watt received nevertheless provided important countercurrents in the gathering wave of adulation that came his way. In 1839 the occasion of a negative review of Arago's eloge of his father caused Watt Jr. to observe that "there has at all times, within my Memory, been a spirit of jealousy among this class [the engineering profession] against the whole race of the Soho engineers, who in return have treated them with due contempt."[71] Watt Jr. recalled that during his father's lifetime, engineering opposition had "stirred up Olinthus Gregory a Professor at Woolwich to stand forward as their champion; to claim for them sundry inventions originating with my father or ourselves, & to vituperate others": "That *roused* my choleric and with the assistance of Professor Playfair and the connivance of Jeffrey, I put forth the *Olynthiad* in the Edinburgh Review. . . . That proved a settler for some years."[72] The incident to which Watt Jr. refers here was sparked by the publication of Olinthus Gregory's *A Treatise of Mechanics* in 1806, which included a history of the steam engine from Savery onward contributed by Jabez Carter Hornblower.[73] This account by a scion of the Hornblower family was highly critical of Watt. It gave credit to Watt for his improvements upon the engine of Newcomen: "First, that the elasticity of the steam itself is used as the active power in this engine; and secondly, that besides various other judicious arrangements for the economy of heat, he condenses the steam, not in the cylinder, but in a separate vessel."[74] But it also entered various criticisms. Hornblower suggested that the economics of Watt's early engines were not so advantageous to their proprietors as to Boulton and Watt themselves. The new engines, though saving money on fuel, "opened other channels of expenditure" in the expense of erecting them; the payment of wages to those engineers who were required to maintain these newly accurate, but also newly sensitive, machines; and the considerable consumption of oil and tallow compared with the old engines. He noted that some proprietors had been left wondering whether old engines with the more accurately bored cylinders might have been just as economical overall.[75]

Hornblower also had the temerity to suggest that too much credit had been given to Watt himself: "Much of the merit ascribed to the fertile mind of Mr. Watt really originated with Mr. John Wilkinson, who was always foremost in the improvement of any thing which related to the iron foundery. . . . Nor must it be forgotten, that some sterling acknowledgments are due to Mr. Watt's coadjutors, of which he availed himself in a region of rare talents." The author observed parenthetically that he would not have needed to say this had not the writer of the article "Steam-Engine" in the *Encyclopaedia Britannica* indulged in "partial, yea fulsome, compliments" to Watt.[76] Having sought to distribute more widely the merit appropriated by Watt and his supporters, Hornblower returned to what he considered a complimentary mode: "Mr. Watt took up the subject of improvement [of the steam engine] with a large and important object in view: it was not only to renovate the principle, but to assimilate this most grand assemblage of philosophical and mechanical skill to those machines which, on a smaller scale, had hitherto engrossed the talents of our most ingenious artificers, and been the admiration of other nations wherever they came: nor do we mean to withhold the acknowledgment, that through his means it now ranks foremost among the productions of the philosophical or mechanical world."[77] This strikingly modern acknowledgment of Watt as orchestrator of the improvement of the new steam engine was not appreciated in the Watt camp, where any deprivation of his individual credit was interpreted as an accusation of theft.

The more detailed history of the improvement of the steam engine discussed the claims of Gainsborough to prior invention of the separate condenser and to be the source from which the idea leaked to Watt, and also the claim of William Blakey to the first use of the expansive power of steam. Jabez provided an extensive discussion of the merits and the priority of some of Jonathan Hornblower's engines, contradicting the treatment of them by Robison in the *Encyclopaedia Britannica*. The attempt to deny Hornblower credit was, naturally, among the chief complaints. But there was more general criticism of the stifling of competition: "We are by no means disposed to detract one atom from the advantages resulting to the community through the perfection of Mr. Watt's engine: but we see it made use of for invidious purposes; not to elevate Mr. Watt above his inherent merit, but to subordinate every other professional man in the scale of comparison, to repress the energies of his cotemporaries, and give a deadly blow to competition."[78] It was in the face of this challenge to the decades of work that Team Watt had invested in the building of Watt's reputation that Watt Jr. mounted the defense that he dubbed the "Olynthiad." This took the form of an article in the *Edinburgh Review* usually attributed to Professor John Playfair, but, as we have seen, Watt Jr. described Playfair as *his* assistant in writing it.[79]

The Olynthiad criticized the account in Gregory's book for its lack of organization and numerous errors. The account of steam engines was described as "disfigured by the railing and misrepresentation of an author, possessing neither the impartiality nor the knowledge indispensable in a work of the kind"—basically, the work of a defeated and disgruntled competitor. The piece then proceeded to give an account of Watt's steam engine and his steps to its successive improvements, highlighting Watt's scientific approach to show "that this great improvement was not the effect of accident, or of casual observation,—but the result of deep reflection, of great ingenuity, and much philosophical investigation."[80] The rhetorical line of the patent trials was followed here. The suggestions that Watt had been anticipated were tackled, as was the claim that he had depended upon, and taken as his own, the contributions of others. While in a civilized age every man who works among other men of talent and ingenuity must incur debts to others, the Olynthiad claimed that self-sufficiency was "as nearly true of Mr Watt's inventions as it can be of any, in an age where art and science are so highly improved, or so extensively cultivated, as in the present. From the first experiments . . . to the last improvement made on his engine . . . we believe the whole series to have been the work of his own genius."[81] Watt's working independence of Black's heat experiments and doctrine of latent heat was also emphasized.

In a final flourish the Olynthiad argued that these attacks upon Watt's claims not only injure the individual but also "obstruct the progress of those improvements which he [the author] professes to explain. It is of the utmost consequence to that progress, that every inventor should be as much as possible assured of the reward to which his discoveries naturally entitle him." Those rewards were the satisfaction of the accomplishment (which could never be taken away); the consequent fame, honor, and reputation (which could be affected, as in the present case, by envy and misrepresentation); and the more occasional enjoyment of wealth that might follow (which was not the main objective of those most likely to succeed). Watt, it claimed, had attracted an abundant share of envy and misrepresentation. There were few whose claims had been subjected to such rigorous and lengthy trial: "We rejoice that he is now in the possession of the ease and affluence to which the services he has rendered to his country, and to the world at large, do so well entitle him. It is not the business of a friend of science, of one interested only for truth and justice, to trouble the repose of an eminent man, retired from active life, and enjoying the fruits of a well earned and justly deserved reputation."[82] The gall of Watt Jr. in penning much of this article is remarkable! Controversy continued, with a response from Gregory in the *Monthly Magazine* of August 1809 in which he maintained that Watt was "too much actuated by a spirit of monopoly for a genuine philosopher," a retort from Playfair/Watt Jr. in the *Edinburgh Review* shortly thereafter, and a final word from Gregory in 1810.[83] Later editions of Gregory's *Treatise*, subsequent to that of 1809, removed Jabez Hornblower's account, substituting instead a summary of it and of the *Edinburgh Review* article. In a footnote in the 1815 edition at the point where Hornblower's account was excised, Gregory explained that he had been "exposed to much calumny and misrepresentation for admitting this historic sketch into my work." He also explained the circumstances of its admission, conceding that he had included it at the behest of a friend who knew Jabez Hornblower and thought that the latter had received much injustice and deserved the chance to "tell his own story." Gregory sought to take the heat out of the situation, forgiving his *Edinburgh Review* critics and hoping himself for forgiveness for simply allowing "an injured (though perhaps hasty) man to defend his own cause, and that of his family."[84]

A letter Watt wrote thanking Professor Playfair is interesting for the possible light it throws on how Watt and Watt Jr. operated in defending Watt's reputation.[85] One peculiarity of the letter is that Watt thanks Playfair for "the pains you have taken to vindicate my fame," while excusing himself for not taking up the pen in his own defense. There is no mention of Watt Jr. If it is true, as Watt Jr. later claimed, that *he* was the primary architect of the Olynthiad, then either Watt did not know that, or he did know and he was feigning ignorance! The former seems unlikely. In fact, Watt wrote to his son in November 1808 with answers to some of the points that Jabez Hornblower had made.[86] The planning and construction of the "Olynthiad" was thus a joint effort. So why did Watt engage in the pretense in his letter to Playfair? Probably because he saw himself as writing a letter that might become public as relevant to a public controversy. The reason Watt gave for not entering the lists himself is surely for public consumption: "I have always had a great repugnance to be the trumpeter of my own fame & I considered that magna est veritas &c & that judicet qua posteritas would do me the justice which my adversaries were disposed to deny me at present.

I am far from being insensible to just pain, but the conviction that I had in some cases deserved it has always been with me a superior sensation."[87] It appears to me that father and son were agreed that in public controversy Watt Jr. would take the active part and Watt would play the retiring character resigned to the judgment of history and seemingly ignorant of the strenuous policing of his reputation mounted from within his own family. Once again, we see a side of Watt very different from the young man concerned about swearing a false oath.

Together with a number of other critiques of Watt, most stemming from the patent trials, the contents of early editions of Gregory's *Treatise* provided material for continuing countercurrents in the years after Watt's death.[88] As A. P. Woolrich noted, in the 1820s and 1830s writers such as Thomas Tredgold and Elijah Galloway echoed the critiques of Gregory and Bramah, while John Farey took a middle path between such willing critics and those for whom Watt could do no wrong, such as Robert Stuart and Charles Partington.[89] But the latter were in the ascendency, thanks to the Olynthiad and to Watt's "corrections" to Robison's articles on steam and the steam engine in the *Encyclopaedia Britannica* discussed earlier.

The often ruthless business tactics of Boulton & Watt had not endeared them to their competitors in the engine business. One suspects that first Boulton and then Watt Jr. were most often the architects of such practices, but it would be a mistake to think that Watt himself was a naïf in such matters. At the very least he was aware of the measures taken. In some cases those measures were questionable in that they stifled innovation and involved dirty tricks and harassment. Whoever was responsible in what degree, these perceived characteristics of the Boulton & Watt enterprise certainly reflected on Watt himself and helped to create countercurrents in the construction of the living legend. Watt, as we have seen, had also often exhibited a harsh, unforgiving attitude toward workmen during his active working life. He rarely wore the velvet glove with which Matthew Boulton disguised the iron fist. Boulton reveled in delivering speeches to the workers at feasts and celebrations; the message was the same as Watt's, concerning the divinely ordained ranks of society and the contribution of each in his station to the work of the enterprise, but it was swallowed more easily with the roast beef and ale than without it. Watt, however, disapproved of such pointless and profligate revelry.[90]

Most business competitors and workmen suffered the harsh approach of the Soho enterprises in silence, but a few spoke out. A case in point, so far as business competitors were concerned, is that of Matthew Murray (1765–1826). Originally from Newcastle-on-Tyne and apprenticed as a millwright or engineer, Murray moved to

Leeds, where he worked for the flax spinners John Marshalls and became chief mechanic there.[91] He began to patent inventions; his first (No. 1752), taken out in 1790, concerned a machine for spinning flax and other fibers. Marshalls purchased a steam engine from Boulton & Watt in 1793, and this enabled Murray to become familiar with its operation and to begin to develop minor improvements. Subsequently, he worked for Fenton, Wood & Lister, a machinery and engine building and repair business, where he took charge of the engine-building operations. He took out further patents on steam engine improvements and began to build and supply various forms of engine. It was at this stage that he attracted the attention of Boulton, Watt & Co. An apparently friendly visit by William Murdoch and another engineer to Leeds resulted in a successful challenge to the validity of two of Murray's patents on the grounds that Murdoch and Boulton & Watt had engaged in prior use of those inventions. *The King against Murray* was initiated by the attorney general to repeal Murray's patent of August 1801. The *Leeds Intelligencer* reported, "The Prosecution was carried out at the instance of Messrs Boulton, Watt & Co of Soho, principally to expose the defendant's unfair conduct and groundless pretensions, by proving that all those parts of the invention in question . . . were invented and practiced at their works for a considerable time before the date of this patent, and that the defendant had obtained a knowledge thereof from some of their workmen." Although scheduled for trial, the defendant withdrew his plea and accepted judgment against him by default. The patent was cancelled and repealed.[92] Effectively, Boulton, Watt & Co. had used its inside track with the patent legal system to blacken Murray's name, and he had been unable to fight an expensive legal battle, thus losing his patent by default.

Murray did not give up, however. He responded in the press "to vindicate my Character as an ENGINEER" against the "foul insinuation" that he had stolen ideas from Boulton, Watt & Co. Quite the reverse, he asserted. He confirmed that he had declined to defend his patent because of the expense of contending against "such rich and powerful opponents" and denied that those opponents had made any improvements since Watt Sr. retired from the management. Murray claimed that during a visit by Storey and Murdoch to his Leeds works, ostensibly of a friendly disposition, the visitors were given free access and "upon their return to Soho, many of our Improvements were immediately adopted." On a return visit, Murray was denied access to the Soho works. To settle the matter of who could make the better engines, Murray issued a challenge, similar, as he put it, to one issued by Boulton & Watt to Hornblower during their dispute with him. They would each build a one horsepower engine, each party also depositing one hundred guineas, the total sum to be claimed by the maker of the best engine as judged by "Twelve Practical Engine Makers" chosen half by Boulton, Watt & Co. and half by himself.[93] Needless to say, the challenge was not taken up.

Prior correspondence between Watt Jr. and Robinson Boulton shows that before making a legal challenge to Murray's patents, Watt Jr. had pursued other strategies. In June 1802 he was in Leeds with William Murdoch to see a customer, Benjamin Gott, but once done was looking at buying land next to the Fenton, Wood & Lister works to try to prevent them from expanding their business, and he seems to have done so.[94] Both sides were engaged in espionage. A worker called Halligan was a focus of attention. He had formerly worked at Soho and now worked for the Leeds firm, having apparently communicated much about the Birmingham operations to his new employer. Watt Jr. and Murdoch visited Halligan and his wife, and they were persuaded to shift their allegiance once again to Soho. Recounting the agreement at which he had arrived with Halligan, Watt Jr. observed that it was a question of policy "whether we should take him back or leave him for some time as a spy upon Murray & Dixon's proceedings."[95] With the help of Halligan's wife, Watt Jr. searched the trunk of another workman called Hughes and found "a roll of drawings of various parts of our Machinery & Engines." Because a warrant for the constable to detain Hughes had to be issued where the "crime" had been committed, that is at Soho Foundry, Watt Jr. recruited his father back in Birmingham to obtain from Mr. Lane a warrant for Hughes's arrest. Having decided that there were "no hopes of getting at Murray's men, as all of the least consequence are engaged and at high wages," Watt Jr. decided that "the only effectual way to harass them seems to be by destroying the basis of their illgot fame and setting up a competition which will diminish their orders." The attack on Murray's patents was no doubt directed at the first of those objectives.

One begins to see why those who were in the same business as Boulton & Watt and Boulton, Watt & Co. would have been disinclined to celebrate the career of the engineering figurehead of those two concerns. As we will see in chapter 8, the rival machine builders and engineers beyond Soho were conspicuous in their absence, from the supporters of the memorial to Watt in Westminster Abbey.

Even some independent thinkers who were otherwise admiring of Watt's achievements found themselves unable to stomach the rampant hyperbole that so often surrounded him. This was certainly true of David Brewster, who, as we have seen, in many ways assisted Watt in "setting the record straight" over John Robison's writings on the steam engine. But there were limits to Brewster's tolerance, as revealed in an exchange with Maria Edgeworth in 1824 when he was advising her on the manuscript of *Harry and Lucy Concluded*. Responding to Maria's quotation in that work from Sir Walter Scott's particularly "purple" characterization of Watt in *The Monastery*, Brewster observed: "I am almost afraid to put down in writing another observation, and yet it would be a weakness to omit it. The passage you quote from the Scotch novel relative to Mr Watt is quite overstrained, and in every respect, utterly incorrect. No man ever admired Mr Watt more than I did. . . . I admired him, however, for what he did, and not for what he never thought of doing; and I confess that I have been guilty, under the influence of personal attachment, of writing of his labours in very exaggerated strains."[96] Great as Watt's improvement of engines through the separate condenser was, Brewster was convinced that steamboats and steam engines would have been as advanced as they then were had Watt never lived:

> I maintain that the high-pressure engine . . . which Mr Watt and all his friends (including myself) reviled as of inferior utility, is in every respect superior to the low-pressure engine, and would have accomplished all the great operations of modern times, even if the low-pressure engine and a separate condenser had never been heard of.
>
> Nay, I am disposed to think that the obstinate adherence of Mr Watt and his friends to the low-pressure engine, long after accurate experiments, made and recorded in Cornwall, had demonstrated the superiority of high-pressure ones, has done much to retard the progress of invention respecting this engine.[97]

Brewster noted that the art of the author of Waverley, like that of other poets and orators, required that "any great national invention" be associated with "one great name" and did not allow for the "sub-division of praise."[98] Once we add the requirement of the patent system tending in the same direction as poetry and oratory, we have here a fine description of the individualistic cult of heroic invention so typical of the nineteenth century and beyond. The counters to Watt's heroic status that were registered in Cornwall; among engine competitors elsewhere like Murray; and among those such as Brewster, who thought soberly about the processes of invention, were to continue in the nineteenth century. But their voices would be drowned out by the chorus of hero-worship.

Capturing a Likeness and Mobilizing It

From the 1790s until Watt's death, there were recurrent efforts to represent through art the important figure he had become and to capture his likeness, not only for the edification of his friends and family but also for posterity. Images of Watt gained wide circulation in the early nineteenth century.

We might think that Watt would be the first of his family to be captured in portraits, but this was not the case. There are oil portraits of Watt's parents, James and Agnes, and of his paternal grandparents, Thomas and Margaret, by an unknown artist, which presumably are still held by the Watt family.[99] The very existence of these portraits is perhaps indicative of a greater level of prosperity than that usually attributed to the family in those early years. If these were all painted at the same time, judging from the apparent ages of the sitters and bearing in mind that Watt's paternal grandparents both died just before he was born, we might tentatively date them to the late 1720s or early 1730s. In October 1785, when Annie was visiting Scotland and went to Greenock to clean and clear out Watt Sr.'s house (he had died three years before) she asked for Watt's instructions on various matters including "some pictures viz. your Father & Mother Grand Father & Grand Mother an Uncle one picture of charity another of some old Don all in gilded frames."[100] An inventory of the contents of Heathfield House taken in 1791, not long after the Watts moved in, reveals that his father's pictures had been given pride of place there. In the stair case and landing were five pictures in gilt frames—"Mr Watt's Father Mother Grand F & M & Uncle"—joined by Lord Napier.[101]

Given this family background it is perhaps surprising that Watt and Annie showed little interest in portraiture prior to the 1790s. Annie did encourage Watt in having his silhouette produced, and he apparently did so. Mary Wray, wife of Daniel Wray FRS and a well-known silhouettist in her time, is recorded as having made one of Watt. Watt mentions to Annie from London in 1787 that he was going "to call on Wessel to get my profile taken" and then a month later that he had "sat 3 times to Mr Wessel & have done with him. I shall bring one impression down with me if [liked?] have ordered 5 more."[102] Such profiles were typically for the consumption of family and friends, and sure enough, in the 1791 inventory of Heathfield House we find in Jessy's closet four profiles of "Mr & Mrs Watt, Gregory and Jessy" in small gilt frames, as well as two profiles of "Miss DeLuc & Mrs Miller"—Miss De Luc being J. A. De Luc's daughter Frances (Fanny), who was a close family friend, and Mrs. Miller possibly being Peggy, who married James Miller in June 1791. The library contained a small profile of "Mr Herschel," presumably William Herschel, with whom Watt was friendly and who was to testify in the patent trials. So, the Watt family and friends participated in the silhouette craze of the period.[103] But more portentous, and more public, art was in the offing.

Carl von Breda (1759–1818) painted the first major portrait of Watt at about the time that Gregory was learning in Glasgow about his father's celebrity.[104] Von Breda was a Swede who had arrived in England circa 1787. He established a studio at 70 St. James Street, London, and gained a significant clientele. In 1792 he was induced to visit Birmingham, where he met members of the Lunar Society and painted portraits of three of them: Boulton, Watt, and William Withering. Joseph Priestley's wife also sat for him, though in London. The recruitment of von Breda was the result of the request for a portrait by Boulton's children and by his friend John Rennie.[105] Rennie wrote to Boulton in June 1792 that he had long wanted a portrait of him and also of Watt. He suggested that they might sit for Mather Brown, an American artist then operating out of Cavendish Square, London, who had already painted John Smeaton and Richard Arkwright. Boulton replied that his children were keen for a portrait and he was considering sitting for Thomas Lawrence instead and would be happy to provide a copy to Rennie. But an intended visit to Birmingham by von Breda that had been anticipated the year before came to fruition in July 1792. No doubt the convenience of having the artist visit them was welcome, for Boulton and Watt were very busy. Stimulus for von Breda's portraits was thus not lacking and certainly received impetus, like the silhouettes, from family and friends who long held the originals. But the timing of this commission, and particularly the nature of the images produced, tempt us to the view that the decision to procure these portraits was perhaps also a business consideration.

Boulton had already sat for a number of portraits and had a keen sense of what their wider display could achieve. If portraits were placed in the Royal Academy Exhibition that was held each year, they would attract public attention. Indeed, the academy exhibitions were in part an opportunity for artists to show their wares by depicting people who were in the news or otherwise gaining celebrity and whom the public would be interested to see.[106] Boulton and Watt were beginning their legal battles in defense of Watt's patent at this time and expected a long contest. Two of the potent tropes they deployed in that cause to try to influence judges, juries, and public opinion in their favor were captured nicely in von Breda's paintings. One trope concerned their supposedly public-spirited contribution to industrial development. This is represented strikingly in von Breda's image of Matthew Boulton, in which he

FIGURE 7.1: *James Watt* by Carl Fredrik von Breda (1792). Watt's pose, and the drawings on the table before him, helped to market him as a "philosophical" engineer at the beginning of the patent contests. © National Portrait Gallery, London.

sits before the vast enterprise of the Soho Manufactory. (No matter that in its detail it referred more to Boulton's minting work than to steam engines; the overall impression was the key thing). Watt's portrait features the other key trope—Watt as philosopher. Von Breda obliges here with an image of a contemplative Watt in classic pose as thinker, with drawings of the output of his ratiocination (steam engines) on the table before him.[107] The drawings depicted have perhaps the additional significance of signaling Boulton and Watt's supposed openness in their dealings as opposed to the secretive anti-modern craft practices. The firm commonly prepared "presentation" drawings for actual and potential clients that had this function.[108] Both pictures were sent to the Royal Academy Exhibition by von Breda in 1793, which would in itself gain them a wide audience among influential people. Boulton's portrait was rendered as a mezzotint engraving by Samuel W. Reynolds in 1796 (see figure 3.2), as was von Breda's portrait of Watt. Engravings enabled prints to be produced in some quantity that would allow an even wider spread of the images to friends, acquaintances, and people of influence.[109]

Watt had been reluctant to sit in the first place, and von Breda seems to have dealt with Boulton when the issue of exhibiting at the academy was broached, leaving him to persuade Watt to allow it.[110] In this, as in so many other areas of marketing, Watt was shown the way by his partner, and if he did not learn the lesson then Watt Jr. certainly did.

Watt appeared a number of other times in the precincts of the Royal Academy during his lifetime: in the portrait of him by Sir William Beechey in 1802; in the wax relief miniature by Peter Rouw in 1803; in a copy of the Beechey portrait by the famous enamelist Henry Bone in 1804; in Sir Thomas Lawrence's effort, exhibited there in 1814; and the next year in the form of Sir Francis Chantrey's bust of him. A pencil drawing of Watt by John Jackson that Watt had sat for was exhibited in 1819 at the academy, though not well liked by the man himself.[111] The Beechey portrait was reputedly Watt's own favorite and later came to be that preferred by his son. Thomas Telford also considered it a good likeness.[112]

To be painted by Beechey and Lawrence, and to be sculpted by Chantrey, was to have "arrived" in late Georgian society. The images of Watt created by these great artists are rather different from that created by von Breda. The Watt in the earliest portrait looks thin, strained, and troubled, as he no doubt was in the early 1790s. This is the working Watt, and the portrait is itself being set to work in a very specific sense also, as we have indicated. The later portraits are of a more serene figure. They present the prosperous, more settled Watt of his retirement, with an eye on posterity and his reputation, not the Watt who as recently as the early 1790s had still been "on the make." A notable exception here is Henry Raeburn's portrait of Watt, which seems to present something of Watt's anguished character. Watt and Watt Jr., perhaps perennially unsatisfied with prior portraits, had decided that Watt should sit for Raeburn when in Edinburgh in 1815. Watt observed that although the work did not "come up to my ideas of my own face" it was "more comfortable to them than any of the others, and by my friends is said to be a good likeness." He also wrote of this portrait that it was "thought like, only it frowns too much."[113] The roll of Raeburn's subjects places Watt firmly among other figures of the Scottish Enlightenment, many of whom were close to him, including John Robison and even Watt's cousin Mrs. James Campbell, the source of so many of the anecdotes about his youth.

There are a number of other images of Watt that were produced by direct sittings. In 1810, sitting for a Miss Andras for yet another profile, Watt observed, "my face seems doomed not to have an exact copy taken of it."[114] He was urged by Rennie in 1816 to sit for "Mr Skirvin," who worked in chalk and crayons and according to Rennie "seems to possess the art of catching the favourable moment for taking the likeness and delineating the character beyond any painter I have seen. I am convinced he is the artist of all others that would do justice to you." Watt replied that he would be happy to sit for Skirvin "either for profile or front view."[115] Watt even sat for copies of existing works, thinking in at least one case that the copy was a better likeness than the original.[116] The sheer number of occasions on which Watt was willing to sit for artists in his later years is perhaps surprising and notable in itself. He was certainly concerned with propagating his image and making sure it was a good one. The perfect likeness seemed elusive.

It was not unusual for clients to be active critics of the efforts of those who sought to capture their likeness. But Watt and Watt Jr. were probably more active than most. The efforts of the mezzotint engraver Charles Turner came in for particular criticism, Watt becoming very frustrated with Turner's efforts to engrave Sir Thomas Lawrence's portrait of him, telling Watt Jr. that the engraver was "averse to make any even the least variation from the painting." Watt thought he saw a way to pressure him into a sitting that might encourage such variation: "I did not make a condition of his coming down to finish the plate from my person, but I think he will easily be persuaded as he wants to see Birm^m and wants to try some experiments of getting his plates of better metal & better rolled in which we can assist him."[117] Turner's second attempt still did not impress:

[It is] very little improved in point of likeness since I saw it in London. . . . I have shown it to all my friends who I thought likely to judge properly of it and I am sorry to say without meeting with any approbation, some thought there was a likeness but not a striking one, but the greater part did not see any & some of the most knowing condemned it as unworthy of Publication both my Son & myself are of the latter number. . . . I think it cannot be mended because some of its greater faults are copied from the picture and Turner *dare* not improve upon that.[118]

The engraving was, however, ultimately published in 1815 and became one of the most widespread images of Watt in the nineteenth century.

Not the least example of Watt's concern with propagating his image was, of course, the extent to which he used his sculpture copying machines in the garret workshop to produce copies of the Chantrey bust of himself for distribution among his friends. Watt wrote to Thomas Thomson in 1818 that he was hoping to be able to produce a reduced copy of Chantrey's bust "fit for a chimney piece, as I do not think myself of importance enough to fill up so much of my friends houses as the original bust does."[119] A bust suitable for a chimneypiece, like an engraving of a por-trait, would above all facilitate the wider circulation of Watt's image. When Chant-rey asked whether he could display his bust of Watt in the academy exhibition, Watt agreed, thinking that it would be hard to deny the artist, given the trouble he had taken on the project. But Watt added that he was not "over fond of exhibiting myself either in person or by proxy."[120] Nevertheless, three months later Watt was asking Chantrey not only for a good cast of the bust but also for "a half head & neck for the purpose I mentioned to you (It should be a little more than half)."[121] This was exact-ly what he needed for employing his sculpture machine in turning out copies! Once again we might convict Watt of a rather strained modesty here, given the enthusiasm with which he aided and abetted the capturing and reproduction of his image, and the active critical interest he took in the process in his later years.

It is true that family and friends held the originals of most of these images of Watt, and the painted copies made of them. The von Breda, Lawrence, and Beechey portraits were, however, engraved during Watt's lifetime, the latter by Charles Picart in 1809 from a drawing based on Beechey by William Evans. Christine MacLeod is correct in saying that the public visual celebration of Watt during his lifetime was limited, certainly when compared with the riot of it after his death. But strategic

FIGURE 7.2: *James Watt* engraved by Charles Turner (1815) after Sir Thomas Lawrence. © National Portrait Gallery, London.

uses had been made of his image, and careful preparations were made to ensure its availability to posterity and to distribute it widely through engraving.

Watt's Death

Eventually posterity arrived: Watt died on 25 August 1819. He had been ill earlier in the year but had recovered reasonably well. He and Annie made their habitual visit to London, but by early July Watt was back at Heathfield consulting with his solicitors about a new version of his will. Muirhead's account of Watt's decline suggests that he recognized the onset of an illness in the autumn as one that he was unlikely to survive: "It was soon recognised by himself, with devout resignation, as the messenger sent to summon him away."[122] Given Watt's long interest in medical affairs and his obsessive concern with his health, it is certainly possible that he recognized a significant change in his condition. In mid-August Watt Jr. told John Rennie, "I found my father much reduced and exceedingly languid. He feels no inclination for food and the little that he takes is frequently rejected with bile. He is free from pain and his mind quite easy and composed." Watt Jr. reported that the physician (Dr. Johnson) thought that in a younger man a cure could be expected, but as things stood he viewed his father's condition with "extreme apprehension."[123] So, the reports continued in the following days with Watt getting weaker but free from pain. Then, for a couple of days there were signs of a rally, with Watt sleeping better, taking some nourishment, and digesting his food, but by 23 August he was going backward again.[124] He was attended close to the end by Watt Jr., Annie, John Tuffen, and Robinson Boulton. The latter somehow managed to report Watt's demise to John Rennie on 24 August before writing the next day to say that Watt had died "this morning at 12 o'clock." Watt Jr.'s "agony of mind" had misled Robinson Boulton the previous day to think that the event had already occurred.

Writing to John Rennie the day after, and for the first time signing himself simply "James Watt," Watt Jr. acknowledged that the loss left a void he strongly felt. But "I have cause not to repine such as seldom accompany such a calamity":

> My father has died in the full possession and enjoyment of the fame and admiration his talents have earned, he has outlived envy itself, and every friend of science has borne a willing tribute to his merit. He has enjoyed to the last the society and regard of the ablest men of the age, and the affectionate esteem of all who knew him. Contented with his circumstances and happy in his domestic circle, his powerful mind has remained clear and tranquil until within a

few hours of his death, and that death has taken place with little pain and in the manner he could himself have wished for.[125]

There is no reason to doubt the general accuracy of this sanguine view of his father's situation at the time of his death, although, as we will see, the claim that Watt had "outlived envy itself" was rather optimistic.

In April 1820 Maria Edgeworth visited Heathfield again and found Watt's image, and his ghost, at every turn. She described the visit in a letter to Honora Edgeworth: "How melancholy to see places the same—persons, and such persons, gone! Mrs Watt, in deep mourning, coming forward to meet us alone in that gay trellis, the same books on his table, his picture, his bust, his image everywhere, *himself* nowhere upon this earth."[126] The next day Maria visited Watt Jr. for breakfast at nearby Aston Hall, where Joseph Priestley's son joined them. Fine Flemish pictures hung in the breakfast room, with portraits of Boulton and Watt at either end. They toured the house. Watt Jr. had fitted out half of it, and Maria remarked on the strange carvings and elegant furniture including fine tables by Bullock made of Brazilian wood. Then they entered the great gallery, which was 136 feet long: "At the furthest end we came to a sort of oriel, separated from the gallery only by an arch, and there the white marble bust of the great Mr. Watt struck me almost breathless. What everybody went on saying I do not know, but my own thoughts, as I looked down the closing lines of this superb gallery, now in a half-ruined state, were very melancholy, on life and death, family pride, and the pride of wealth, and the pride of genius, all so perishable."[127] In the case of Watt, family pride and the pride of genius had some way to go yet.

Eight

Afterlife

⚜ ⚜ ⚜

A Man for All Causes

fter his death in 1819, James Watt had a big future ahead of him. There **A** were many ways in which the legend built in Watt's lifetime was perpet- uated and extended. Prominent among these was the "filial project," the fervent and enduring effort of Watt Jr. to perpetuate his father's memory and to promote his reputation as an engineer and natural philosopher. This project had a number of dimensions involving the construction of memorials, a literary "campaign" to tell a particular story of Watt's life, inventions and discoveries, a concern with material preservation of Watt's possessions as heirlooms, and even the fortune of the company with which his name remained associated. As we have seen, the filial project began in the early years of the nineteenth century but in some ways originated long before, when Boulton and Watt themselves had built and used Watt's reputation for innova- tion as a way to promote their commercial ventures.

However, the filial project was only one aspect of Watt's afterlife. Public com- memoration was also quick, extensive, and enduring. Watt, the heroic inventor of the machine widely seen to underwrite industrial progress, could not fail to be extensive- ly obituarized, memorialized, and celebrated in a country that was transformed by industrial change. Stories about Watt, and his image, were everywhere, assisted not least by the impact of the steam press on the world of newspapers, magazines, and book publishing. Images such as Marcus Stone's *Watt Discovers the Condensation of*

Steam of 1863 and Robert Buss's *Watt's First Experiment on Steam* of 1845 were engraved and widely distributed as representations of the child genius prefiguring the mature engineer's accomplishments. James Eckford Lauder's painting *James Watt and the Steam Engine: Dawn of the 19ᵗʰ Century* of 1855 loaded into the image not just the kettle but other popular symbols: the dividers of the geometrician, the basic tools of the craftsman, and the model Newcomen engine of the experimentalist. The viewer could interpret accordingly.[1] Watt became a flexible resource for many groups in British society, including engineers, scientists, and skilled workers, who employed characterizations of him to legitimate and extend their own status and concerns. Nationalisms as well as civic and professional pride competed to collar him as theirs. Educational objectives and the promotion and defense of the patent system were other popular concerns with which he was routinely associated. Anniversaries of Watt's death and birth, in 1919 and 1936 especially, drew wide attention to the legacy and gave a focus for interplay of many of the ongoing uses of Watt. In that sense he became "a man for all causes."

Until recently, the overall tone of commemorations of Watt, however divergent they might have been in character and substance, was generally positive and celebratory. However, with growing doubts in the late twentieth century about the future of an industrial civilization based on high fossil energy consumption, and concern about its environmental consequences, invocations of Watt's name have begun to take a different turn. He has been identified as a key progenitor of fossil capitalism and of the Anthropocene, a putative geological era in which humanity's pursuit of technological mastery began to leave its indelible mark on Earth. We will also briefly explore this latest, nascent phase in Watt's afterlife.

The Filial Project

We have seen that in the 1790s the troubled relationship between Watt and his eldest son was repaired when Watt Jr. returned from revolutionary France, at least partially renounced the politics that he had espoused there, and knuckled down to work in the family business. Though it took a little while for Watt to gain full confidence in his errant son, and somewhat longer for the relationship with Annie to be reestablished, the bond between the two men became very strong. Watt came to admire James's capacities and achievements, while James's devotion to his father became almost pathological. This devotion expressed itself in everyday ways. He was at his aging father's shoulder during most of the older man's business transactions, especially those relating to the Welsh estates. He attended to his father's letters and other business when Watt was away from Heathfield, and he assumed considerable responsibility for the welfare of the Scottish branch of the family and friends. In these ways he sought simply to make life easier for the old man.

Given Watt's ambivalence about matters of reputation, Watt Jr. also early assumed responsibility for monitoring, responding to, and where possible controlling what was written about his increasingly famous father. Thus he had been the prime mover in the Olynthiad, and also helped in Watt's dealings with David Brewster about editing Robison's writings on the steam engine.

In August 1819 Watt Jr. kept a vigil at his dying father's bedside and recorded the vicissitudes of Watt's final illness in a series of letters to close friends, including James Weston and John Rennie. On the day after Watt's death, James wrote: "I have indeed lost a father whom I deeply reverenced and dearly loved, and whose affection I enjoyed in return. Long years of intimate communication had sunk the father in the friend, and the loss of such a father and such a friend cannot but leave a void that I must strongly feel. . . . It remains for me now to pay due honours to his memory."[2]

Most immediately this meant attending to the funeral, which took place at nine o'clock in the morning on Thursday, 2 September. Watt had specified in his will that he be buried "in the most private manner, without show or parade as soon after my decease as may be proper."[3] We might question whether this wish was honored since the event reputedly cost £700. But local press reports suggest that it was conducted "in conformity with the simplicity and retiredness of his character": "In accordance with his own wishes, his funeral was private, but highly respectful; his hearse having been followed to the grave by a numerous assemblage of his most attached friends, among whom were many Gentlemen, eminent in science and literature, from various and distant parts of the Island, all anxious to pay the last sad tribute of respect, friendship and affection to this great and excellent Person."[4] Watt was interred in Handsworth Parish Church, St. Mary's, near the remains of his partner Matthew Boulton. In asking for simplicity Watt had had in mind a point of contrast between his own funeral and Boulton's. On the occasion of that ostentatious affair, Watt had written to his son, "Considering the publick life Mr B had led I think the whole a highly proper respect to his memory; but I pray that I may not be buried with so much parade, I have all my life hated show and ceremony, let it not follow me to the grave."[5] The Church of England buried the Presbyterian after all.

Handsworth Church was under repair at the time of the interment; in July Watt had given a gift of fifty guineas toward its rebuilding. Among the consequences of that partial rebuilding and enlargement was that Watt's grave, originally in the church grounds, was brought eventually inside the building.[6] A more immediate consequence was that there could be no funeral service in the church on the day of Watt's interment. Instead, the rector of Handsworth, the Reverend Thomas Lane Freer, preached a sermon at Aston Church on Sunday, 5 September, in which he marked the occasion. It is hard not to see Watt Jr.'s influence, and most likely his pen, behind Freer's remarks that day:

> [Watt's] character and course of life were of no *ordinary* cast. . . . His youth was devoted to such intense study, as in some measure injured his health, but his exertions were at length crowned with almost unparalleled success. His transcendent abilities, invigorated by habits of *application*, were particularly employed in *scientific* pursuits. Not however contented with understanding in theory the *known* principles of science, he urged his way, by experiments, to *fresh* discoveries, and he applied his discoveries, with inventive genius to the most *practical* results. By these results he put into the hands of men, such an increase in mechanical power . . . that in commerce and manufactures, a new epoch was produced. . . . They who had the happiness of knowing him in the private and social circle were alone able to appreciate fully, the strength of his intellect, the compass of his attainments, and the moral excellence of his character. . . . His knowledge was no less various than profound . . . [and was] preserved for use by the most faithful and retentive memory. Nor was the manner in which he imparted these *vast* treasures of his understanding, the least portion of his praise. . . . He appeared to follow up the remarks of *others*, than to lead the way with his *own*. . . . The *suavity* of his temper, the *benignity* and the *benevolence* of his heart—these qualities, accompanied by the strictest integrity, influenced his domestic and public life, and they rendered him kind as a neighbour—faithful as a friend—candid and patient respecting the faults of others—*habitually* desirous of increasing the comforts and relieving the sorrows of all around him.[7]

Of course, many of those attending would have remembered a more complex and flawed character than the paragon that the wranglers of his legend, and the requirements of a funeral oration, had created.

After Watt's death, James carefully monitored the published obituaries and other writings about his father. He described to Rennie Francis Jeffrey's "most capital delineation" of his father's character published in *The Scotsman* (Jeffrey's piece was to be an enduring source of panegyric concerning Watt).[8] He also knew from Rennie that John Barrow was contemplating writing an article on his father, and he offered copies of the Jeffrey piece and also a copy of the Freer sermon. Watt Jr. also recommended a short notice of the leading facts of his father's life that was annexed to Cadell and Davies's *British Gallery of Portraits*.[9] Watt Jr. was less sure about another "biography" of his father that Rennie had intimations of, which was to be published in the *New Monthly Magazine:* "I presume that the pupil of my father and brother of the late Professor Playfair who proposes to undertake the Biography . . . is not James, but William Playfair. How far he may be a proper person for this undertaking I know not, but probably he may from his own knowledge be able to supply some Anecdotes and some Remarks on his Character & Talents at the time of his [Watt's] first entering upon the Engine business, which may be interesting."[10]

William Playfair (1759–1823) had a long history of entanglement with Boulton & Watt and their associates.[11] He was indeed the younger brother of Professor John Playfair, who looked after him when their father died. William was apprenticed to Andrew Meikle at the same time as John Rennie, and he was then recommended by Robert Small (the brother of Dr. William Small) to Boulton & Watt for a position as draughtsman and assistant. He worked in that position alongside Watt for a few years, leaving in 1781. Watt was not impressed by him and called him a "blunderer," but he was "close to the action" of the early engine business and also had a lot to do with the copying machine venture. He witnessed Watt's signature on the patent for that machine. Playfair also worked for James Keir when the latter managed James Watt & Co, the copying machine company, and also at Keir's alkali manufacture at Tipton.

When Playfair left Boulton & Watt, his varied career combined real achievement (notably his pioneering works on graphical representation) with dodgy dealing: he variously stole others' ideas in brazen fashion (including some of Keir's) and patented them; engaged in land fraud in France, where he also participated in the storming of the Bastille before fleeing the country as the revolution fragmented; and in 1816 attempted to extort money from a Scottish lord on the basis of his supposed discovery of documents undermining an old legal settlement of the lord's title and estates.

Playfair had not been well regarded by Watt, or by Boulton's family. Watt was very cool when Playfair sent him a copy of his *Commercial and Political Atlas* in 1785, describing it to Annie as "mere plummery."[12] Boulton's wife Ann accused Playfair of "scandalous conduct" with women other than his wife. It is unclear just how much of Playfair's colorful career Watt Jr. was aware of, or whether he had had any wind of the recent extortion attempt, but he was clearly suspicious.

At first Playfair did the right thing, sending Watt Jr. his essay for comment. Watt Jr. detected serious errors in the essay, which stated that his father had served an apprenticeship in Greenock and made out Watt's family to be obscure and poor. Watt Jr. was compiling corrections when Playfair wrote again sending an already printed memoir of Watt from the *Monthly Magazine* that contained the same mistakes. Playfair was pressing Watt Jr. for pecuniary reward. Then Playfair sent him a copy of a letter that he had purportedly received from a "John Smith."[13] This letter, supposedly a response to having read Playfair's memoir, criticized it for suppressing "any thing that is not advantageous to the Hero of the Tale." It went on to portray Boulton as ruthless in the way he had garnered the whole fortune of the Robinson family by marrying both sisters, as cheating his partner John Fothergill, and as in deep, desperate financial trouble when he met Watt. It further alleged that Boulton tricked Fothergill out of a third share in the engine partnership and left him a ruined man, and also that Boulton used menaces when chased for a large debt, seeing off the bailiffs by threatening to bring in his workmen. The latter event occurred the same day, the letter claimed, that "Mrs Boulton drowned herself . . . from shame, grief & despair." Coming to Watt, Smith's letter charged that he knew of all this but did nothing, that he was mean toward Murdoch, and that Murdoch—not Watt—improved the engine. The letter concluded: "Should I find time to do this subject Justice as I intend your Memoir will look a bad piece of work."

Watt Jr. realized that he was the victim of threats and attempted extortion, and that Playfair had forged the Smith letter: "I could not for an instant doubt who the author of the letter signed John Smith was, for I verily believe there does not exist another person in the world artful and malicious enough to have written it. There is a basis of facts in what is stated about Mr Boulton, which could only have been known to some person in his confidential employ, but every thing is distorted and misrepresented and the statement as it stands is wholly false. What relates to my father is too wretched to require contradiction."[14] Watt Jr., with Rennie's agreement, replied civilly to Playfair, effectively inviting him to do his worst. The article appeared in the *New Monthly Magazine,* but with the editor's cooperation it was headed by a statement that it had been under correction by "a friend of Mr Watt" when another memoir appeared in the *Monthly Magazine,* despite the "handsome remuneration" that Playfair had received from the *New Monthly.* The editor left readers to "form their own opinion on the conduct of Mr Playfair."[15]

This strange affair confirmed for Watt Jr. how easily "misrepresentations" could occur, and the importance of controlling what was written because evil forces were abroad that might challenge his father's reputation. He had an immediate exchange with the publishers of *Annual Obituary*, having learned that their obituarist had relied on Playfair's memoir. Watt Jr. demanded that his own composition be substituted for the intended, tainted piece.[16] The Playfair business also stimulated Watt Jr. to conduct research about his father's early life. He solicited recollections from his father's friends and associates, especially from the early days, having resolved that a proper biography should be produced. He told Rennie that he intended "to devote all my spare time to the collecting and arranging materials for a history of my father, as soon as I can set about it."[17]

In May 1823 Macvey Napier approached Watt Jr. to write a brief biography of Watt for the supplement to the fourth, fifth, and sixth editions of the *Encyclopaedia Britannica*, of which he was editor.[18] Napier was keen to obtain a definitive account "from having been informed, that some inaccurate allusions to some part of your father's life, in an earlier portion of the work, had given pain."[19] This was a reference to the articles "Steam" and "Steam-Engine" by Robison in the *Encyclopaedia Britannica* that Watt had "corrected." It was not until 22 December that Watt Jr. sent the memoir to Napier. There followed over Christmas and New Year a fussy flurry of amendments. The piece was overlong, but rather than delay further Napier offered to edit other articles down to accommodate Watt Jr. Napier wanted to name the author, since named authors were commercially valuable to him, but Watt Jr. successfully insisted on anonymity. He felt that writing in brief he had been forced into claims without being able to supply evidence, and he wanted to avoid the accusations of "partiality & prejudice" that revealing his name would inevitably elicit.[20] Watt Jr. was expecting trouble, and he also gave it. He had learned of the intended contents of the biography of John Rennie (1761–1821) as written by John Barrow that was to appear in the *Britannica*. He and Robinson Boulton were unhappy with some of the things said about Rennie's work with Boulton & Watt and about Boulton and the mint. Watt Jr. insisted on a rewrite. Barrow was upset and Napier was aghast at the delay, but the Rennie family decided to comply. This is a good illustration of Watt Jr.'s powers of insistence where his father's history was concerned. Barrow thought him "the most obstinate & wrong-headed man I ever met with—different, very different, in all respects from his father."[21]

For a time, the initial flurry of literary interest in Watt after his death subsided. Watt Jr. was caught up in other sorts of commemorations that will be discussed later in this chapter. The literary dimension of the filial project gained new impetus when in 1834 Watt Jr. learned that François Arago, perpetual secretary of the Académie des Sciences, intended to deliver an eloge of his father later that year. Why did it take so long for the academy to memorialize one of its distinguished foreign members? There was a long backlog. But there were also political reasons. Arago's predecessor, Georges Cuvier, had used the eloges to highlight natural philosophers who had been relatively disengaged from the problems of social and economic change. Arago had very different political objectives and saw in Watt an exemplar useful in arguing for the greater engagement of natural philosophy with the solution of practical problems and the creation of wealth for the French nation.[22] Arago decided to attend the meeting of the British Association for the Advancement of Science to be held in Edinburgh that year. Watt Jr. learned of Arago's intended visit from a letter written to the Reverend William Buckland, the geologist, by a close associate of Arago's, J. B. Pentland. Pentland announced: "Arago intends to visit Birmingham to see your friend Mr Watt, as he proposes to read his Eloge of the Illustrious Inventor of the Steam Engine, at the next Publick meeting of the Institute in November and will be glad to talk over many circumstances with your friend & Mr Bolton, as well as to visit their superb Establishment. Will you therefore write a line to Mr Watt on the subject."[23]

Watt Jr. was anxious to see Arago and to let him have all the information he needed. It would not be possible, however, for Arago to visit the Soho works. Such visits were banned, with no exceptions, because of problems of industrial espionage.[24] In the event, Arago visited Aston Hall on 1 September 1834. As Watt Jr. recalled to George Rennie, the Frenchman "was in such a hurry to get to the Edinburgh Meeting that I could only get him to give me one day, during which I communicated all that time permitted, and above all completely satisfied him by an inspection of my fathers private correspondence that he had had foul play between Cavendish & his friends, with regard to the priority of publication."[25]

Watt Jr. was clearly preoccupied in this exchange with the question of Watt's priority in the discovery of the composition of water. He had come to see this, along with his father's independent heat and steam experiments, as central to promoting the character of Watt as a philosopher. The *Encyclopaedia Britannica* memoir had brought this to the fore without being able to discuss all the evidence. Watt Jr. saw the eloge as an opportunity to provide a fuller account.

It was to be five years before that happened. In the interim Watt Jr. continually prodded Arago to finish his work and plied him with materials. Those materials included the recollections of his father's early years by his cousin Marion Campbell that had been sent to him by her daughter Miss Jane Campbell. Watt Jr. advised Arago to "obliterate what you have already written of this period of my father's life, and to subscribe, if you please, this entire narrative in its place.".[26] Arago had provided a draft of the eloge for Watt Jr.'s inspection by June 1835. After further delays, eventually, in March 1837, Watt Jr. went to Paris for the express purpose of meeting Arago to finalize the manuscript. Then other material intervened—Henry Brougham sent Watt Jr. the results of his research in the Cavendish papers, and they went to the Royal Society together to examine materials there that convinced them that Cavendish's paper had been surreptitiously altered to make it appear that he had priority. It was agreed that Brougham's memoir would be appended to the eloge when it was published. Another visit to Paris saw Watt Jr. simultaneously poring over the proofs with Arago and suggesting further additions. Finally drawing a line under the process, Watt Jr. showed some self-knowledge: "I will not answer that if I should remain here a month longer, fresh matter may not occur to me. I therefore beg you to suggest your adopting the expedient proposed respecting a Gentleman we talked about yesterday and applying forthwith to Louis Philippe for an order for my banishment from Paris."[27]

Arago's eloge, and more specifically the reaction to it, brought Watt to the very center of arguments about the nature of science and the industrial arts, and their relationship in Victorian Britain. The water controversy was kicked off by the Reverend Vernon Harcourt's provocative rejection of Watt's character as a chemical discoverer in his address to the British Association meeting in Birmingham in 1839. This polarized the nascent scientific community around Watt on the one hand and Cavendish on the other. Controversy rumbled on for twenty years, generating a large literature,[28] raising in a sharp form the question of Watt's philosophical credentials and achievements. It fed for some a view of Watt as a scientifically driven engineer, and for others a view of him as chemist and philosopher manqué whose real strengths and achievements lay in his profound engineering insights. These contrasting views were important in the scientific politics of the period. The scientific elite of the British Association and the Royal Society, composed in significant part of Cambridge-trained mathematical physicists, sided with Cavendish as a model of scientific practice consonant with their own. That model in turn supported their preferred conception of their own kind of science as the ultimate source of material progress. For them Watt's untutored engineering genius was certainly to be praised, but his credentials as a natural philosopher were dubious at the least.

Watt Jr., it must be said, did not adopt either of these polarized views of his father, preferring to claim all these capabilities for him. Thus one of his struggles with Arago was to try to have material concerning his father's craft skills and civil engineering work inserted into the eloge. Arago resisted this, being much more inclined to see civil engineering as a waste of Watt's talents and being intent upon holding up an image of a scientific engineer. Watt Jr. did not live long enough to see the full flowering of another characterization of his father from within the scientific and engineering elite. We might call this the "thermodynamic Watt" that was promoted by the so-called "North British" philosophers led by William Thomson (later Lord Kelvin) and Macquorn Rankine, who were central figures in arguing for, and exemplifying in their own careers, the development of "engineering science" in the mid-nineteenth century.[29] These North British philosophers rejected the hierarchy that the Cambridge Network maintained in which abstract, mathematical science sat above and informed practical engineering. A central nexus of North British engineering science was thermodynamics, which sprang from the unification of heat theory, more generally energy physics, and steam engineering. An important symbol and instrument of this unification was what Rankine called the "diagram of energy," which, as we have seen, he identified with the indicator diagram produced by "Watt's" recording steam indicator.[30] Here the esoteric theory of thermodynamics, among the most fundamental insights into the physical world ever accomplished, was unified with the practical management of steam engines, the latter being facilitated in entirely new ways by the former.

Rankine was happy to promote this engineering science by identifying Watt as one of its key progenitors. In his improvements of the steam engine and his—supposed—invention of the recording steam indicator, Watt had contributed crucially, it seemed, to the foundations of thermodynamics. More broadly, Watt had pointed the way to what might be achieved when artificial boundaries between sophisticated natural philosophy and engineering were broken down. This gave Watt a central place in the scientific transformation of nineteenth-century engineering, a special role in the hearts and the teachings of the North British philosophers, and a consequent longevity in the engineering and shipbuilding community of western Scotland in particular. It was no accident, as we will see, that the statue of Watt erected in the early twentieth century in his birthplace was unique in holding a recording steam indicator in its outstretched hand.

For a long time Watt Jr. intended to write the definitive biography of his father. But this was never to be. He decided that the filial project required an assistant, whom he found in the person of James Patrick Muirhead (1813–1898). Muirhead was the son of Lockhart Muirhead, Regius Professor of Natural History at Glasgow University, and his wife Anne Campbell, the daughter of Watt's cousin Marion.

(Anne Campbell was the young woman who had visited Heathfield during Annie's absence and received so much interesting instruction from Watt, much to Annie's consternation). Educated at Glasgow College and Balliol College Oxford, Muirhead was admitted advocate in Edinburgh in 1838 and practiced law there for eight years, including work as a patent lawyer. In 1844 he married Katharine Elizabeth Boulton, the second daughter of Robinson Boulton, and so became doubly, if not triply, linked to the Boulton and Watt families.[31] During his time at Oxford, Muirhead visited Watt Jr. at Aston Hall, where he worked in the library and fell into the role of literary aide-de-camp to Watt Jr. While still practicing the law, he undertook the English translation of Arago's eloge, and so was drawn decisively into Team Watt. His next project was publication of Watt's correspondence on the composition of water, which in the mid-1840s Watt Jr. finally decided to publish to try to settle the water controversy once and for all. While Watt Jr. provided a prefatory letter for the work, it was Muirhead who edited the correspondence and wrote a substantial introduction to it, offering a carefully crafted argument for Watt's priority.[32] Watt Jr.'s health was failing by this time: his eyesight was especially bad and he suffered severe pain in his legs, which affected his mobility. Major changes were afoot in Muirhead's life too. He had married Katharine Boulton in 1844 but she was unhappy in Edinburgh and so Muirhead abandoned his career as an advocate, and in 1847 the couple went to live at Haseley Court, Great Haseley, in Oxfordshire. That magnificent house belonged to Katharine's family, having been bought by Matthew Piers Watt Boulton, the grandson of Watt's partner, that same year.

Watt Jr. died on 2 June 1848. Even in death he sought to perpetuate his father's name and promote his reputation. His estate devolved upon a young member of the Gibson family, James Gibson,[33] the son of Watt's granddaughter Agnes, but only when he turned twenty-five years old. Muirhead became, in effect, Watt Jr.'s literary executor. He saw through the publishing plans that they had worked out together, duly producing the three volumes of *The Origin and Progress of the Mechanical Inventions of James Watt* in 1854 and *The Life of James Watt* in 1859. The former contained a long account of Watt's life in volume one, extracts from the correspondence, which was heavily edited to burnish Watt's halo, in volume two, and the mechanical drawings and other technical material in volume three. Thereafter, Muirhead considered his job essentially done and moved on to other literary pursuits of a lighter nature. However, he did not escape entirely because as a trustee of Watt Jr.'s will, he had been charged with various duties to perpetuate the memory of Watt the father. He was to devote some of the rent raised from the lower meadow of the Heathfield

Estate "to keep and maintain the Chapel or Cemetery in the Parish Church of Handsworth . . . wherein my late honoured father is buried and the Monument there created to his memory in good and perfect repair order and condition."[34] A key section of the will established a number of items as "heirlooms" in the technical legal sense of items that could not be bequeathed away from the estate:

> I give and bequeath unto the said Trustees of this my will their executors administrators and assigns the portrait of my said late Father by Sir William Beechey his Bust by Sir Francis Chantrey his portrait in Enamel by Bone and all my family portraits busts and prints and the portraits of the late Mrs Watt and Matthew Boulton Esquire and as well all my said late Fathers private letters and papers of every description as also all my own together with all my watches trinkets plate and plated goods all my books in the several libraries elsewhere at Aston Hall . . . all the oak furniture . . . and such other of my furniture articles and effects at Aston Hall . . . as my said executors may think proper to select and remove to my several Mansion Houses at Doldowlod and Heathfield . . . my Maps and the iron safe in my study . . . and all the books of my said late father now at Heathfield . . . to go along with and be held and enjoyed with and attached to the several mansion Houses at Doldowlod and Heathfield aforesaid in the nature of heirlooms sofar and so long as the Law will permit.[35]

Thus it was Watt Jr.'s clear intent that the houses at Doldowlod and Heathfield should contain these heirlooms as long as possible. Those who would in future inherit the houses containing the heirlooms would "upon taking possession of such personal effects sign a script for the same at the foot of an Inventory thereof in the usual manner."[36]

So it was that the houses at Doldowlod and Heathfield continued to hold these various relics and memorials of Watt and Watt Jr. When Annie Watt died in 1832, Heathfield became the home of Watt's granddaughter Agnes (née Miller) and her husband James Gibson MD, who practiced as a medical man in Birmingham. Gibson himself died in 1835.[37] It is well known that Watt's garret workshop at Heathfield was long preserved as a memorial.[38] Watt Jr. locked up the room on his father's death, and though it was, as we shall see, disturbed on various occasions, it was claimed to be substantially intact until Heathfield was demolished in the 1920s, when the workshop was removed to the Science Museum. What has perhaps not been realized is that Watt Jr. had larger plans that involved the houses themselves being living memori-

FIGURE 8.1: Watt's garret workshop, Heathfield House. Photograph taken on the occasion of the contents being moved to the Science Museum in December 1924. © Science & Society Picture Library.

als by virtue of the heirlooms that they contained. The Pemberton family, to whom Heathfield House was leased in 1857 by Gibson Watt, found other heirlooms as well as the garret awaiting them. The son of the house, Thomas Edgar Pemberton, described their arrival: "It was a cold and cheerless February day, and as we drove up to the deserted house it had a forlorn and melancholy appearance. It was in the hands of caretakers, and, save for a quantity of old and very dusty furniture placed and piled up anywhere and anyhow, was empty."[39]

Beginning to explore, young Pemberton found a newspaper from 1819 containing an account of Watt, and then entered a ground floor room where he saw "a vividly striking portrait (it was the only picture on the walls of the somewhat damp-stained house) of a handsome man." Elsewhere he found a marble bust of what was clearly the same man, confirmed by his father to be James Watt. These were, I suggest, the heirlooms that had been left at Heathfield. Presumably, apart from the garret workshop, much of the rest had been removed to Doldowlod, where most of them remained in the care of generations of Gibson-Watts until disposed of by Sotheby's in *The James Watt Sale* after the death of David Gibson Watt, Baron Watt in 2002.[40] The James Watt papers bequeathed by Watt Jr. (along with his own papers) as an heirloom were held at Doldowlod until a large portion of them were purchased from Lord Gibson-Watt in 1994 and housed in Birmingham Central Library.

Watt's garret workshop was long treated as a kind of shrine not to be desecrated by intrusion, certainly during Watt Jr.'s lifetime. But it was opened on occasion. J. P. Muirhead saw it in 1853 and commented that "the presence of some reverend dust silently announced that no profane hand, forgetful of the "religio loci," had been permitted to violate the sanctities of that magical retreat."[41] The dust was still there, "two inches deep,"[42] when the Pemberton family received a visit from their landlord, James Gibson Watt; a key was turned in the rusty lock; and the room was opened and entered. The fact that, at least initially, Gibson Watt was required to be present for the room to be opened (it seems probable that he had the key) suggests that the garret was indeed regarded as an heirloom. Thomas Pemberton recalled, however, that thereafter the room was frequently opened, and it appears that as a young boy he examined its contents.

There were other visitors over the years. Some, like Samuel Smiles, who visited in 1864 or 1865 as he was writing the lives of Boulton and Watt, were concerned with somehow getting in touch with the departed Watt, or finding a stirring conclusion. In his description of the room, which did indeed conclude his book, Smiles evoked the feeling of such proximity by emphasizing the extent to which the room and its contents were as Watt left it, down to the piece of iron in the lathe, the ashes in the grate, and the withered bunch of grapes in a dish. There were the tools and the sculpture machines, the dried up chemical substances, the trunk containing the young Gregory's belongings. But amid the dust and the decay, Smiles was inspired to contemplate that which endured: "The spirit of his work, the thought which he put into his inventions, still survives, and will probably continue to influence the destinies of his race for all time to come."[43]

Others were on a more worldly mission, seeking to rescue the workshop from its isolation and open its inspirations to a wider public, and perhaps in so doing achieve some of their own objectives. Such was Bennet Woodcroft, clerk to the Commissioners of Patents, who visited on 5 April 1864. Woodcroft (1803–1879) was instrumental in establishing the administrative "apparatus" that systematized Britain's patent affairs. He also believed that he and others would find instruction and inspiration from historical information and artifacts. To this end he collected numerous items for the Patent Museum in collaboration with Francis Pettit Smith, the museum's cura-tor.[44] They got wind that "the Trustees of the late Mr Watt contemplate opening the room at Heathfield Hall" and saw a chance to acquire the workshop for the museum. This was the context of their visit in April 1864. As discussions with James Gibson Watt continued, Woodcroft optimistically began construction of a special room in his London offices to receive the workshop. The law of heirlooms achieved its impact by prohibiting the separation of the house and the heirloom by bequest. It did not, strict-ly, preclude the sale of the item during the lifetime of an inheritor of the estate. In this case, then, Gibson Watt may have been attracted by the financial arrangement that Woodcroft proposed. But the plan faltered because Woodcroft's superiors baulked at the deal and the potential ongoing expense involved.[45]

There were still others with designs on the workshop, notably Samuel Timmins of Birmingham, who was inspired in part by civic pride and concern that such an important relic of Birmingham's industrial past should not be lost to London.[46] But Timmins's efforts were frustrated in 1876; even as he began removing items from the workshop into the courtyard below, he received a telegram from Gibson Watt stating that "they were left as heirlooms in the family, and . . . he had no power to have them placed in a public Museum."[47] It was subsequently revealed that at about the same time the owner himself removed a large number of items from the room (perhaps with a view to taking them to Doldowlod), thinking that the new lessee of the prop-erty, George Tangye, would need the garret for his own purposes. But Tangye, who also showed himself eager to conserve Boulton and Watt artifacts and papers in other ways, said he was happy to preserve the garret as it was.[48] Timmins was joined in his concern for the Watt workshop at Heathfield and its artifacts by others including members of the Institution of Mechanical Engineers. At their quarterly meeting in Birmingham in November 1883, a large attendance and a number of visitors, includ-ing Timmins, heard a paper from Mr. E. A. Cowper on "The Inventions of James Watt."[49] In the discussions after the reading, Timmins said that he hoped the insti-tution's engineers would join those in Birmingham who wanted to ensure the preser-vation of the Watt workshop. As reported by the *Birmingham Daily Post*, Timmins described "the body of mechanical engineers" as the "lineal descendants of James Watt. They were the heirs-at-law; a sort of apostolic succession of one of the greatest inventors of our time."[50] He noted that many people in Birmingham felt that the relics should be preserved in their city. Now that Aston Hall, Watt Jr.'s longtime residence, was public property, he suggested that the solution was to preserve them there.

Shortly thereafter, however, suggestions began to appear that Heathfield Hall

and estate itself might be purchased by Birmingham Corporation to serve as a museum and public park, so that the Watt relics might be kept permanently and securely
"where they had hitherto remained as they were left by the great inventor."[51] House
builders were nibbling away at the estate, and there were fears that Heathfield House
itself was threatened. An approach was made to the owner, J. W. Gibson Watt, to
purchase the hall and grounds, but it came to nothing.[52] The hall remained under
lease to George Tangye and the relics effectively under his care for another thirty-five
years.

   Tangye lived in the house until his death in 1920. It is clear from the long sequence
of intrusions upon the workshop that the idea that it had been undisturbed since
Watt's time was a fiction. "Foreign objects" had appeared in it from time to time.[53]
During Tangye's tenancy there were periodic visitors to the room, including a number
of the engineers from the 1883 meeting just mentioned. In 1899 a visiting journalist,
Arthur Lawrence, who wrote an account of George Tangye and of his home at Heathfield for the *Cornish Magazine,* recounted a late-night tour of the garret given to him
by Tangye, during which they rifled through drawers and Lawrence tried some of
Watt's snuff, which was "dead."[54] Clearly, the display of the relics extended beyond
interested engineers. In 1910 Tangye entertained local Liberal party leaders at the
hall, and after the political speeches the men were shown the Watt relics. In 1912 a
group of German students inquiring into social conditions in Birmingham were entertained at Heathfield Hall, the visit terminating with "an inspection of the interesting Boulton, Watt and Murdoch relics assembled in the museum" there.[55] Among the
last visitors that Tangye welcomed to Heathfield and the garret were attendees at a
garden party in 1919 for the Watt centenary celebrations.

   In the wake of those celebrations and also of Tangye's death, the fate of Heathfield was once again uncertain. Its owner by then was Major James Miller Gibson
Watt, and according to newspaper reports, he made an offer to Birmingham Corporation detailing the terms upon which he was prepared to dispose of the house and
grounds.[56] The Public Works and Town Planning Committees of Council met with
the Watt Centenary Committee to discuss the proposal and entered into negotiations
with Gibson Watt. The relics themselves were not for sale. If the sale of the hall and
grounds was clinched, the relics, as heirlooms, would be loaned to the city on condition that they were kept at Heathfield Hall.[57] The hope appears to have been that the
Watt Centenary Fund, which had launched an appeal to raise £150,000 for various
memorials, would fund the venture. But a year later that fund stood only at a disappointing £11,000, and there were disagreements about whether Heathfield would

be suitable as a museum, creating fears that the city of Birmingham would lose the hall entirely.[58] Two years later the Gibson Watt trustees, presumably tired of wait, ing, had sold the Heathfield estate and the hall to a local syndicate who intended to develop it as a miniature garden city. The corporation approved plans for fifty-one houses. The new owner was willing to sell the hall as a memorial on condition that the land around it was also purchased. The asking price was £10,000, which was seen by most as beyond the reach of the ratepayers through their council. There were hopes for a while that the World Power Conference, then meeting in London, might help, but that too fell through. A few locals persisted in efforts to save the hall, even after Major Gibson Watt announced that he had presented the relics to the muse, um at South Kensington.[59] By March 1925 the removal of the relics, the fireplace, floorboards, doorway, window frame, and other fittings was complete. The hall was demolished in 1927, closing a long chapter opened by the heirloom clause in Watt Jr.'s last will and testament.[60]

Watt Jr.'s explicit literary accounts of his father, and those that he sponsored, stressed Watt's philosophical character and the public benefits his work had pro, duced. Additionally, the material that Watt Jr.'s stipulation of heirlooms helped to preserve, especially the workshop at Heathfield as shrine, provided fodder for that other major interpretation of Watt as at bottom a craftsman. An example is H. W. Dickinson's biography of Watt, which was published in the bicentenary year of Watt's death and subtitled "Craftsman and Engineer," with a photograph of the garret workshop as frontispiece.[61] But such interpretations were, and are, highly la, bile.

As the demise of Heathfield seemed ever more certain, descriptions of the Watt relics housed there became, in public discussion at least, more and more hyperbolic (and historically inaccurate). Now "it was in the garret that [Watt] conducted the ex, periments which were the genesis of the 'steam age'; and which later revolutionised industry," and it was also "a rendezvous for many of the leading scientists of Great Britain and the Continent" as well as being, supposedly, the site of meetings of the Lunar Society. Such hyperbole failed to save the relics for Birmingham. Interestingly, ten years later, as the two hundredth anniversary of Watt's birth was celebrated, in Birmingham the headline was "James Watt Plus Birmingham." H. Hopkins told readers of the *Daily Gazette* that it had been said that "more than any other single man [Watt] was the author of our civilisation. But left to himself, James Watt would never have been the author of anything, except perhaps a few unfinished scientific pamphlets. It was James Watt plus Birmingham that did the trick."[62] Watt was now described as a "pure scholar," timid and diffident, in action "pathetically ineffectu-al." It was only "a succession of honest and able Birmingham-bred men"—John Roe-buck but preeminently Boulton and William Murdoch—who had made the engines possible and made Watt's fortune for him. Once that fortune was established, Watt "flew off with a great sigh of relief to spend the rest of his days pottering profound-ly—beyond the reach of worldly care or wifely retribution—among the test-tubes and glue-pots of the garret of Heathfield Hall." Civic pride was a powerful thing as, in the face of Birmingham's loss of the garret, it was transformed from the cradle of industrial civilization into merely a place of profound pottering!

For Watt Jr. literary memorials and heirlooms were only part of the picture. He regarded Soho Foundry, and James Watt & Co., as memorials to his father as well as ongoing concerns. He strove to run a successful business, and as his own powers began to fade, he tried to ensure the business's future by bringing in other directors who might do the same. It has been a matter of some contention whether he succeeded. Ironically, precisely because of the luster that Watt's name acquired and the heroic reputation of his engineering feats, the nineteenth-century history of Soho Foundry has often been seen as rather humdrum. On this view, as an engineering business it was like many others of the period, and certainly by Watt Jr.'s death (perhaps before that) its best days were behind it. Watt Jr. has also been criticized for a fussy pre-occupation with administrative matters and as lacking strategic vision.[63] However, a rather different account of these issues has been offered in Laurence Ince's care-ful study.[64] He shows that Watt Jr. in particular worked extremely hard from the mid-1790s into the 1830s, reorienting the production of steam engines from the large beam engines typical of the early years of Boulton & Watt to compact engines for small industrial uses. He also pioneered the production of marine steam engines. Prof-its were substantial, but in the 1830s problems emerged and losses were incurred. When the largely inactive Matthew Robinson Boulton retired from the business in 1840, Watt Jr. became the sole owner. A few years before that, Watt Jr. had confided to Arago:

I did hope that a Firm which has now lasted for more than 60 years, might have continued to the end of our respective lives . . . But . . . I have given way to his wishes. At my period of life, and with a fortune ample to my wants; it may seem to you an Act of folly, that I should continue to expose myself to the exertions which the management of such a business as ours requires, and I might add to its possible vicissitudes; but so it is, that I have an insuperable aversion to desert a concern, founded by my father, and which has contributed not a little to his reputation, and has placed myself in a forward rank among the Manufacturers of this Country. It has been the pride & pleasure of my life to continue & keep up that business.[65]

By the time Robinson Boulton retired formally, Watt Jr. was over seventy years old and sought other partners. Henry Wollaston Blake became the chief support of the firm, now renamed—surely with an eye to ongoing posterity—James Watt & Co. Blake was a son of a Fellow of the Royal Society, and a Fellow himself as well as a Cambridge-trained mathematician, long-term director of the Bank of England, and director of a number of other companies that brought business to James Watt & Co. On his watch there was a revival in the 1850s and 1860s, but then decline set in again. It seems that Blake propped the firm up with his own money through the bad times, since he was once a very wealthy man and died with only just over one hundred pounds to his name. The firm was closed in 1895.[66]

There was another small, but very significant, way in which Soho Foundry and Soho Manufactory contributed to preserving the Watt (and Boulton) legacy. This took the form of what was known as the "Watt Room," which was devoted to the history of the company and its illustrious founders. In it were preserved for many years the Boulton & Watt company correspondence and papers as well as numerous other documents and artifacts. Among those artifacts was a steam indicator that had been dated, quite erroneously, to 1785 by persons unknown, but probably by someone who curated the collection. For some years and on multiple occasions, this was used to promulgate the idea that Watt had invented the recording steam indicator in that year. This encouraged the misleading impression that Watt had somehow intuited the indicator diagram and its modern thermodynamic meaning at the same time as he perfected his improved steam engine.[67] The contents of the Watt Room passed into the hands of George Tangye and then mainly into the care of Birmingham Corporation, becoming the basis of the marvelous collections concerning Boulton & Watt and its chief protagonists now held in the Library of Birmingham. The Boulton & Watt Collection, as donated by Tangye in 1911 to Birmingham Corporation, was opened by the mayor of Birmingham on 22 October 1915 in the reference department of Birmingham Public Libraries. Perhaps more than any other factor, the simple survival of that documentary record has preserved and promoted Watt's reputation. The differential survival of documents inevitably shapes the stories that historians can tell.

FIGURE 8.2: Watt statue by Francis Chantrey in St. Paul's Chapel, Westminster Abbey, London, late nineteenth century. Photograph by Frederick Henry Evans. © Country Life/Bridgeman Images.

Public Commemorations

The largest, and arguably most significant, public memorial to Watt was the giant statue of him by Francis Chantrey that was installed in St. Paul's Chapel, Westminster Abbey, in 1832. It carried subsequently on its base (see figure 8.2) an inscription written by Henry Brougham in consultation with Watt Jr.:

> Not to perpetuate a name / Which must endure while the peaceful arts flourish / But to shew / That Mankind have learnt to honour those / Who best deserve their gratitude / The King / His Ministers and many of the Nobles / And Commoners of the Realm / Raised this monument to / JAMES WATT / Who directing the force of an original genius / Early exercised in philosophic research / To the improvement of / The Steam Engine / Enlarged the resources of his country /Increased the power of man / And rose to an eminent place/ Among the most illustrious followers of science / And the real benefactors of the World.[68]

Christine MacLeod has argued convincingly that this commemoration was a major turning point for the status of inventors in nineteenth-century Britain.[69] This "arrival" of the inventor in the form of Watt in the ancient bastion of Westminster Abbey caused much disturbance. The installation of the massive monument in St. Paul's Chapel disrupted its very fabric when the paved floor cracked, revealing rows of gilded coffins beneath. The statue dwarfed everything else in the chapel. Public notices concerning the installation circulated by its supporters sang its praises: the pedestal was "of a design in harmony with the architecture of the place"; the likeness was "considered perfect and the look is intellectual and serene"; "the work may well take a place among the best portrait statues of ancient or modern times."[70] Not all agreed. There was much criticism on aesthetic grounds. One critic of Chantrey's contributions to the abbey's monuments convicted him of an "utter want of architectonic feeling," a lack demonstrated by "his master-piece" in St. Paul's Chapel. That chapel, the critic observed sarcastically, "should rather be called 'Mr Watt's Chapel,'" its "beautiful lancet forms" having been sacrificed to "a huge white image of the square-built mechanician . . . made, against all ecclesiastical decorum, to 'rump' the high altar itself."[71] Such aesthetic concerns were, as MacLeod argues, underwritten by deep anxieties about social and economic change.

From a few decades of perspective, Arthur Penrhyn Stanley, longtime Dean of Westminster, assessed the intrusion:

Well might the standard-bearer of Agincourt, and the worthies of the Courts of Elizabeth and James, have started from their tombs in St Paul's Chapel, if they could have seen this colossal champion of a new plebeian art enter their aristocratic resting place, and take up his position in the centre of the little sanctuary, regardless of all proportion, or style in the surrounding objects. Yet, when we consider what this vast figure represents, what class of interests before unknown, what revolutions in the whole framework of modern society . . . there is surely a fitness even in its very incongruity.[72]

When we look at the origins of the Westminster memorial, we do indeed find the marks of those professionally and politically supportive of the transformative effects of science, engineering, and industry, which Watt symbolized. As MacLeod shows, the push to memorialize Watt was redolent with a challenge of the industrial and commercial classes to an aristocracy that was enjoying renewed buoyancy, having been rejuvenated by military victory over the French. There was no shortage of com-memorations of military heroes, but there was a widespread sentiment that those behind the increasingly industrial sources of the nation's burgeoning wealth should also be celebrated.

The initial idea, as broached by Edward Littleton, MP, in a letter to the prime minister, Lord Liverpool, in May 1824, was raised on behalf of "several of the prin-cipal manufacturers of the kingdom." He sought the government's agreement to fund a public monument in Westminster Abbey in Watt's memory. Littleton suggested that this would be "a most popular act with the manufacturing body of the kingdom, who justly regard Mr Watt as the greatest promoter of the manufacturing skill and pre-eminence of the country that has ever lived."[73] Liverpool was reluctant to set a precedent in terms of government funding but agreed that a public meeting, attended by himself and many members of his administration, to launch subscriptions for a monument would be a good idea. In fact, it is clear that Watt's friends (including Rob-inson Boulton, Charles Hampden Turner, Henry Brougham, Humphry Davy, Francis Jeffrey, and the industrialist G. A. Lee) were closely involved from the beginning, and probably the real initiators.[74] Though Watt Jr. publicly kept a respectful distance, he was privately very much involved. Writing to Wedgwood on 12 June, Watt Jr. was clearly fully apprised of what was to happen at the proposed public meeting, adding, "I am of course supposed not to take part in these proceedings, although my wishes have been consulted throughout by all the parties to them, in the handsomest manner possible."[75]

This was the context in which a meeting was held in Freemason's Hall, Westminster, in June 1824 to promote a lasting memorial to Watt. For some of those attending, this was simply a chance to elevate an old friend to a status they felt that he deserved. For others, like liberal Whigs led by Henry Brougham and Francis Jeffrey, it was a chance to galvanize reformist constituencies in the provinces and in their native Scotland. Jeffrey's obituary in *The Scotsman* had staked a bold claim that it was the "improved steam engine that has fought the battles of Europe" and now fought the economic contest of the peace.[76] The architect of that improved engine deserved most of the credit, Jeffrey suggested, for his country's successes, in war and peace. A number of speakers at the 1824 meeting relied upon similar ideas. They were of the view that Watt was being celebrated as a rival figurehead to the Duke of Wellington.

Prime Minister Lord Liverpool headed those attending the meeting, added his own panegyric, donated £100 to the appeal, and communicated a pledge of £500 from King George IV. Other members of Liverpool's cabinet were also present, including Robert Peel, Chancellor of the Exchequer, as was Sir Humphry Davy, president of the Royal Society and an old friend of the Watt family, whose members had helped to set him on the road to fame and fortune. The presidents of the Royal Academy and of the Society of Antiquaries also attended, along with a number of other politicians, men of science, merchants, and the like. Those pursuing public identification with the cause raised a subscription of £6,000 that was used to commission the work from Chantrey.

MacLeod's analysis of those involved shows that many were members, like Chantrey himself, of the Athenaeum club, which had been founded the previous year and catered specifically for men of science and literature, and those who patronized and supported them.[77] Many, too, were Fellows of the Royal Society of London, whose president spoke at length at the meeting on Watt's scientific accomplishments and their relation to his engineering triumphs. This was a transparent bid by Davy to use Watt, and the occasion, to hammer home the necessity for increased government support for science.[78] Manufacturers, many of whom had been customers of Boulton & Watt and continued to use their engines, were also prominent subscribers to the appeal.

Notable by their absence among supporters of the Watt memorial, however, were the engineers and the machine builders with whom Watt, and Soho, had had a tense competitive relationship. They did not share the monument committee's desire to celebrate him as a national hero. Their countercurrents saw him rather as a ruthless monopolist, stifling the inventions of others, shaping history to his own purposes, and capable of anything in search of competitive advantage. As we have seen, the experience of the Hornblowers, Edward Bull, Richard Trevithick, Matthew Murray, and others with the temerity to engage in competition with Boulton & Watt had left many of those most closely concerned with the steam engines of the early nineteenth century disinclined to celebrate the memory of Watt. National commemoration proceeded without them.

Watt's memorial remained in Westminster Abbey with those to Newton and other "Abbey Scientists" until 1960 when, finally, a decision was made that it should be replaced by a more modest likeness.[79] After the coronation of Queen Elizabeth II, efforts mounted to restore the abbey, both its crumbling fabric and its internal arrangements, to former glories. The Dean of Westminster, Alan Campbell Don (1885–1966) initiated a funding campaign,[80] and large sums were raised and invested in the process, which often involved tidying up after what were seen as the misplaced enthusiasms of previous ages. The chapter and dean of the abbey echoed the old complaint that the Chantrey statue of Watt was simply too large for St. Paul's Chapel, and they decided that it should be removed elsewhere and replaced by a smaller representation of the great engineer.

The result was that a new Dean of Westminster, Eric Symes Abbott (1906–1983), wrote to the chairman of the British Transport Commission (BTC) wondering if the colossal statue might find a home in one of its public buildings. The response was that it might be accommodated in the BTC's museum of "historical relics" at Clapham. Once the sharing of the cost of transporting it was sorted out, the statue was delivered to Clapham on 14 December 1960.[81] It sat there rather incongruously among transport relics, as many noted.

The move was unpopular in some quarters. Deep historical and familial resentments surfaced: for example, R. E. Trevithick, president of the Cornish Engines Preservation Society, wrote to J. H. Scholes, curator of historical relics for the BTC, questioning the legitimacy of the statue's presence in a transport museum. Trevithick observed with more than a hint of sarcasm: "Surely this [is] not the most suitable place for this to rest, as the Great Man said publicly 'that Trevithick deserved hanging for introducing high pressure steam.' Without high pressure steam neither rail nor road steam locomotion would have been possible. I should be interested to hear that you agree." Scholes's enigmatic reply advised that he could not comment since "on a question relating to historical relics we should at all times record the true facts to the best of our ability."[82] The Watt family were also unhappy and surprised, for partly similar reasons. David Gibson-Watt (1918–2002), later Baron Gibson-Watt, recorded his reaction some years later. Recalling the occasion when "the Dean & Chapter of Westminster removed the statue . . . from the Abbey & dumped him without ceremony in your excellent museum," Gibson-Watt observed: "What Watt had to do with railways only the Dean knows."[83]

However, most correspondence concerning the possibility of moving the statue elsewhere came from those keen, as they saw it, to honor Watt more appropriately or usefully. The town clerk of the Council of the Borough of Smethwick asked if the statue might be moved there, where Soho Foundry had been located, but was batted away by the offer of a photograph. The Education Committee of the County of Renfrew, which wanted the statue to be placed in a new technical college to be built soon in Greenock, received a similar response. The Institution of Mechanical Engineers' suggestion that the statue might be placed in its garden in Birdcage Walk, not far from the abbey, met with more interest, but faltered on the impossibility of placing a porous marble statue in the open air.

A particularly concerted approach came from the president of the Institution of Engineers & Shipbuilders in Scotland, Iain M. Stewart, who argued for the location of the statue at what was to be Glasgow's "fifth university." Stewart's plan was interesting for a number of reasons. He had the backing of David Gibson-Watt now that the latter had been unsuccessful in an attempt to have the statue reinstated in the abbey. Stewart, himself a Scottish industrialist, became widely known for his enterprising attempts to revitalize the Scottish shipbuilding and engineering industries in Glasgow.[84] The Watt statue had an important place in his plans, as he explained to Dr. Beeching himself.[85] The Institution of Engineers & Shipbuilders was instituting a "James Watt Day" and an exhibition on Watt's life and works, intended as an annual event to revive the memory of Watt in the country of his birth and stimulate the members of the institution "to find new solutions to some of our very real difficulties in Scotland." The institution, from its foundation in 1857 with Rankine as a prime mover, had used Watt as an emblem on its printed transactions and in other ways.[86] Stewart thought that "there is really much to be gained for us in Scotland through the correct handling and siting of this important statue." A television program on Watt's life had been developed and was to be shown, and articles prepared for the press. The hope was that the statue could arrive by mid-October 1962 as a climax of these events.

Faced with these competing claims on the statue, it was decided within the BTC to refer the matter to its Consultative Panel for Preservation of British Transport Relics, which, having considered all the suggestions, concluded unanimously that the statue ought to remain in London.[87] It did so on the grounds that the original meeting of 1824 called to raise subscriptions had resolved for erecting a monument "in the Metropolis of the British empire" and that the location reflected its status as "a tribute of national gratitude" erected by "the King, his Ministers and many of the Nobles and Commons of the Realm." The committee also noted that its research showed that there were at least twelve statues of Watt in the country in Glasgow, Birmingham, Greenock, Manchester, Edinburgh, and Leeds, so that applications that would deprive London of a statue originally intended for the metropolis were misplaced.

The committee appears to have been particularly unimpressed by the strenuous and creative efforts of Iain Stewart to secure the statue.[88] Stewart had made strategic use of the media, engaged in concerted political lobbying, and struck an attitude presumptive of success to try to achieve his objective. The Consultative Panel clearly found his efforts infra dig. To Stewart's suggestion that Watt was "a forgotten man in Scotland" it responded that if this was so, "it was not for want of statues (please see the foregoing list, in particular one colossal bronze in George Square, Glasgow, and others in Scotland)." It rejected his claim that this was a national monument that should be "returned to Scotland" with the observation that the statue had "of course, never been there and was never intended to be."[89] It was almost certainly also unimpressed by the article "James Watt Broods in a Bus Depot" planted in *The Sunday Times* earlier in the year, probably by Stewart, at the beginning of his campaign.[90]

So the statue stayed at Clapham until the early 1970s, when another crisis arose. It had been decided that the museum at Clapham would close and its contents be transferred to a new national transport museum to be built at York. This in itself was a complex and controversial move.[91] The Science Museum was taking responsibility for the contents of the new museum at York but did not want the Watt statue. So a new hunt began for a site. The Science Museum stood firm against it despite opinions that it would be by far the best place. The Victoria & Albert Museum considered it briefly but declined it because of its "colossal" scale. The general manager of the Scottish Region of British Rail rejected the idea that it might sit in the concourse of Glasgow Central Station. By early January 1974, Eric Merrill, chief public relations officer, was consulting the British Rail legal adviser about options, describing Watt as "alone and forlorn" in the museum at Clapham. "Nobody wants James Watt," he lamented. The dispassionate reply suggested that the statue might be sold as a statue "or for scrap"![92] Merrill knew that public sale, probably to America, might secure a great deal of money but that it would lead to a lot of adverse publicity for British Rail: the symbolism of offloading this great symbol would not be positive. He and

FIGURE 8.3: "James Watt Broods in a Bus Depot," *The Sunday Times*, 17 June 1962, 16. Reproduced by permission of The Times/News Licensing.

many others, including David Gibson‑Watt, who had been kept informed, were re‑lieved when it emerged that the dean and chapter of St. Paul's Cathedral were willing to house the statue in their crypt.[93] There was a frantic "whip around" to raise money to fund the transfer and reinforce the floor that was to receive the monument. The removal of the statue from Clapham and its installation in the crypt at St. Paul's was completed by April 1977.

Watt's "bus depot" period had come to an end, but his place in St. Paul's turned out to be temporary too. In 1996 the dean and chapter of the Cathedral donated the statue to Heriot Watt College in Edinburgh. As a supposedly temporary measure, it was removed first to the Scottish National Portrait Gallery in Edinburgh, where, covered with a St. Andrew's Cross flag, it was unveiled by the Lord Chancellor, Lord Mackay of Clashfern, on 22 August 1996. As *The Sunday Herald* newspaper head-lined it, "A Great Scot Comes Home after 150 Years in Exile."[94] The Consultative Panel for the Preservation of British Transport Relics would have had something to say about that!

We have seen that the newfound mobility of the Watt National Monument from the early 1960s stirred competing civic interests. But, as the Consultative Panel noted in its report, the competitors had long been supplied with their own public commem-orations of Watt.

Contemporary with the installation of the Chantrey statue in Westminster Abbey was the erection of a bronzed version of the same colossal figure in George Square, Glasgow, in 1832. From a very broad cross-section of Glasgow's trades £3,000 had been quickly raised, suggesting that Watt enjoyed the status of a more popular hero among the mechanic class in Glasgow. But there were some who suggested that a more appropriate memorial than a statue would have been a new building for the Mechan-ics' Institute.[95] In fact, in 1824 the Glasgow Mechanics' Institute commissioned a life-size statue by John Greenshields of Watt leaning on a steam cylinder. It was in-tended for their premises on Shuttle Street. By 1831, when the institute moved to new premises on North Hanover Street, the Watt statue was exhibited on the main street frontage. Glasgow proved prolific in its memorialization of Watt. Another ver-sion of the Chantrey statue, a gift from Watt Jr. to the University of Glasgow, took its place in the Hunterian Museum. Private commissions produced a number of others.[96]

Later in the century other important Glasgow relics of Watt faced destruction. The university left its old site in 1870, when many of the buildings, including that which had housed Watt's workshop, were demolished to make way for the Glasgow Union Railway. The *Glasgow Herald* pondered the supposed ironies of the situation: "Deep-brooding Watt, sitting in his academic shop, studying great physical powers, evoked from his brain the very spirit the progress of which is about to lay the walls of his student-cell in ruins. It is to the railway that the University is about to yield up its ancient dwelling-place, and in a few months there will sweep over the spot where the great philosopher sat the very spirit which he was then quietly chaining to the car of civilisation."[97]

Though in this case Watt is depicted as destroying his own relics, his name and ac-complishments, particularly the connection between science and wealth that his story could conjure, were instrumental in justifying the significant expenditures on the university's new buildings and featured prominently in the speeches given on their opening. Much historical violence was done. As Crosbie Smith puts it, "Glasgow College and its reformers were too shrewd to allow historical accuracy to stand in the way of their claim to the legends of Watt."[98] In Glasgow, mobilization of Watt to promote educational objectives was perhaps most prominent among his uses.

Watt's birthplace, Greenock, also raised a good deal of money quickly for its own memorial, accumulating over £1,700 by 1826. At a meeting in the assembly rooms at Greenock on 30 August that year, local worthies gathered to discuss the commemoration of Watt and, in a carefully staged move, were advised that important visitors were in town, who were then called to the meeting.[99] The visitors were Watt Jr. himself and the president of the Royal Society, Sir Humphry Davy. The chairman of the meeting, Sir Michael Shaw Stewart, and Davy both gave speeches about Watt and the "triumph of philosophy, as applied to practical purposes."[100] Watt Jr. announced his donation of £2,000 to build a library, in which a mere life-size version of the Chantrey statue of his father was placed in 1838, where it remains. This completed a process begun by Watt himself in 1816, when he had donated money for the purchase of books for Greenock. The town's commemoration stressed the importance of measures to encourage emulation of their great predecessor by the mechanic class. In this way the Watt Monument Library was established, becoming a focus for regular remembrances of Watt, which were plentiful in his birthplace.

One might have thought that this was enough, but a plan was hatched when the Greenock cemetery grounds were acquired in the 1840s to establish there a "universal memorial" to Watt. This was an anti-parochial cosmopolitan venture. The key idea was that stone and marble materials would be donated from around the world, transported to Greenock, and there constituted into a memorial, the result of a "united effort of the whole civilised world." It was apparently inspired by the example of the Washington Monument in the United States. The design, by the architect David McIntosh, was for an Italian campanile tower rising to almost three hundred feet on a forty-foot square base. It would have been visible from Glasgow.[101] But even in Greenock, on this occasion reality fell short of ambition. Donations of material produced only a ramshackle collection of rocks that were consolidated into a Watt cairn. The project got no further.

By the 1880s pressure continued for a truly public monument to Watt in Greenock, and there was a strange sense of shame that all that had been managed was the naming of a recently completed wet dock after him. One possibility was to use the supposed site of Watt's birth on William Street to house a truly national memorial to which those regular visitors searching for the birthplace might flock.[102] Not until 1902 was a plan developed for that site as a result of the visit to Greenock by the US steel magnate Andrew Carnegie to open the Public Free Library, which he had funded. When Carnegie, who was immensely proud of his Scottish origins, saw the derelict site of Watt's birth at the corner of William and Dalrymple Streets, the idea of a memorial was discussed. This eventually took the form of the Watt Memorial Engineering and Navigation School. In a niche at the corner of that school was displayed a full-length bronze statue of Watt by Henry Charles Fehr, holding in its right hand a recording steam indicator. Carnegie had encouraged an attempt to raise money from around the world through small donations. He considered this appropriate, given what he took to be the lowly origins of Watt himself as a practically self-made man.[103] But little was subscribed, and both building and statue were funded almost entirely by Carnegie.

On 1 June 1908 Carnegie unveiled the statue at the opening of the school. A large audience of top-hatted gentlemen and of workers wearing their bunnets (flat caps) was present to hear Carnegie's speech. The statue itself was a copy of one by Fehr erected in 1898 in Leeds, except that instead of the dividers in the hand of the Leeds statue, in Greenock Watt held the recording steam indicator. For the university-educated, top-hatted types present, the indicator would symbolize Watt as a philosophical progenitor of the science of thermodynamics. It would have had significance too for the practical engineers and mechanics in their bunnets as a familiar device used in the business of testing, regulating, and adjusting the performance of the engines of steam ships out on the Clyde.

Once understood, memorials of Watt could have such intensely local significances. While memorials in Western Scotland (at least outside of academic circles) usually conveyed, and were interpreted through, a view of Watt as practical improver, those in Edinburgh tended more to the high-blown picture of him as Enlightenment philosopher. From the beginning, the impetus behind the public commemoration of Watt in Edinburgh had been driven less by civic pride in industrial achievement (Watt was, after all, a Glasgow man) than by a more diffuse, but powerful nationalism among the educated legal and political elites of southeast Scotland. The Whigs' Henry Brougham and Francis Jeffrey set the tone with their high-flown characterisations of Watt as philosopher, intent upon thrusting him forward in the Edinburgh pantheon against the local Tories' celebrations of Scotland's military contributions. But in Edinburgh the tension between nationalism and unionist sentiment meant real uncertainties about whether the appropriate course of action was support of the

Figure 8.4: Watt statue by Henry Fehr outside Watt Memorial Engineering and Navigation School, Greenock, showing the steam indicator in Watt's right hand. Author's photograph.

Westminster Abbey monument subscription or a local effort. Francis Jeffrey and Sir Walter Scott, who called the initial Edinburgh meeting in 1824, were focused upon mobilizing Edinburgh behind the Westminster subscription. But other voices, notably Professor James Pillans, argued for a local commemoration; a subsequent meeting attended by large numbers of artisans batted the opposing ideas back and forth, adding the suggestion of raising money for a new building for the School of Arts in which a memorial might be housed. In the end none of these plans came to immediate fruition for the simple reason that Edinburgh did not raise much money.[104] Only in 1851 were accumulated funds used to buy a building for the Edinburgh School of Arts, renamed the Watt Institution and School of Arts, and a statue of Watt by Peter Slater, which was placed outside it.[105]

Other cities that commemorated Watt in public statuary tended, like Edinburgh, to achieve success in doing so in the mid- to late nineteenth century. Birmingham and Manchester are of particular interest. Although Manchester had no direct connection with Watt, it was obviously a place that had benefited enormously from the fruits of his labors. As early as 1836 a subscription list had been opened for a Watt statue, but the idea was not revived until 1855, with the Manchester Literary and Philosophical Society leading the push.[106] A Watt Memorial Committee was established at a public meeting in which commemorations beyond a statue were considered, including a Watt Institute to promote the application of mechanical science to industry. But the focus was upon a statue to sit in the newly developed Piccadilly Esplanade that would form a pair with the statue of local scientific hero John Dalton, a bronze copy by William Theed of a Chantrey marble statue. The Watt statue, also by Theed after Chantrey, was unveiled in June 1857 by the engineer William Fairbairn. The lack of originality in the commemoration was criticized in some quarters, it being observed that Manchester was "dealing in the old clothes of Art."[107] Nevertheless, the pair of statues was a powerful statement about the place of science and engineering in industrial development, Watt's role being clearly that of mechanician.

In Birmingham, thanks to Watt Jr.'s drive in the cause, and thanks also to his purse, in September 1827 the first of the Chantrey marble statues of Watt was installed in St. Mary's Church, Handsworth, on a pedestal in a special chapel designed by the architect Thomas Rickman. Busts of William Murdoch and Matthew Boulton joined it there subsequently. But of course this statue of Watt was a private memorial, and there were rumblings over the next decades about the lack of a public memorial to Watt in Birmingham. In 1865 the Institution of Mechanical Engineers approached the mayor of the city about a memorial and the city council approved its commissioning. A public subscription attracted support from Matthew Piers Watt Boulton, Boulton's grandson, and from the workers of Soho Manufactory and Soho Foundry as well as local industrialists. The artist commissioned was Alexander Munro, who had previously sculpted a statue of Watt for Oxford University's Museum of Science, and the result was unveiled in October 1868 in Ratcliffe Place. It was a naturalistic depiction of Watt and regarded as a remarkable likeness. Mayor Thomas Avery acknowledged that it was long overdue recognition of one of the greatest benefactors of the city.[108]

The other major commemoration of Watt in Birmingham in the form of statuary was unveiled in 1956. This is a group statue of Boulton, Watt, and Murdoch, cast in bronze, finished in gilt, and mounted on a stone pedestal in front of the registry office on Broad Street. Watt is the central figure, and Boulton and Murdoch hold a drawing of mechanical devices between them, which the figures are evidently discussing. The sculptor was William Bloye, and the sculpture is unusual in being explicitly intended by the city council, which commissioned it, to represent "the teamwork of these men."[109] The council's emphasis on teamwork reflected an ongoing ambivalence in Birmingham about stressing Watt's individual accomplishments, an ambivalence that we saw expressed in the wake of the removal of Watt's garret workshop to London when the opinion was voiced that industrial transformation had been the product of "Watt Plus Birmingham."

The meanings of public statuary are not contained within the objects themselves. Conventions of representation, the significance of poses and accoutrements, and so on all have to be read into statuary by those observing it. The Watt statue in Greenock that holds the recording steam indicator is a case in point. Modern observers of it would likely have no idea what it is. Even when it was unveiled, the meanings read into it would have varied, as we saw, with the education and occupation of the observer. We can in the end only presume that such statuary would be read in ways that reinforced existing knowledge, preoccupations, and prejudices. It was undoubtedly read at various times as inspiration to the working mechanic as well as an affirmation of the philosophical basis of Watt's achievements; as an embodiment of the virtues of technical education or as an endorsement of the importance of the patent system; as a celebration of local industry or of national achievement. I have argued in detail elsewhere that one concerted message of the statuary was that Watt was a "mechanician." There was virtually no possibility of reading Watt the chemist into it, and in that sense it conveyed a historically slanted view of the nature and basis of his achievement.[110]

Watt, the Patent System, and Industrial Civilization

We have concentrated on the varied sites for commemorating Watt, and we now turn briefly, in a different focus, to two particular reasons for doing so. One was to promote the patent system and defend it against its critics. Another was to advance a particular view of the development of industrial civilization.

Given what we know about Watt's reliance on the patent system and the criticisms, even during his lifetime, of his use of it, it is no surprise that Watt lived a good part of his afterlife as a key figure in the to-and-fro between patent reformers and abolitionists. The reformers were concerned with strengthening the patent system so that the inventiveness of individuals might be encouraged to promote national prosperity. Patent abolitionists, by contrast, considered the system a pernicious and counterproductive creator of monopolies of no use to the development of new industrial arts and machinery, and in many ways an obstacle to that.[111]

For those seeking to improve the situation of the struggling inventor, the example of Watt was an important one. Surely, they argued, the enormous benefits that flowed from Watt's inventions would not have been forthcoming without the period to develop and exploit his invention that the 1769 patent and its extension had given him. More generally, patents were defended as a just reward for individual originality. As Christine MacLeod has shown, the burst of biographies of engineers and inventors in the middle decades of the nineteenth century was closely tied up with the attempted reforms of the patent system. Numerous stories of the inventive struggles, successes, and failures of individual inventors of all kinds reinforced the point about the vital importance of patents.[112] One focus of reform was making patents easier to obtain by reducing the fees charged and streamlining the labyrinthine procedures involved. Much was achieved by the Patent Law Amendment Act of 1852, carried in the wake of the Great Exhibition. It appears that the legal environment for the defense of patents also improved from the 1840s.[113]

While heroic biographies of inventors were promulgated by the likes of Samuel Smiles, John Timbs, and Henry Dircks,[114] Bennet Woodcroft, as head of the patent office, promoted record systems and instituted a museum that reinforced the individualistic view of the process of invention. Watt, of course, was always a prominent example in these discussions, and we saw that Woodcroft made a serious, if unsuccessful, effort to obtain Watt's garret Workshop as a centerpiece of his Patent Office Museum. That would have been a true jewel in the crown of the patent promoter's cause. William Siemens, in addressing the Mechanical Science Section of the British

Association in 1869 as abolitionist sentiment mounted, put the case clearly using the example of Watt:

The greatest illustration of the beneficial working of the Patent Laws was supplied, in my opinion, by James Watt when, just 100 years ago, he patented his invention of a hot working cylinder and separate steam-engine condenser. After years of contest against those adverse circumstances that beset every important innovation, James Watt, with failing health and scanty means, was only upheld in his struggle by the deep conviction of the ultimate triumph of his cause. . . . Without this opportune help Watt could not have succeeded in maturing his invention. He would in all probability have relapsed into the mere instrument-maker, with broken health and broken heart, and the introduction of the steam engine would not only have been retarded for a generation or two, but its final progress would have been based probably upon the coarser concep-tions of Papin, Savery, and Newcomen.[115]

There were powerful forces, however, assembling under the banner of free trade that argued for the abolition of the patent system. This was a Europe-wide phenomenon, and in some countries abolitionism made great headway. The Netherlands did actu-ally abolish its patent system in 1869, and it was not reestablished until 1912. In Britain, Robert Andrew McFie (1811–1893), a Scottish sugar refiner and member of Parliament for Leith, led the abolitionists. He wrote a number of works on the question and kept the abolitionist cause before the Parliament and the press.[116] Many manufacturers who paid out large sums in patent royalties found their pocketbooks and their principles carrying them in the same abolitionist direction. Other oppo-sition to patents came from some prominent engineers. I. K. Brunel was among the best-known opponents; much of his opposition derived from the frustrations of pur-suing large-scale projects that were constantly crisscrossed by rent-seeking patent holders.[117] Sir William Armstrong, the inventor and arms manufacturer, was anoth-er prominent opponent, as was Sir William Cubitt. When Cubitt, then president of the Institution of Civil Engineers, had been pressed during his evidence before the House of Lords Select Committee on Patents in 1851 about the supposed importance of Watt's patent to his success and his ability to recover the cost of developing it, he did not concede the point but rather emphasized that Watt himself had told him that "it cost him a great deal to defend his invention in the courts of law."[118]

Abolitionists were not without material to support a non-individualistic, collec-tivist, or even determinist account of the process of invention, which, if accepted, challenged both pragmatic and moral arguments about the need to provide the patent incentive to individuals if inventions were to flourish. In their accounts we see treatments of Watt's life and accomplishments that played down the heroic character of his inventive achievements.[119] We have seen that there was ample material to do this: the countercurrents to Watt's heroic reputation provided stories of other inventors who might be seen as anticipating Watt, or coming up with "his" ideas at much the same time. This could feed the arguments about simultaneous invention that abolitionists mounted, taking a deterministic view that economic demand was the primary stimulus of inventive effort. There was no surprise, then, that multiple candidate inventors of a given machine might be found. The corollary was, of course, that if one person had not come up with the invention then others would have stepped into place. The stories that floated around about Watt's reliance upon others and his individual appropriation of what were really others' inventions were also liable to surface in abolitionist texts. While the hyperbole that attributed to Watt all the nineteenth-century miracles of steam could be useful to the supporters of the patent system, it could also be an easy target for abolitionists intent upon revealing the excesses of individualistic accounts of invention.

These more balanced representations of Watt were, however, very much in a minority. The overwhelming and continually reinforced view of him was as an inventor of heroic proportions. But there were variations on this theme also. The images of Watt propagated by and for the skilled labor aristocracy were also formed in their own image. Watt's heroism in this case was that of a skilled workman, a man whose hard work and perseverance as much as anything else were presented as the crucial ingredients of his success.

Beyond Watt's utility in arguments about the patent system was the connection drawn between him and his exploits and the whole process of industrial change. Indeed, as Christine MacLeod has argued, long before the idea of an "industrial revolution" centered upon the transformation of manufacturing through machines officially took root in the late nineteenth century, "veneration of Watt helped to identify the British industrial revolution increasingly with the steam engine and cotton factories."[120] The commemoration of Watt through statuary and in other ways, and his use to promote the patent system and to raise the profile of the individual inventor generally, gave a new positive impetus in peoples' minds to the transformative role of invention. The totemic status acquired by the steam engine as producing a revolution in power production, and hence manufacturing, and in human affairs was underwritten by Watt's reputation. Thus the idea of transformative industrial change and what had driven it was already well established in public affairs before Arnold Toynbee published his famous *Lectures on the Industrial Revolution in England* in 1884.[121]

Through the later nineteenth century and well past the middle of the twentieth century, Watt retained an important place in this basic story of Britain's rise to industrial might. But he became part of a bigger picture. The Festival of Britain celebrations in 1951 were many and varied. In Glasgow the big event was the Exhibition of Industrial Power in Kelvin Hall, which treated the machines and the people who developed and used them, and more generally the impact of British inventiveness on the world. Watt had his place in all this, but the subject matter was laid out in two parallel sequences of "coal" and "water," leading to a grand finale with a presentation of the presumed power source of the future, atomic energy.[122] The eighteenth-century inventor had a place of honor still, his craft remembered though his natural philosophy largely forgotten, but the science-based power of the atom was the star of the show. Earlier in the century Watt had been somewhat overshadowed by the 1931 commemoration of the scientist Michael Faraday, estimated as the father of modern electrical generation.[123] Now in the atomic age, Watt as historical figure was eclipsed by Rutherford and other exponents of atomic energy.

Anniversaries

Watt became one of those figures whose significant anniversaries were celebrated, and we can see over time and across different sectors of British society the way that quite different "Watts" were put to work.

The earliest and most regular anniversary celebrations of Watt were those held by the Watt Club in Greenock.[124] The club traced its origins back to the winter of 1813, when it began its social meetings in the Greenock Tavern. Watt did not become the central excuse for these meetings until 1820, when the members noted the passing of their famous townsman and decided to name their club after him. They also began an effort to record details of Watt's life, beginning with the location of his birth. They discovered that Watt had been born in a house that had previously occupied the site of their meeting place, the Greenock Tavern itself, which duly changed its name to "The James Watt Tavern." As a public recognition of Watt, the club decided that it would hold a dinner to celebrate Watt's life on his birthday, 19 January, the first of these meetings taking place in 1821. In its regulations, which were formulated a few years later, the basic structure of the club was established, as was the practice of awarding testimonials "to encourage native genius and merit." So began the regular reports in the local press at that time of year of the anniversary dinners at which

Watt's memory was toasted. So too began the inquiries into Watt's early life that in the hands of one of the club's Presidents, George Williamson, became a valuable source of anecdote and information on that topic.

The Watt Club communicated the fact of their existence and the nature of their celebration to Watt Jr., who returned his gratitude and that of Mrs. Watt. Watt Jr., in return, expressed the wish to present a bust of his father by Chantrey to the town of Greenock; as we have seen, this resulted in the construction of the Watt Monument Library, which housed a large Chantrey statue of Watt. Eventually the Greenock Philosophical Society was established in 1861, absorbed the Watt Club, and held Watt Anniversary Dinners that became the occasion for a distinguished lecture series; in 1865 James Joule became the first in a long line of national and local worthies who gave the lectures. Collectively, the lectures remain a very valuable source on the various interpretations in the later nineteenth century of Watt's work and significance.[125]

Glasgow also held annual Watt Anniversary Dinners, jointly mounted by the Philosophical Society of Glasgow and the Institution of Engineers and Shipbuilders in Scotland from the late 1880s, and previously held under the auspices of a Glasgow-based James Watt Club, also known as the Association of Foreman Engineers, beginning in the 1840s or early 1850s.[126] In the 1890s these dinners were attended by three or four hundred people, but there were some signs of "Watt fatigue." In 1896 the chairman of the anniversary dinner, John Inglis, lamented the fact that the chairman had the duty to offer some observations on Watt's work and career, an "item in the bill of fare, what is apt to become an indigestible hash of scraps from a biographical dictionary."[127] He simply proposed the toast. Even that was overlooked the following year, only to be reinstated in 1897 when Chairman Sir William Arrol concisely observed that the achievements of Watt spoke for themselves, as "the fruits of his genius and labours are to be seen everywhere."[128] But the contest over which Watt was to be celebrated kept him alive. A recurring theme was the tension between the practical and the university-trained engineer. In 1902 the chairman of the dinner, William Foulis, put it this way:

> We sometimes heard it said Watt's great achievement was the conception of the separate condenser. To his mind that was only one incident in his career. He thought his great achievement was the result of those long series of years in which he strove with imperfect tools, and under many disadvantages, and under much discouragement, to perfect the whole details of that engine, and to

make it a workable machine. Without all his labour and all the detailed invention which he brought to bear upon it, the engine might have been little more than a scientific toy.[129]

Foulis, while he did not deny the value of technical education, was skeptical about the frequent claims that other nations were outstripping Britain because of their superiority in it. He believed that, as in the case of Watt, it was individual exertion and qualities of perseverance that would keep British engineers at the forefront of technological progress. Unsurprisingly, Professor Archibald Barr, on behalf of the University of Glasgow, stood up to offer a rather different estimation of the relative importance of scientific and technical knowledge and qualities of perseverance in the engineers of the future, as, by implication, in the most prominent of those in the past.

In 1911 the dinner was held in St. Andrew's Hall, Glasgow, the largest venue in the city, so that the entire company of more than six hundred people could dine together. Numerous engineering, shipbuilding, and military dignitaries attended, and toasts were raised to Watt and to "the Imperial Forces." The chairman read a cablegram from "the Steam users of Japan" who were "holding a James Watt dinner today." The reply had hoped for the long continuation of the "cordiality of the feelings existing between the two societies."[130] Preparations for war and hopes for peace were frequently mentioned at the Watt Anniversary Dinners before 1914.

There was some recognition among other scientific and engineering bodies of the 1869 anniversary of the granting of Watt's first patent one hundred years before, but it was not until 1919, the hundredth anniversary of Watt's death, that a major series of celebrations was planned and held around the United Kingdom. This may seem rather late in the day, but it must be remembered that the "cult of the centenary" did not begin to gain a strong foothold in our imaginations in Britain and Europe until the second half of the nineteenth century, though its origins lay somewhat earlier.[131]

In Glasgow in 1919 it was clear that the technical college and university educators' view of appropriate commemoration of Watt had won out. In Scotland he was remembered as a celebrated natural philosopher and civil engineer, and the use of him was determinedly forward-looking. This was undoubtedly part of the legacy of the North British philosophers, for whom Watt had long been symbolically integrated into an active program of engineering science that operated at the innovative edge of modern industrial arts. The Institution of Engineers and Shipbuilders in Scotland, though it fell short of an ambitious target, still raised subscriptions of more than £30,000 as an endowment for two chairs of engineering at the University of Glasgow. There were hopes to establish a James Watt Institute of Engineering in Aberdeen and distinctly up-to-date ideas about what a modern-day Watt educated there might be concerned with: "Now that coal is become a precarious commodity owing to industrial strife, and when the discovery of oil in this country bids fair to furnish a vast supply of other fuel, a Watt who can invent new machinery or adapt existing forms and secure economy of consumption is a genius much to be desired."[132]

As MacLeod and Tann have shown, this lively centenary commemoration in Scotland contrasted with a full but comparatively sclerotic and rather inchoate commemoration in Birmingham.[133] Perhaps Birmingham had too much of the past to focus upon—too many relics. At the invitation of a group of Birmingham engineers, a conference was held on 27 February 1919 at the Birmingham Chamber of Commerce, and a committee was established with William Mills as chairman.[134] The committee decided on an ambitious threefold project: the collection of funds to support the erection of a Watt Memorial Building in Birmingham, the raising of funds to endow a Watt engineering chair at the University of Birmingham, and the publication of a memorial volume about Watt. When this scheme was put before a subsequent meeting in Birmingham—chaired by the Lord Mayor of Birmingham and attended by more than two hundred men representing various engineering, scientific, and public organizations—it became even more ambitious. It was now to include a week of commemorative meetings and to be "international" in character.[135]

The proceedings began on Tuesday, 16 September, with a civic welcome at the university buildings and with the presentation of a number of papers about Watt. In the afternoon a memorial service was held at St. Mary's Church, Handsworth, where wreaths were laid in the memorial chapel, and visitors then proceeded to Heathfield, which (as we have seen) was opened for the occasion by George Tangye. Many took the opportunity of a peep into the garret workshop. In the evening, a mayoral reception was held and guests strolled around an exhibition of Boulton and Watt relics. The second day began with more lectures, one with Sir Oliver Lodge looking forward to "New Sources of Energy," and others polishing up the relics, such as the paper by two professors from Glasgow on the "Model of the Newcomen Engine repaired by Watt." In the afternoon, cars whisked guests to various locations to inspect Boulton & Watt engines that survived in the area. The oldest was the Smethwick Engine, built originally in 1777 for the Birmingham Canal Navigations but in the early twentieth century located at Ocker Hill, Tipton, where it was put under steam and indicator diagrams were given as souvenirs for the guests. (It is now in the Think Tank in Birmingham).[136] The evening of the second day saw the commemoration dinner held at the Grand Hotel, an evening of toasts and telegrams for more than three hundred guests, including members of the Watt family and the ambassador of the United States as well as guests from Australia, France, and Japan. The third day of commemorations involved visits to the Soho Foundry, then used by W. & T. Avery Ltd. but still exhibiting structures and machines surviving from the old foundry. The university then held a degree ceremony in which various honorary degrees were awarded to visiting dignitaries, and those still keen were given the chance to visit other surviving Boulton & Watt engines.

The Saturday following saw a workers' procession through the streets of Birmingham led by a lorry on which a half-size model of the Boulton & Watt "Lap Engine" was carried. Marchers carrying the banners of the major trade unions as well as more exhibits from local engineering firms followed behind. More public lectures ensued, and the Boulton & Watt relics at the art gallery were made available for viewing once more to the wider public.

As MacLeod and Tann observe, this commemoration was certainly very full, but the fund-raising efforts were much less successful. While the target had been in excess of £100,000, only about £11,000 was raised and expenses had amounted to £4,000. A few thousand pounds were given to the university to establish a fellowship in engineering, but the grand plans for endowment of a Watt Chair and the construction of a Watt Memorial Building were abandoned. The memorial volume on Watt did appear, and was an outstanding book, but it wasn't published until 1927.[137]

The marketability of the Watt who was emphasized at Birmingham, that is Watt the craftsman rather than Watt the natural philosopher, seemed much lower at this period than the philosopher engineer who was conjured by the Glasgow commemoration. This was a time when the ideology of applied science was being developed by an active scientific community that proved much more successful at garnering resources than an engineering community still divided on the value of scientific and technical education and research.[138]

In 1936, when the bicentenary of Watt's birth came around, there was an altogether less strenuous feel to the various commemorations. Perhaps a lesson had been learned in 1919 about what the market in commemorations and fundraising would bear. There was a service in Westminster Abbey on 19 January, at which it was claimed that "every qualified engineer in Britain was represented" through the presence of about twenty presidents of engineering institutions and societies. Three wreaths were laid at the foot of the statue in the abbey: by the Institution of Mechanical Engineers, the Institution of Naval Architects, and by a representative of all the remaining engineering institutions.[139] An exhibition at the Science Museum in London inevitably featured the garret workshop not long installed there. In Glasgow the Watt Club of Heriot-Watt University was revived and student prize Watt medals introduced. In Edinburgh the Scottish National Portrait Gallery put on a display of Watt portraits. New technology made its contribution when the radio play *James Watt and the Romance of Steam*, written by Alastair Dunnett and John Gough, was broadcast in Scotland.[140] An air of levity reigned when the New Theatre Cinema in Cambridge was the scene for a performance entitled "Watt's Bust" by the music hall comedy duo Raymond Bennett and Fred McNaughton. The manager of the theater said that after the act, a pro-proctor of the university sent for him and "appeared to be very angry about a particular gag—a joke with an application to the James Watt centenary [sic] celebrations, the pro-Proctor's objection being to a play upon the word 'bust.'"[141] In Greenock itself, a "Pageant of Greenock" was performed with the life of Watt as centerpiece. With considerable historical license, this documented his early struggles; his "adoption" by the university, which was presented as the scene of his "important improvement" in the steam engine; and the formation of his partnership with Boulton. The story climaxed in Cornwall in a scene where, incongruously, the Watt rotative engine (a model constructed by students and staff of the Watt Technical School) sputtered to life on stage. At that point the audience was stirred into hearty applause.[142] In more sober vein, Lord Rutherford gave the annual Watt Lecture, his "The Transformation of Energy," in which he argued that Watt's association with Glasgow University was the turning point of his career, "for not only did it give him a favourable opportunity to begin his experiments on the steam engine, but what was more important, provided him with an ideal environment for acquiring that breadth of scientific outlook and knowledge which so markedly distinguished him from the engineers of his time."[143] Rutherford was an astute judge of the land of the North British philosophers. Watt relics were exhibited in the Watt Memorial Library, and, not to be left out, "The Children's Corner" column of the *Greenock Telegraph* devoted itself to the story of Watt's mother as the nurturer of his genius, exhorting the children to "Honour thy father and mother."[144] On even higher moral ground, in Birmingham, St. Mary's Church, Handsworth, once again held a commemoration service on 1 March, attended by the Lord Mayor, members of the Watt family, a representative of the citizens of Greenock, and a representative of the Institution of Mechanical Engineers. The latter read one of the lessons, Sir Herbert Austin read the other. A vice-chairman of Messrs. W. & T. Avery Ltd. of Soho preached a sermon.

The emphasis was very much upon Watt as a symbol of the continuing vitality of local engineering industry.[145]

During the 1936 commemorations, then, the images of Watt as philosopher and as engineering craftsman were both alive and doing duty, but if anything the craftsman was in the ascendant, as in Dickinson's biography of that year. Watt was becoming a multimedia phenomenon, even a subject of music hall jokes. His uses were being generalized, a process that we see in his deployment in advertising, where he became a general purpose symbol of innovation.

The twentieth-century iconography of James Watt as expressed in advertisements carried a number of the nineteenth-century tropes forward, particularly the notion of Watt as inspired by the boiling kettle. The anecdote that began with Marion Campbell and was variously rendered artistically by Robert Buss and Marcus Stone has been used in the twentieth century to sell numerous goods and causes, including televisions, Johnny Walker whisky, Dewar's whisky, Toshiba semiconductors, and, perhaps most innovatively, kettles![146] The streets of Greenock and along the River Clyde have seen the procession and transport of a giant yellow kettle in the cause of local industry, innovation, and information. Other images and stories of Watt have been deployed in the advertising service of Lloyds Bank, Whitbread Brewery, Smith's "Glasgow Mixture" pipe tobacco, Siemens Nixdorf IT systems, Newbro's Herpicide for Dandruff treatment, Isaac Black's tailor-made clothing, and the *Encyclopaedia Britannica*.[147] My favorite Watt advertisement, one that avoids the simple, cringe-worthy historical inaccuracies in the majority of them, is an advertisement placed by United Technologies in *The Times* in 1985 that, in trying to be clever, manages much more complex inaccuracies. The advertisement stated that Watt

didn't actually invent the steam engine. Nor did he get the idea watching his mother's teakettle, as we were told. For one thing his mother was dead, and for another the Newcomen engine already existed. . . . In 1764 Watt came up with the idea of using two chambers, one cool and one hot. He also applied steam to both sides of the piston for extra speed. He invented the device that converted the piston's movement into the turn of a wheel, and the Industrial Revolution was born.

The companies of United Technologies didn't invent the jet engine, either. Or printing, or lifts, or helicopters. But like Watt, we've made some dramatic improvements here and there.[148]

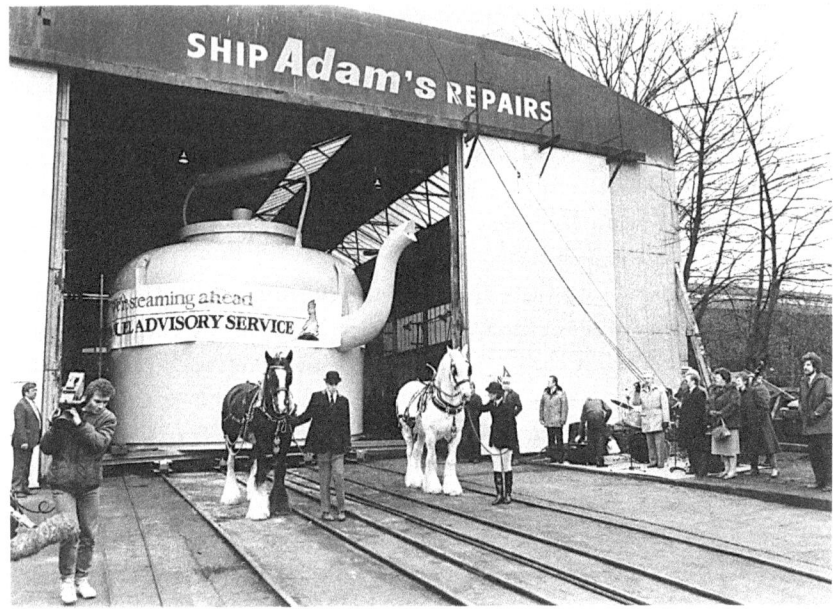

FIGURE 8.5: The Big Yellow Kettle prepares for an outing in 1988. Watt and his kettle became a general symbol for innovation in the late twentieth century. Courtesy of the *Greenock Telegraph*.

But my point here is not the fact of the inaccuracies of advertisements deploying the legends of Watt, but rather the observation that through the public culture of the twentieth century, the live contested symbol that Watt was in the nineteenth century fades away into the generic legendary inventor. It is this same generic movement that has significant numbers of people believing in the twenty-first century that James Watt invented the light bulb.[149]

Watt and the Anthropocene

We saw that in the Festival of Britain celebrations Watt as the Great Steamer was overshadowed by the prospects of atomic energy. The history of human generation of industrial power was there conceptualized in terms of the sources of energy: coal, water, and the atom. In recent decades, as the potential of nuclear energy has been questioned and the prospects of renewable energy sources have brightened according to many, Watt has been identified by those "in the know" not so much with steam as with coal. There is clearly sense in this: the great economic value of his invention lay in the saving of coal that it made possible. But of course this was the kind of "saving" of coal that led to its ever-escalating exploitation. Paul J. Crutzen, in his widely cited paper suggesting the idea of the Anthropocene, stated: "It seems appropriate to assign the term 'Anthropocene' to the present, in many ways human-dominated, geological epoch, supplementing the Holocene—the warm period of the past 10–12 millenia. The Anthropocene could be said to have started in the latter part of the eighteenth century, when analyses of air trapped in polar ice showed the beginning of growing global concentrations of carbon dioxide and methane. This date also happens to coincide with James Watt's design of the steam engine in 1784."[150] Crutzen's nomination of a coincidence did not claim a univocal causal relationship, but it has often been taken as doing so in popular treatments of the question. We often read that it was Watt's "invention of the steam engine" that marked the beginning of the Anthropocene. But others have taken up alternative accounts, some wanting to drive the beginning back to the 1600s and the Columbian Exchange, during which disease and the decimation of human populations had important knock-on effects on climate, and some bringing it forward to the mid-twentieth century, when human-produced high levels of radioactive isotopes joined carbon emissions as crucial markers.[151] Geologists are closely wedded to locating the "Golden Spike," the stratigraphic marker to define the commencement of the Anthropocene, while advocates of Earth system science regard that as a limited and mistaken perspective.[152] The latter favor linking the Anthropocene to the "great acceleration" in carbon emissions conjoined with the large-scale radiation emissions of the nuclear weapons era, and so suggest a nominal year of 1945 as the beginning of the Anthropocene. Still others caution against seeing the destructive activities marking the Anthropocene as a generalized product of "humanity," and instead see it as a result of inequality. They are keen to attribute the "blame" not to Watt, or any individual, but to the industrial, coal-consuming paradigm of development pioneered in Britain and the United States from the eighteenth century onward and the social groups that controlled those means of production.[153]

So Watt's afterlife continues into the present. While he might not be blamed for the state of the Earth system, except implicitly in the most unthinking popular treatments of the Anthropocene concept, it does seem likely that our perspective on him will change substantially in the decades to come as a result of these debates.

Concluding Reflections on the Great Steamer

We have said that Mr. Watt was the great *Improver* of the steam-engine; but, in truth, as to all that is admirable in its structure, or vast in its utility, he should rather be described as its *Inventor*.

—FRANCIS JEFFREY, "The Late Mr James Watt," *The Scotsman*, 4 September 1819

I do not think I merit much praise. I am in fact a mere compilator. I have tacked and sewed together the seraffo and remnants of many a man's ideas, altered and fitted them to purposes he never thought of, but I have invented or discovered little.

—JAMES WATT TO ERASMUS DARWIN, 31 January 1790

So, what in the end do we make of James Watt, known to some in the nineteenth century as the Great Steamer, as we mark the two hundredth anniversary of his death?[1] The dimensions of his life and work that we need to consider include: the nature and extent of his technical and economic achievement; the relationship between his engineering and his natural philosophical experimentation; the connection of Watt with the spectacular industrial development of the late eighteenth and nineteenth centuries; his qualities as a man and his relationships with family, friends, and collaborators; and his ongoing symbolic significance after his death, including for us today.

Watt's low-pressure steam engines, which he thought of in terms of a chemistry of the substance of heat, had more in common with Newcomen's era than with the steam engines of high Victorian times. Watt's work was in many respects a culmination of eighteenth-century industrial art in its relations with natural philosophy. In this sense his critics in the early nineteenth century were essentially correct. It must be agreed that Watt to some extent caused a brake on innovation in high-pressure steam, but not in a major way. It does not automatically follow, as some like to argue, that the patent system should be blamed for this. In a related point, we have to say that Watt in many ways remained a Scottish improver rather than became an industrial entrepreneur. Boulton and other members of Team Watt did more to supply the entrepreneurship than Watt did himself. Having made what he regarded as a sufficient fortune, Watt (unlike Boulton, who pursued new business ventures to the end of his life) was happy to leave business behind and turn himself into a landed gentleman—if a rather unusual, and unusually active, one. It was in the nineteenth century that Watt was retrospectively converted into a central figure in, and progenitor of, the Industrial Revolution. This process involved a good deal of hyperbolic extension of Watt's achievements, both by a heroic treatment of him that neglected the collective dimension of what he accomplished and by a thoughtless presumption that the marvels of steam power of all sorts in the nineteenth century could be laid at his door. They could not.

To gain such perspective on Watt's achievement is not to devalue it, but to honor it properly. It is, as David Brewster put it, to admire Watt "for what he did, and not for what he never thought of doing." In the epigraphs that I supply for this conclusion, Francis Jeffrey drew too long a bow in his obituary of Watt, but Watt sold himself a long way short in his despondent response to Erasmus Darwin's request for an account of his achievements. We should find a middle level between the heights of Jeffrey's hyperbole and the depths of Watts despond. Wherever the separate condenser idea came from—and there are grounds, as we have seen, to doubt the inspiration on Glasgow Green—it was a great invention, and Watt's persistence and inventive fecundity in devising a multiplicity of ways to save steam might in itself qualify him for immortality. So too would the way he tackled the use of steam in expansive working in rotative applications, and the mechanical insight and skill exhibited in inventions that increasingly rendered his engines "self-acting." Others may have invented "Watt's governor," but inventive capacity lies as much in seeing applications to new purposes as it does in being "first." Watt's engineering and experimental skills and inventiveness lay at the base of numerous achievements that seeded industrial development through steam power.

We have seen that Watt was not only an inventive engineering craftsman but also an experimental natural philosopher, if not a conventional, systematic one. His experimentation was usually stimulated by the requirements of a practical project. But it did not fragment entirely along the separating lines of those projects; I have shown the evolution of his ideas across and between those practical ventures. With the exception of his work on water, however, Watt was not inclined to publish his experiments or his conclusions from them, with the result that recovering that work requires us to examine carefully his private papers and correspondence. In the tradition of improvement, Watt was ever alert to the potential commercial value of his ideas, and this may well be the reason why he guarded them closely. Although Watt, and Boulton, moved in Royal Society circles and enjoyed the friendship and indulgences of the Society's president, Sir Joseph Banks, and other Fellows, their involvement with that institution seems always to have had a commercial motive. Watt's first major dealings with the Society were concerned with trying to give it a role in the adjudication of patents, obviously a key commercial concern of his. Apart from the publication of his work on water, his interactions with the Society were underwritten by commercial concerns to promote his engines, his copying machines, and after his retirement, the Boulton, Watt & Co. venture into gas illumination. With the possible exception of the informal, and unrecorded, Lunar Society meetings, the pursuit of natural philosophy in an open fashion was not high on Watt's, or Boulton's, agenda. The Lunar meetings were, significantly, private to the group. Completely open sharing of the fruits of his intellectual and experimental labor was, for Watt, not only a commercially unwise option but also a strange and puzzling concept. He believed that only those of independent fortune who did not live by their wits could, or indeed *should*, behave in that way. Future exploration of the constellation of economic, social, and moral ideas that underwrote this kind of thinking will perhaps enable us to understand this stance better.

There were significant tensions at the heart of what modern scholars call the Industrial Enlightenment that are also exhibited in Watt's life, and they constitute a central enigma. We have seen Watt inhabit a conflicted moral universe in a number of respects. On the one hand, there was his cultural Presbyterianism and high moral propriety—the concern about swearing oaths and truth telling when taking out his patents, the threat of hell and damnation that he inflicted on Watt Jr. as a little boy caught stealing, the insistence on proper accounting by his children to the point where he alienated them. On the other hand, there was his clear duplicity in business and in the closely related matter of promoting and defending his reputation. Certainly he hid much of the time behind Boulton and Watt Jr. in this regard, but he could be ruthless himself. Playing the ever-virtuous philosopher behind key members of Team Watt was, in any case, itself an example of the same duplicity.

In some ways this enigmatic character is no surprise: it inhabits us all and the imperfect world that we live in. We might attribute to Watt a Manichean character and so privilege the consistent pursuit of self-interest above all else. Or, perhaps more satisfactorily, we might recognize hard-to-resolve tensions between aspects of his announced moral precepts. One of the key tensions was perhaps between the imperative of ensuring just desserts for the inventor or discoverer and the question of how far down the road of duplicity one should go to try to ensure this. Managing these tensions, which appeared also between the openness encouraged in the Republic of Letters and the need to guard commercial secrets, was a central dilemma of the Industrial Enlightenment.

We can look too hard, however, for elaborate philosophical justifications, and might consider that Watt, like the rest of us, was simply trying to get by as best he could in a practical sense. Pragmatism was Watt's hallmark in many ways across the range of his concerns. In his experimental inquiries, which bore in some cases on major theoretical points of contention in chemistry at the time, he was both tentative about drawing broad theoretical conclusions and eclectic in using promising ideas, whatever their theoretical context. Certainly his approach to understanding airs and steam had this pragmatic quality. He was drawn to what "worked" in both a practical and theoretical sense. Although he clung to phlogistic chemistry, he adapted his ideas and incorporated modified versions of the ideas of the likes of Lavoisier and Berthollet. He merrily transplanted and combined concepts that others would see as theoretically incompatible because he found potential explanatory value in them.

As a mechanician, too, Watt was certainly inventive, but there are deep continuities at a thematic level across many of his projects. Thus the issue of linkages of various sorts, progressively more complex, unites a range of his inventions, from his perspective machines during his merchant days; to the piston, parallel motion, sun-and-planet gear linkage, centrifugal governor, and other aspects of the steam engines; to his steam indicator; and finally to his sculpture machines during his retirement. In each of these cases, he was returning to old mechanical themes and issues and developing them to apply to newly conceived purposes.

In politics Watt combined a temperamental conservatism with a fear of the mob and a distrust of elites. Politics for him was not, or should not have been, a doctrinaire business but rather a matter of adopting measures that in his judgment would work. Watt pitted a rationalist political outlook against both the ignorant violence of the mob, which he equated with democracy, and the unthinking incompetence of entrenched elites, especially hereditary aristocracy.

In religious matters, attributing a pragmatic approach to Watt makes sense of the few clues we have about his religious attitudes and practices. Certainly after he left Scotland, Watt was not observant in religion, despite the strong and deep Presbyterian traditions of his family and upbringing. I have suggested that his cultural Presbyterianism, especially the insistence upon careful accounting in all the business of life, remained strong. Watt believed that this discipline worked in the sense that it kept a person on the straight and narrow path and discouraged extravagance that was the route to ruin. In Birmingham, Watt appears to have made an early practical decision that on political grounds it would be a mistake to attend a dissenting meeting house. His already established political conservatism would not have sat well with the strong Whig and radical commitments of his fellow dissenters in Birmingham. He became identified, like Boulton, with the parish of Handsworth and was buried in its sacred ground, and he was memorialized within the walls of St. Mary's Church. When it was convenient, however, as in the case of escaping service as high sheriff of Staffordshire or Radnorshire, Watt had no compunction in declaring that he was not of the Church of England.

Watt's pragmatic character might also be appealed to in understanding his family life. He was content in most respects to be led in domestic affairs by his second wife Annie, often siding with her against the children of his first marriage, James and Peggy. But in extremis he could lose his temper with what he saw as Annie's tendency to stick with her prejudices. Judging from Annie's reaction, Watt's outbursts could be frighteningly severe, but they may well have been calculated. In the case of his son James, in particular, Watt was inclined to judge him by his actions rather than by his presumed nature. This willingness to take people as he found them probably saved his relationship with Watt Jr., whom he persuaded to do the same, and also carried him through often stormy relationships with other members of Team Watt. But he did not so readily apply it to the women in his circle; with them Watt could be domineering and uncompromising. He treated his daughter Peggy very badly during her short life. Nor would he brook opposition from the sisters of his first wife, who spent long years looking after his children, especially Peggy. He shared the common view of his time that women were, and should be, dependent, and that their role was to provide for his ease and comfort in return. Apart from occasional bursts of ill temper, Watt seems to have been very tightly contained emotionally through much of his life. In his later years his mood lightened and his behavior changed somewhat. He became close to his sons and was hit especially hard by Gregory's early death, which almost destroyed his habitual advocacy of resignation and trust in Providence in the face of loss. That advocacy is remarkable given that Watt lost at least one member of his immediate family in every decade from the 1750s: his mother, his brother, his first wife, his father, two daughters, a son, and a grandson.

Watt's much remarked upon timidity in face-to-face business is borne out, for the most part, by a close study of him at work. But in his dealings with subordinates and employees, he was often severe and even ruthless. This may have been a result of overcompensation on his part for a lack of confidence in handling such relationships. This timidity perhaps sits strangely with Watt's often-exhibited compulsive, rationalist control in many aspects of his life, though it might be observed that people, as distinct from problem-areas, often resist the reasonable and defy control. He studied, and sought to master, everything. He became for a time perhaps the leading practical authority on the law of patents in Britain, partly to serve his own interests but also to serve those of inventors and improvers generally. The pneumatic medicine project was only the tip of the iceberg of Watt's medical studies. Incidental remarks dotted through his correspondence show that he confidently dispensed for himself, his family, and his friends from his store of medical knowledge and tested therapies and remedies. Annie, his children, his father-in-law, his cousins and their families, his grandchildren, and his friends all received advice and sometimes treatment from "Dr. Watt." When the time came to invest his fortune, Watt made a similarly thorough study of alternative strategies and, having decided on investment in land, became a font of information and firm opinion, always justified by extensive factual evidence, on which land was the best investment. The planting of his estates—the world of horticulture and land management generally—was another area in which Watt became a self-made authority.

In retirement, even the recreational and therapeutic possibilities of the towns and spas of southwest England became a subject of ardent inquiry. Here too, Watt spilled out advice to family and friends about the best tourist routes, accommodation, facilities, and attractions. A simple sitting for an artist or sculptor often devolved into inquiries about, and advice upon, the practitioner's materials and methods. We have to conclude that Watt could not help himself. The pursuit of investigations, large and small, weighty and trivial, was just what he did. Watt was not only well read, but also compulsively self-instructed on all matters that his life's progress brought before him. Contemporaries remarked on the encyclopedic quality of his knowledge, a result of this compulsion to inquiry combined with his unusually retentive memory.

We have seen a great deal of the symbolism that settled on and surrounded Watt in his later years and in his afterlife. Almost from the beginning the Boulton & Watt partnership made commercial use of the idea of Watt as philosopher taking an informed experimental approach to the improvement of steam engines. Propagating this image was one way the partnership sought to display its achievement and distinguish its product. This approach imbued the push for a patent extension and the strategy used in the patent trials of the 1790s. It carried on in the nineteenth century and was extended by the filial project, especially in attempts to gain Watt recognition as discoverer of the compound nature of water.

Watt became a symbol of varied sorts, pressed into the service of numerous causes such as the promotion of engineering science and the defense and improvement of the patent system. Stories about his life, especially his early life, entered into popular culture through print and images; images of the young Watt playing with the kettle are endlessly reproduced in many variations in public places, museums, and in print. Major anniversaries provided occasions for bursts of memorialization that reflected the preoccupations of the time.

Watt is no less a symbol for modern-day historians than he was for engineers and others in the nineteenth century. Because of his accumulated symbolic importance, there is a great eagerness to place him in the histories of science-technology relations and of technological and industrial change. Watt was the single most important and ubiquitous figure in Musson and Robinson's accumulation and exposition of evidence for their claim about the close relations of science and "technology" in the British Industrial Revolution.[2]

One of the most penetrating interpretations of Watt's life and family offered in recent years has been that given by Margaret Jacob, most notably in an essay titled "The Watts, Entrepreneurs."[3] Within a broad cultural interpretation of industrialization as a product of Enlightenment, Jacob gives a perceptive social-psychological and cultural account of the Watt family, with James Watt at its center. The key notion that Jacob deploys is that of "entrepreneur." The application of this term to Watt is a bold move since, in commonsense terms, the dominant historical view of Watt has emphasized his relative lack of entrepreneurial qualities. Surely it was Watt's deficiency in that department that made Boulton a necessity and rendered their partnership such a productive combination? It is certainly true that much scholarship in recent years has argued persuasively that the usual view of the Boulton & Watt partnership is a caricature of their real capabilities and relationship: Boulton was more of a natural philosopher and inventor and Watt was more of a businessman than either is usually granted credit for.[4] I am doubtful, however, whether "entrepreneur" is the appropriate descriptive term for him.

The Watt who has emerged in this book does conform in many ways to the stereotype created of him. He was very often timid and diffident, and he was, compared with Boulton, risk averse. He was uncomfortable, or at least professed himself to be, with the ostentation and self-promotion that was involved in Boulton's marketing of their steam engines. Certainly some of this self-presentation as diffident was a guise: Watt could take risks (the move to Birmingham, though prepared for, was difficult), he could drive a hard bargain, he could be ruthless in competition. But to consider him a "natural" or "cultural" entrepreneur is in my view misleading. Not only his behavior in particular situations but also the whole arc of his career points in a rather different direction. While Boulton pursued new large-scale, innovative industrial ventures through his later years, even when crippled by chronic illness and right up to his death, Watt was happy, indeed anxious, to retire from the fray to his sculpture machines, recreational travel, and consumption of light fiction, with an occasional public-spirited foray into civil engineering projects.

The term "entrepreneur" is in any case only uneasily applicable to the eighteenth century. Economic historians have found this twentieth-century, Schumpeterian term historically useful, but it seems to me that in this case the actors' category is more cogent. Watt was an "improver," and "improvement" was his game. More particularly, he was a "Scottish improver" and, I think, remained so through his golden days in Birmingham, returning to the role north of the border in a minor way during his retirement. Margaret Jacob also uses the term "improver"; in fact, she uses it interchangeably in many ways with "entrepreneur." But there are, I think, subtle and important differences. An "improver" was a developer and perhaps also implementer of a new technique or process with a focus upon advancing the economy or productivity of the enterprise to which it related. We would never call Joseph Black, for example, an entrepreneur, but we could call him an improver. Watt was an improver in a constitutional sense, to the heart of his being: he could not take up a subject, a device, or an issue without contributing to it—without, in some respect, adding value to it. This was a compulsive behavior on his part, a heritage of generations, and the main thing that Watt brought to the party. He was a "saver of steam" in his engines and had good ideas about how to measure and charge for those savings. At a strategic level he was less productive. He often had to be dragged along by others—the move from designing and producing large pumping engines to smaller rotative engines being a crucial case in point. His resistance to high-pressure steam is another.

Watt certainly stood up for and sought to promote what was increasingly called the "manufacturing interest," even when, during the last twenty years of his life, he became essentially a landed gentleman. Earlier in his career Watt took a keen interest in patent law and sought to defend the rights of fellow industrialists under that law. But this does not make him an entrepreneur. Jacob makes the entirely valid observation that the Watt family was imbued with the ethos of "getting ahead," but again this does not make him, or them, entrepreneurial. Theirs was an ethos of material self-improvement, imbibed in Watt and Annie's case in the west of Scotland. We have certainly seen that financial success was important to Watt and his headaches disappeared when he achieved it, although his rheumatism worsened.

In the end Jacob's treatment of Watt is most concerned with presenting his shift to secularism, his emphasis on science and its value. She sees him as living the Enlightenment as part of what Peter Jones, following Joel Mokyr, calls an "Industrial Enlightenment." This much can readily be granted, although Watt's secularism had its limits. In times of stress and loss, he was still inclined to invoke a providential God whose mysterious ways had to be accepted. Crucially too, as Jones himself shows, for Watt and Boulton there were strict limits on the open sharing of information that Enlightenment sensibilities enjoined.

The notion most clearly redolent of Watt's desired career path was, I think, the term "ingenious indolence" that George Jardine coined for him, or possibly derived from Watt himself, on the eve of Watt's move to Birmingham. The freedom to explore, to invent, to follow his nose, with due regard to economic considerations but without the everyday pressure of business, was (after the amassing of fortune) Watt's most ardent wish. This he saw as his strength and his role in a world of "improvers."

Peter Jones and Ursula Klein have suggested other important characterizations of Watt.[5] These have been directed specifically at the issue of the relationship between science and engineering in Watt's career and, more generally, in the era during which he worked. Jones suggests the descriptor "savant-fabricant," which solves the problem of the relationship of science and industrial art in a career by harnessing them together, by seeing them as capacities embodied in the one person. Jones argues that this term applies readily to both Boulton and Watt and also to other members of their circle, as also, presumably, to numerous individuals in France who represented a culture that routinely merged high engineering and natural philosophy in substance and intent. Ursula Klein has suggested that we should see eighteenth-century people and practices as exhibiting "hybrid" character, and she certainly finds in Watt a prime example of such hybridity. While Jones hyphenates science and the art of making things in a way designed to capture the character of an historically contingent European

Industrial Enlightenment, Klein's hybridization is much inspired by the modern idea of "technoscience," which unites science and technology as a matter of deep and universal philosophical principle.

Both of these concepts—savant-fabricant and hybrid—are certainly helpful in directing us away from linear conceptions of relations between science and the industrial arts and the prior presumption of them as clearly separate realms that the linear model requires. If this biography has shown nothing else, it has supported this view in the case of Watt. With regard to steam and his inventive life more generally, it was not the case that Watt conducted separate philosophical and experimental inquiries and then applied them to his engineering. Their mutual dependence was such that it had an organic or holistic character of the sort that Klein, and perhaps Jones, want in their different ways to suggest. I have shown that Watt's inventions in steam and other fields had this sort of mutually permeating relationship with his evolving chemistry of heat. I have suggested that one consequence of this is that the worlds of laboratory, workshop, and industrial manufactory should all be seen as sites of philosophical insight for him.

Yet when it comes to labeling Watt in a way that reflects his historical realities, the term "improver" has much to recommend it. It was part of the language of Watt's youth and early maturity, of his formative period. While "improvement" was widely deployed throughout the eighteenth-century Anglosphere, it acquired a characteristically insistent and urgent tone in post-Union Scotland that has made it a central theme of that country's history. It also connoted a kind of relationship between experimental natural philosophy and the practical arts (whether agricultural or industrial) that Watt embodied. Robert Anderson has recently highlighted what he describes as a "telling statement" about the "interlacing of the roles of academic chemist and industrial practitioner" made by Watt's mentor, supporter, collaborator, and lifelong friend, Joseph Black:

> I call every man a Philosopher who invents anything new or improves any
> business in which he is employed—even the Farmer who considers the nature
> of different soils or makes improvements on the ploughs he uses, I must call a
> Philosopher, though perhaps you can call him a Rustic one. Nor am I inclined
> to give much credit to those men who shut up their Closets in study and retire-
> ment have obtained the appellation of Learned Philosophers they in general
> puzzle more than they illustrate, they are wrapt in a veil of Systems and of
> Theories and seldom make improvements or discoveries of Use to Mankind.[6]

By this standard, Watt was certainly a philosopher, even a learned philosopher, but one who illustrated more than he puzzled—a man whose pragmatic temperament was inclined to avoid the entanglements of "Systems and Theories" and who certainly qualifies as an improver of "Use to Mankind." Our current preoccupations increasingly cast him as a major architect of the fossil capitalism that many now consider a great threat to our planet. As that threat materializes, the extension of the use of steam engines by increasing the efficiency of their use of fossil fuels will, perhaps, come to be seen as a dubious achievement. Even though the rise of steam was a collective accomplishment and its use as much the responsibility of the consumers of industrial civilization as of its producers, the "great improver" may lose his luster and fade into history. That would be unjust since, in this connection also, we should remember Watt "for what he did, and not for what he never thought of doing."

Notes

Introduction

1. See, for example, Naomi Klein, *This Changes Everything: Capitalism vs. the Climate* (New York: Simon & Schuster, 2014), 171–74; Andreas Malm, "The Origins of Fossil Capital: From Water to Steam in the British Cotton Industry," *Historical Materialism* 21 (2013): 42–45, 58.

2. "Technology" is an anachronistic term in the eighteenth century, though of growing popularity in its current sense from the middle of the nineteenth. See Thomas Misa, "The Compelling Tangle of Modernity and Technology," in *Modernity and Technology*, ed. Thomas J. Misa et al. (Cambridge, MA: The MIT Press, 2003), 7; Ben Marsden and Crosbie Smith, *Engineering Empires: A Cultural History of Technology in Nineteenth-Century Britain* (Houndmills: Palgrave Macmillan, 2005), 3–4. I will strive to use "industrial arts" or "engineering" in preference, except where I am deploying someone else's usage.

3. On experimental "natural philosophy" see Steven Shapin, *The Scientific Revolution* (Chicago: The University of Chicago Press, 1996), 65–117; on its increasingly public character in the eighteenth century, see Margaret C. Jacob and Larry Stewart, *Practical Matter: Newton's Science in the Service of Industry and Empire 1687–1851* (Cambridge, MA: Harvard University Press, 2004). On "scientist" see Sidney Ross, "*Scientist:* The Story of a Word," *Annals of Science* 18 (1962): 65–85.

4. Richard L. Hills, *James Watt, Volume 1: His Time in Scotland, 1736–1774* (Ashbourne: Landmark Publishing Ltd., 2002); Richard L. Hills, *James Watt, Volume 2: The Years of Toil, 1775–1785* (Ashbourne: Landmark Publishing Ltd., 2005); Richard L. Hills, *James Watt, Volume 3: Triumph through Adversity, 1785–1819* (Ashbourne: Landmark Publishing Ltd., 2006).

5. Ben Marsden, *Watt's Perfect Engine: Steam and the Age of Invention* (Cambridge: Icon Books, 2002).

6. Ben Russell, *James Watt: Making the World Anew* (London: Reaktion Books, 2014).

7. H. W. Dickinson, *James Watt: Craftsman and Engineer* (Cambridge: Cambridge University Press, 1936).

8. Maxine Berg, *The Age of Manufactures 1700–1820: Industry, Innovation and Work in Britain*, 2nd ed. (London: Routledge, 2005); Celina Fox, *The Arts of Industry in the Age of Enlightenment* (New Haven: Yale University Press, 2009).

9. David Philip Miller, *Discovering Water: James Watt, Henry Cavendish and the Nineteenth-Century "Water Controversy"* (Aldershot, UK: Ashgate, 2004).

10. David Philip Miller, *James Watt, Chemist: Understanding the Origins of the Steam Age* (London: Pickering & Chatto, 2009).

11. I have benefitted in this enormously from the work of Christine MacLeod, especially her *Heroes of Invention: Technology, Liberalism and British Identity 1750–1914* (Cambridge: Cambridge University Press, 2007).

12. Margaret C. Jacob, *Scientific Culture and the Making of the Industrial West* (New York/Oxford: Oxford University Press, 1997), especially 116–30.

13. Peter M. Jones, *Industrial Enlightenment: Science, Technology and Culture in Birmingham and the West Midlands 1760–1820* (Manchester: Manchester University Press, 2008); Ursula Klein, "Hybrid Experts," in *The Structures of Practical Knowledge*, ed. M. Valleriani (Cham: Springer, 2017), 287–306.

Chapter 1: The Making of a Scottish Improver

1. T. M. Devine, "Scotland," in *The Cambridge Economic History of Modern Britain, Volume 1: Industrialization 1700–1860*, ed. Roderick Floud and Paul Johnson (Cambridge: Cambridge University Press, 2004), 393.

2. Christopher A. Whatley, *Scottish Society, 1707–1830: Beyond Jacobitism, Towards Industrialisation* (Manchester: Manchester University Press, 2000), 66.

3. Alan McKinlay and Alistair Mutch, "'Accountable Creatures': Scottish Presbyterianism, Accountability and Managerial Capitalism," *Business History* 57 (2015): 241–56.

4. The following details on Watt's family rely heavily on George Williamson, *Letters Respecting the Watt Family* (Greenock: W. Johnston & Son, 1840) and George Williamson, *Memorials of the Lineage, Early Life, Education, and Development of the Genius of James Watt* (Greenock: Printed for the Watt Club, 1856).

5. Williamson, *Letters*, 19–20.

6. Daniel Weir, *History of the Town of Greenock* (Greenock: Privately printed, 1829), 16.

7. H. W. Dickinson and Rhys Jenkins, *James Watt and the Steam Engine* (Oxford: The Clarendon Press, 1927), 78, quoting Watt to John Roebuck, 16 March 1769. Even the writing of letters on Sunday was a problem in the strict Sabbatarianism of many Presbyterians.

8. For example, in 1803–1804 consideration was given to using his allegiance to the Church of Scotland to escape recruitment as sheriff of Staffordshire. See chapter 8.

9. Hills, *James Watt, Volume 1*, 40.

10. Thomas Ahnert, *The Moral Cultures of the Scottish Enlightenment, 1690–1805* (New Haven: Yale University Press, 2014); Richard B. Sher, *Church and University in the Scottish Enlightenment. The Moderate Literati of Edinburgh*, 2nd ed. (Edinburgh: Edinburgh University Press, 2015).

11. Alistair Mutch, "Religion and Accounting Texts in Eighteenth-Century Scotland," *Accounting, Auditing & Accountability Journal* 29 (2016): 926–46.

12. McKinley and Mutch, "'Accountable Creatures,'" 243–46.

13. See Crosbie Smith, *The Science of Energy. A Cultural History of Energy Physics in Victorian Britain* (London: The Athlone Press, 1998).

14. See, for example, James Watt Sr. to Watt, 3 April 1766 and 20 February 1769, Archives of Soho, Birmingham City Libraries, James Watt Papers, MS3219/4/1.

15. Williamson, *Memorials*, 154–55.

16. James Watt Sr. to Watt, 28 March 1763, M3219/4/1. For the unfolding story, see James Watt Sr. to Watt, 5, 7, 9 March 1763, MS3219/4/1. Jockey was apparently chief mate to Captain Alexander Buchanan on a ship that sailed from Havana to Jamaica, from there intended for Bristol. The ship was wrecked on "Abbott Keys" off the Bahamas, and all but three of the crew drowned.

17. Samuel Smiles, *The Lives of Boulton and Watt: Principally from the Original Soho MSS* (London: John Murray, 1865), chapter 5, gives a readable account of Watt's schooling. On the mythology of the kettle, see Eric Robinson, "James Watt and the Tea Kettle: A Myth Justified," *History Today* 6 (1956): 261–65, and David Philip Miller, "True Myths: James Watt's Kettle, His Condenser and His Chemistry," *History of Science* 42 (2004): 333–60.

18. Williamson, *Letters*, 17.

19. Williamson, *Memorials*, 138.

20. "Sundries Belonging to Town of Greenock in Mr Marrs Schoole House," MS3219/3/118. "Mr Neal" may well be the watchmaker (Neale) for whom Watt worked temporarily during his time in London. The Marrs, father and son, had dealings with Neale over his patent globes (see Hills, *James Watt, Volume 1*, 56).

21. J. P. Muirhead, *Life of James Watt, with Selections from his Correspondence* (London: John Murray, 1858), 15–16; James Watt Jr. to Francois Arago, 22 January 1835, MS3219/6/106.

22. John Sinclair, "Killearn, County of Stirling," in *The Statistical Account of Scotland*, 21 vols. (Edinburgh: William Creech, 1791–1799), XVI: 102–103.

23. "Memoranda of the early years of Mr Watt, by his cousin, Mrs Marion Campbell (born Muirhead, daughter of his mother's brother), who was his companion in early youth, and friend through life; dictated to and written down by her daughter Miss Jane Campbell in 1798." An original version of this document was supplied to Watt Jr. (who had solicited it) in 1834 by Jane Campbell; from there its contents were copied and populated the biographical works of Arago and J. P. Muirhead on Watt. I have used a copy of the version that Watt Jr. supplied to Arago on 22 January 1835. This version is at MS3219/6/219. The version supplied to Watt Jr. was sold by Sotheby's, with a number of other related papers, to an unknown buyer in 2003. Part of the document is depicted in the sale catalogue: Sotheby's, *The James Watt Sale: Art and Science* (London: Sotheby's, 2003), 22–23.

24. This anecdote was the basis for Robert W. Buss's famous genre painting *Watt's First Experiment on Steam* (1845). See Miller, *James Watt, Chemist*, 21–2.

25. The English translation by J. T. Desaguliers of the original Latin edition published in Leiden in 1720, was Willem Jacob's Gravesande, *Mathematical Elements of Natural Philosophy, Confirmed by Experiments, or, an Introduction to Newtonian Philosophy*, 2 vols. (London: W. Innys, 1747).

26. In the version of the recollections I have used (see note 23), Watt Jr. adds a note here for Arago's benefit that his father remained attached to the medical art, and deeply studied his own constitution and treated his own complaints and those of his family and friends, as "homme de métier," that is as a person skilled in the art.

27. In at least one respect, the account had been "doctored" by Watt Jr. before being transmitted to Arago. See Miller, "True Myths," 342–43.

28. Watt to James Weston, 6 August 1815, MS3219/4/120.

29. Williamson, *Memorials*, 130–31. This less flattering account of Watt was roundly rejected by Watt Jr. when it was first promulgated in Williamson, *Letters*. See Watt Jr. to George Williamson, n.d. [1840], MS3219/6/219, in which Watt Jr. describes the stories told by his father's surviving school-fellows as "absurd legends."

30. Williamson, *Memorials*, 153.

31. Hills, *James Watt, Volume 1*, 47–48.

32. See David B. Wilson, *Seeking Nature's Logic: Natural Philosophy in the Scottish Enlighten-ment* (University Park, PA: Pennsylvania State University Press, 2009), 69–102.

33. Richard B. Sher, "Muirhead, George (*bap.* 1715, *d.* 1773)," *Oxford Dictionary of National Biography*, Oxford University Press, 2004, http://www.oxforddnb.com/view/article/19504.

34. Richard B. Sher, "Commerce, Religion and the Enlightenment in Eighteenth-Century Glasgow," in *Glasgow: Beginnings to 1830*, ed. Thomas M. Devine and Gordon Jackson (Manchester: Manchester University Press, 1995), 312–59.

35. See Roger L. Emerson, "The Philosophical Society of Edinburgh, 1737–1747," *The British Journal for the History of Science* 12 (1979): 154–91; Roger L. Emerson, "The Philosophical Soci-ety of Edinburgh, 1748–1768," *The British Journal for the History of Science* 14 (1981): 133–76; Roger L. Emerson, *Academic Patronage in the Scottish Enlightenment: Glasgow, Edinburgh and St Andrews Universities* (Edinburgh: Edinburgh University Press, 2008); Roger L. Emerson, *An En-lightened Duke: The Life of Archibald Campbell (1682–1761), Earl of Ilay, 3rd Duke of Argyll* (Edinburgh: Humming Earth, 2013).

36. Watt to John Craig, 30 April 1805, MS3219/4/119. This letter is quoted in J. P. Muirhead, *Or-igin and Progress of the Mechanical Inventions of James Watt*, 3 vols. (London: John Murray, 1854), II: 299–303. It is also quoted in John Strang, *Glasgow and its Clubs* (London: Richard Griffin, 1857), 156–57, who identifies the place of meeting as Mrs Scheid's tavern.

37. John Millar, *The Origin of the Distinction of Ranks*, ed. John Craig (Edinburgh: W. Black-wood, 1806), iii. On Millar see also William C. Lehmann, *John Millar of Glasgow 1735–1801* (Cam-bridge: Cambridge University Press, 1960).

38. William R. Brock, *Scotus Americanus: A Survey of the Sources for Links between Scotland and America in the Eighteenth Century* (Edinburgh: Edinburgh University Press, 1982), 19, cited in Richard B. Sher and Andrew Hook, "Introduction," in *The Glasgow Enlightenment*, ed. Richard B. Sher and Andrew Hook (East Linton: Tuckwell Press, 1995), 5. Devine found that between 1728 and 1800 at least sixty-eight tobacco and West India merchants had been students at Glasgow University (T. M. Devine, *The Tobacco Lords: A Study of the Tobacco Merchants of Glasgow and Their Trading Activities c. 1740–90* [Edinburgh: John Donald, 1975], 8).

39. H. W. Dickinson, *James Watt: Craftsman and Engineer* (Cambridge: Cambridge University Press, 1936), 21.

40. A gold mourning brooch long preserved in the Watt family collections in memory of Major John Marr records his death on 19 October 1786, age 64. See Sotheby's, *The James Watt Sale: Art & Science* (London: Sotheby's, 2003), 150.

41. Hills, *James Watt, Volume 1*, 50–58.

42. Richard Sorrenson, *Perfect Mechanics: Instrument Makers at the Royal Society of London in the Eighteenth Century* (Boston, MA: Docent Press, 2013), 13–22.

43. Anita McConnell, *Jesse Ramsden (1735–1800): London's Leading Instrument Maker* (Aldershot: Ashgate, 2007), 25–26.

44. See D. J. Bryden, *James Short and His Telescopes* (Edinburgh: Royal Scottish Museum, 1968).

45. Alison D. Morrison-Low, *Making Scientific Instruments in the Industrial Revolution* (Aldershot: Ashgate, 2007), 140–41; Hills, *James Watt, Volume 1*, 55–56; Watt to Watt Sr., 15 July 1755 and 21 July 1755, MS3219/3/3.

46. Hills, *James Watt, Volume 1*, 51–52. Compare Smiles, *Lives of Boulton and Watt*, 103–05.

47. Nicholas Rogers, *The Press Gang: Naval Impressment and Its Opponents in Georgian Britain* (London: Continuum, 2007).

48. Instrument makers of the mathematical, optical, and philosophical variety had no guild of their own, and there was in that sense no "closed shop." In order to be freemen of the city, men in that trade had to join one of the city companies. Many were members of the grocers' company, for example. See J. R. Milburn, *Benjamin Martin: Author, Instrument-Maker and "Country Showman"* (Leyden: Noordhoff International Publishing, 1976), 85–86. Martin was another, like Watt, who skirted the edges of the trade in making his way.

49. Watt to Watt Sr., 8 June 1756, MS3219/3/93.

50. Hills, *James Watt, Volume 1*, 58.

51. D. J. Bryden, "The Jamaican Observatories of Colin Campbell F.R.S. and Alexander Macfarlane F.R.S.," *Notes and Records of the Royal Society of London* 24 (1970): 261–72.

52. On Robison see Paul Wood, "Robison, John (1739–1805)," *Oxford Dictionary of National Biography*, https://doi.org/10.1093/ref:odnb/23894, and Paul Wood, "Introduction," in John Robison, *A System of Mechanical Philosophy* (Reprint, London: Thoemmes Continuum, 2005), I: v–xxiii.

53. See Muirhead, *Origin and Progress*, I: xliii. Caution is required, given that these recollections date from forty years later and their context was Robison's testimony in the 1796 patent trials, intended to reinforce the picture of Watt as philosopher.

54. David Murray, *Robert & Andrew Foulis and the Glasgow Press, with Some Account of the Glasgow Academy of Fine Arts* (Glasgow: James Maclehose and Sons, 1913), 21. See also John Cleland, *Enumeration of the Inhabitants of the City of Glasgow and County of Lanark*, 2nd ed. (Glasgow: John Smith & Son, 1832), 145.

55. Smiles, *Lives of Boulton and Watt*, 107. Smiles reports (in 1865) visiting the room "some years since," and describes it and its location in detail. It then contained Professor William Thomson's mirror galvanometer for testing the Atlantic telegraph cable.

56. Smiles, *Lives of Boulton and Watt*, 107. Smiles was here relying on the testimony of "Professor Fleming," presumably William Fleming (1794–1866), professor of moral philosophy at the college, who either had evidence of, or himself recalled, this old house, which by the time of Smiles's visit had been demolished. Fleming was an important informant regarding Watt. It was he who communicated to J. P. Muirhead the papers held by the college concerning Watt's famous repair of the model Newcomen engine (see Muirhead, *Life of James Watt*, 93n).

57. Muirhead, *Origin and Progress*, 1: xxxv. The source of this claim was probably Joseph Black. See "Memorial by Dr Black respecting Mr Watt's Invention of his Improvements on the Steam Engine etc. Written in 1796," transcribed in *Partners in Science: Letters of James Watt and Joseph Black*,

ed. Eric Robinson and Douglas McKie (Cambridge, MA: Harvard University Press, 1970), 253–56. Black described Watt as being "molested by some of the corporations who considered him an intruder on their privileges" (253).

58. Muirhead, *Origin and Progress*, I: xliv–xlv.

59. See John R. R. Christie, "The Origins and Development of the Scottish Scientific Community, 1680–1760," *History of Science* 12 (1974): 122–41; Jack Morrell, "The University of Edinburgh in the Late Eighteenth Century: Its Scientific Eminence and Academic Structure," *Isis* 62 (1970): 158–71; Paul Wood, "Science, the Universities and the Public Sphere in Eighteenth-Century Scotland," *History of Universities* 13 (1994): 99–135.

60. Hills, *James Watt, Volume 1*, 72–73. On Anderson (1726–1796), see Paul Wood, "Jolly Jack Phosphorus in the Venice of the North, or, Who Was John Anderson," in *The Glasgow Enlightenment*, ed. Hook and Sher, 111–32; James Muir, *John Anderson, Pioneer of Technical Education and the College He Founded* (Glasgow: J. Smith, 1950).

61. Others have argued that Anderson has been given perhaps too much emphasis in the development of vocational education in Scotland during this period, contending that such education for adults was provided for more generally, including by the openness of classes at the college, even to non-matriculating students. See John A. Cable, "Early Scottish Science: The Vocational Provision," *Annals of Science* 30 (1973): 179–99.

62. Arthur Donovan, *Philosophical Chemistry in the Scottish Enlightenment, The Doctrines and Discoveries of William Cullen and Joseph Black* (Edinburgh: Edinburgh University Press, 1975). See also David V. Fenby, "The Lectureship in Chemistry and the Chemical Laboratory, University of Glasgow, 1747–1818," in *The Development of the Laboratory: Essays in the Place of Experiments in Industrial Civilization*, ed. Frank A. J. L. James (Houndmills: Macmillan Press, 1989), 22–36.

63. On Black see the papers in *Joseph Black 1728–1799. A Commemorative Symposium*, edited by A. D. C. Simpson (Edinburgh: The Royal Scottish Museum, 1982), and R .G. W. Anderson and Jean Jones, "Joseph Black: Life and Work," in *The Correspondence of Joseph Black*, ed. R. G. W. Anderson and Jean Jones, 2 vols. (Farnham: Ashgate, 2012), I: 25–57.

64. Joseph Black, *De Humore Acido a Cibis Orto, et Magnesis Alba* (Edinburgh: G. Hamilton and J. Balfour, 1754). That Black saw himself as a disciple of Cullen is suggested by the unusual dedication of the Edinburgh thesis to the Glasgow Professor (see Anderson and Jones, "Joseph Black," 29).

65. Joseph Black, "Experiments on Magnesia Alba, Quicklime and Other Alcaline Substances," *Essays and Observations, Physical and Literary* 2 (1756): 157–225.

66. Miller, *James Watt, Chemist*, 87–99; Anderson and Jones, "Joseph Black," 30–32.

67. James Watt Sr. to Watt, 9 December 1768 and 20 February 1769, MS3219/4/1.

68. Archibald and Nan L. Clow, *The Chemical Revolution: A Contribution to Social Technology* (London: The Batchworth Press, 1952), chapter 4.

69. Draft of letter, Watt to William Small, 20 October 1769; Watt to Small, 27 October 1769, MS3219/4/62.

70. See Eric Robinson, "James Watt and Early Experiments in Alkali Manufacture," in *Science and Technology in the Industrial Revolution*, ed. A. E. Musson and Eric Robinson (Manchester: Manchester University Press, 1969), 354.

71. Hills, *James Watt, Volume 1*, 158–59. See also, on Keir and subsequent developments, Kristen

M. Schrantz, "The Tipton Chemical Works of Mr James Keir: Networks of Conversants, Chemicals, Canals and Coal Mines," *International Journal for the History of Engineering and Technology* 84 (2014): 248–73.

72. The following section relies heavily on Richard L. Hills, "James Watt and the Delftfield Pottery, Glasgow," *Proceedings of the Society of Antiquaries of Scotland* 131 (2001): 375–420.

73. "Journal," 13 and 14 June 1772, MS3219/4/135.

74. David Bryden, "James Watt, Merchant: The Glasgow Years, 1754–1774," in *Perceptions of Great Engineers: Fact and Fantasy*, ed. Denis Smith (London: The Science Museum, 1994), 9–22.

75. See Michael Wright, "James Watt: Musical Instrument Maker," *The Galpin Society Journal* 55 (2002): 104–29.

76. Hills, *James Watt, Volume 1*, 109–11.

77. Bryden, "James Watt, Merchant," 12.

78. Wright, "James Watt: Musical Instrument Maker," 122–25.

79. The identity of this John Craig, Watt's business partner, remains uncertain. We know that he died in 1765. He was described by John Robison as "a very intimate Acquaintance of Mr. Watt's" (Eric Robinson and A. E. Musson, *James Watt and the Steam Revolution* [London: Adams and Dart, 1969], 28). It is tempting to identify him as part of the same family as the John Craig with whom Watt corresponded in the early nineteenth century about his early recollections of Professor John Millar. That John Craig was referred to by James Strang (*Glasgow and Its Clubs*, 156) as "John Craig of Waterport," which suggests that he was of a family of timber merchants whose timber yard was at the Waterport, the southern entrance to the city of Glasgow near the old bridge on Clyde Street. Watt recalled John Craig's father, also called John Craig, as a member of the informal club that met in the early 1750s. It seems possible that Watt's partner was either the father or the grandfather of Watt's early-nineteenth-century correspondent.

80. Miss Millar to James Watt, "Gourock 18th" (1761/2), MS3219/4/3.

81. "Memoranda of the early Years of Mr. Watt, by his cousin Mrs Marion Campbell," MS3219/6/219.

82. See Margaret Watt to James Watt, 15 November 1766, MS3219/4/3. This concerned a house belonging to Mr. Martin the upholsterer. Its garret Peggy considered unsuitable for Watt because it lacked a fire (Watt to Margaret Watt, 4 May 1767, MS3219/4/3). However, Watt's assistant, John Gardner, recalled steam experiments being done when Watt lived upstairs from "one Martin above the Cross" (reported in Gregory Watt to James Watt, September 1796, MS3219/4/8). Perhaps Watt's experimentation had strayed into the kitchen!

83. Apart from the shop, which was first at the Saltmarket nearly opposite St. Andrew's Street and then on "Buchanan's Land" in the Trongate, from 1765 Watt also kept a workshop near King Street, in a little court at the north end of the beef market (see Cleland, *Enumeration of the Inhabitants*, 272–73; Muirhead, *Life of Watt*, 50).

84. Watt to Margaret Watt, 4 May 1767, MS3219/4/3.

Chapter 2: Improving Ventures

1. Watt to William Small, 20 September 1769, in Muirhead, *Origin and Progress*, I: 71.

2. See R. A. Buchanan, *The Engineers: A History of the Engineering Profession in Britain, 1750–*

1914 (London: Jessica Langley, 1989), 38–45; A. W. Skempton, *John Smeaton F.R.S.* (London: Thomas Telford, 1981). See also Peter Jones, "Becoming an Engineer in Industrializing Great Britain circa 1760–1820," *Engineering Studies* 3 (2011): 215–32.

3. Hills, *James Watt, Volume 1*, 181–88.

4. Mackell was another of Watt's associates who had benefited from the patronage of the Third Duke of Argyll. Interestingly, Mackell made scale models of machine mills for the duke, then Earl of Ilay, that he was also involved in building as working mills. Mackell was also employed by Lord Milton, and the Board of Trustees for Fisheries and Manufactures, to develop and test new machines for the linen trade. See Emerson, *An Enlightened Duke*, 134, 137.

5. Len Paterson, *From Sea to Sea: A History of the Scottish Lowland and Highland Canals* (Glasgow: Neil Wilson Publishing, 2006), VII.

6. Brian Watters, *Where Iron Runs Like Water! A New History of Carron Iron Works 1759–1982* (Edinburgh: John Donald, 1998), 54–55.

7. Hills, *James Watt, Volume 1*, 193–95.

8. Watt to Margaret Watt, 5 April 1767, MS3219/4/3.

9. Hills, *James Watt, Volume 1*, 197.

10. See Watt to Margaret Watt, 18 April 1767, MS3219/4/3.

11. The following draws on George Thomson, "James Watt and the Monkland Canal," *The Scottish Historical Review* 29 (1950): 121–33.

12. Watt to William Small, 24 November 1772, in Muirhead, *Origin and Progress*, II: 34.

13. Watt to William Small, 9 September 1770, in Muirhead, *Origin and* Progress, II: 2, quoted in Hills, *James Watt, Volume 1*, 210.

14. Joseph Black to James Watt, 31 December 1772, in Anderson and Jones, *The Correspondence of Joseph Black*, I: 266–67.

15. Watt to "Sir," 31 March 1770, MS3219/4/200.

16. On Newcomen see L. T. C. Rolt and J. S. Allen, *The Steam Engine of Thomas Newcomen* (New York: Science History Publications, 1977) and Brian Corfield, "Thomas Newcomen the Man," *The International Journal for the History of Engineering and Technology* 83 (2013): 209–21.

17. H. W. Dickinson, *A Short History of the Steam Engine* (Cambridge: Cambridge University Press, 1938), 29.

18. J. T. Desaguliers, *A Course of Experimental Philosophy*, 2 vols. (London: W. Innys, 1744), II: 532.

19. Desaguliers, *A Course of Experimental Philosophy*, II: 536.

20. See Paul A. David, "Path Dependence, Its Critics and the Quest for 'Historical Economics,'" in *Evolution and Path Dependence in Economic Ideas: Past and Present*, ed. Pierre Garrouste and Stavros Ioannides (Cheltenham: Edward Elgar, 2001), 15–40.

21. Rolt and Allen, *The Steam Engine of Thomas Newcomen*, 146–54; Hills, *James Watt, Volume 1*, 297.

22. J[ohn] R[obison], "Attempts Towards the Improvement of a Machine as Useful as the Contrivance of It Is Ingenious," *The Universal Magazine of Knowledge and Pleasure* 21 (1757): 229–31.

23. Muirhead, *Life of James Watt*, 74.

24. Watt's Note Book [MS3219/4/170, 7e] reproduced in Robinson and McKie, *Partners in Sci-*

ence, 434. The fact that Watt says "6 or 8 years ago" may seem to date the writing of this recollection to the mid-1760s. But we know that there was a great deal of subsequent interference with the Watt archive, and so it is possible that the entry was actually made much later to deliberately masquerade as an entry of the mid-1760s.

25. "J. Watt's Recollections of his Friend, Dr J. Robison," April 1805, MS3219/4/119. This document is reproduced in Muirhead, *Origin and Progress*, II: 293–99.

26. On the filial project, see Miller, *Discovering Water*, chapter 5, and this book's chapter 8.

27. "J. Watt's Recollections," in Muirhead, *Origin and Progress*, II: 295.

28. Hills, *James Watt, Volume 1*, 318.

29. Robert Hart, "Reminiscences of James Watt," *Transactions of the Glasgow Archaeological Society* 1 (1859): 3–4. I date the meeting that the brothers had with Watt to 1813 or 1814, not to 1817 as is sometimes stated by others. The "Reminiscences" were read to the Archaeological Society in November 1857 and mention that the meetings with Watt took place "forty-three years since." At one point it is mentioned explicitly that after talking to Watt they went searching for one of the workshops where he experimented (in King Street, Glasgow) and that "this was in the year 1813 or '14." (5)

30. Hart, "Reminiscences of James Watt," 4.

31. See David Philip Miller, "Watt in Court: Specifying Steam Engines and Classifying Engineers in the Patent Trials of the 1790s," *History of Technology* 27 (2006): 43–76.

32. I refer here to Watt's work, at David Brewster's behest, on the recollections eventually published in 1822 in John Robison, *A System of Mechanical Philosophy* (Edinburgh: John Murray, 1822). See this book's chapter 7.

33. See Miller, *James Watt, Chemist*, 37–41.

34. Jim Andrew, "Boulton, Watt and Wilkinson: The Birth of the Improved Steam Engine," in *Matthew Boulton, Enterprising Industrialist of the Enlightenment*, ed. Kenneth Quickenden et al. (Farnham: Ashgate, 2013), 85–100.

35. Hills, *James Watt, Volume 1*, 350ff.

36. "Memorandum made at the request of J. Watt Jun. in Jan. 1808," reprinted in Robinson and McKie, *Partners in Science*, 419.

37. See Brian Watters, "Charles Gascoigne of Carron Company," Falkirk Local Historical Society, 2005, http://www.falkirklocalhistorysociety.co.uk/home/index.php?id=108.

38. Hills, *James Watt, Volume 1*, 355–56.

39. Hills, *James Watt, Volume 1*, 359–60.

40. On Carron Iron Works see Watters, *Where Iron Runs like Water!* and R. H. Campbell, *Carron Company* (Edinburgh: Oliver and Boyd, 1961).

41. John Roebuck to Matthew Boulton, 12 December 1768, cited in Hills, *James Watt, Volume 1*, 160.

42. See especially George Selgin and John L. Turner, "Strong Steam, Weak Patents, or, the Myth of Watt's Innovation-Blocking Monopoly Exploded," *Journal of Law and Economics* 54 (2011): 841–61, and Michele Boldrin and David K. Levine, "2003 Lawrence R. Klein Lecture: The Case Against Intellectual Monopoly," *International Economic Review* 45 (2004): 327–50, esp. 348–49.

43. On the history of the patent system, see Christine MacLeod, *Inventing the Industrial Revolu-*

tion: The English Patent System, 1660–1800 (Cambridge: Cambridge University Press, 1988) and a recent revisionist account, stressing the relative ease of patenting for inventors in Britain as a crucial aid to industrialization, Sean Bottomley, *The British Patent System during the Industrial Revolution 1700–1852: From Privilege to Property* (Cambridge: Cambridge University Press, 2014).

44. On the London Society of Arts, see most recently Matthew Paskins, "Sentimental Industry: The Society of Arts and the Encouragement of Public Useful Knowledge" (PhD diss., University College London, 2014), and on the Select Society see Roger L. Emerson, "The Social Composition of Enlightened Scotland: The Select Society of Edinburgh, 1754–1764," *Studies on Voltaire and the Eighteenth Century* 114 (1973): 291–329.

45. Watt to John Roebuck, 24 May 1768, as quoted in Dickinson and Jenkins, *James Watt and the Steam Engine*, 100. Letter in Muirhead, *Origin and Progress*, I: 25.

46. Hills, *James Watt, Volume 1*, 387. On the process see MacLeod, *Inventing the Industrial Revolution*, 53, and Sean Bottomley, *The British Patent System*, 35–38.

47. This value is obtained by multiplying £170 by the percentage increase in the Retail Price Index from 1769 to 2015. See measuringworth.com.

48. George Jardine to Watt, "Monday," 1768, MS3219/4/59, printed in Muirhead, *Origin and Progress*, I: 25–28. Muirhead dates the letter to May, but Hills (*James Watt, Volume 1*, 388) correctly suggests that it must have been written circa November 1768.

49. George Jardine to Watt, "Monday," 1768, MS3219/4/59.

50. Watt was later to express this same idea to William Small, reflecting back on the beginnings of his serious work on engines from 1765: "I was at that time spurred on by the alluring hope of placing myself above want, without being obliged to have much dealing with mankind, to whom I have always been a dupe" (Watt to Small, 28 April 1769, in Muirhead, *Origin and Progress*, I: 56).

51. MacLeod, *Inventing the Industrial Revolution*, 49.

52. Watt to Roebuck, 9 November 1768, MS3219/4/58, in Muirhead, *Origin and Progress*, I: 34.

53. Watt to Small, 28 January 1769, in Muirhead, *Origin and Progress*, I: 35.

54. Handley's advice appears to have been limited to urging that Watt guard the draft specification carefully to avoid theft of his ideas. See Miller, "Watt in Court," 72, note 16.

55. They are described and analyzed in Hills, *James Watt, Volume 1*, 393–97.

56. Small to Watt, 5 February 1769, in Muirhead, *Origin and Progress*, I: 37.

57. Small to Watt, 5 February 1769, in Muirhead, *Origin and Progress*, I: 40.

58. Watt to Small, 22 February 1769, in Muirhead, *Origin and Progress*, I: 47.

59. Watt to Small, "March" 1769, in Muirhead, *Origin and Progress*, I: 49.

60. See, for example, Kenneth L. Cuthbertson, *The Last Presbyterian? Remembering the Faith of my Forebears* (Eugene, OR: Resource Publications, 2013), especially chapter 11.

61. Small to Watt, 18 April 1769, in Muirhead, *Origin and Progress*, I: 51

62. Watt to Small, 28 April 1769, in Muirhead, *Origin and Progress*, I: 53–54.

63. The importance of tacit knowledge in technology (and scientific research), and more specifically in the replication of experiments and machines, is an important theme of science and technology studies, building initially on insights of Michael Polanyi. See particularly Harry M. Collins, *Changing Order: Replication and Induction in Scientific Practice* (Chicago: The University of Chicago Press,

1992, first published 1985), and Harry Collins, *Tacit and Explicit Knowledge* (Chicago: The University of Chicago Press, 2010).

64. English patents were, rather peculiarly, granted for "England Wales and Berwick-on-Tweed," the latter being the English town closest to the Scottish border. Witnessing of a specification presumably had to be done in England, and Berwick, where a master in chancery would be found, was a favored destination for Scottish patentees.

65. See Hills, *James Watt, volume 1*, 393–413.

Chapter 3: Birmingham, Boulton, and Steam Enterprise

1. Patrick Walcot, *A Sketch of the Life of Dr William Small and His Relationship with Matthew Boulton & James Watt* (Sutton Coldfield: Privately printed, November 2015), 49.

2. See George Demidowicz, "The Origins of the Soho Manufactory and Its Layout," in *Matthew Boulton: Enterprising Industrialist of the Enlightenment*, ed. Kenneth Quickenden et al. (Farnham: Ashgate, 2013), 67–84. See also George Demidowicz, "A Walking Tour of the Three Sohos," in *Matthew Boulton: Selling What All the World Desires*, ed. Shena Mason (New Haven: Yale University Press, 2009), 99–107.

3. "Memorandum concerning Mr Boulton, commencing with my first acquaintance with him," Glasgow, September 17, 1809, printed as *Memoir of Matthew Boulton by James Watt* (Birmingham: College of Arts and Crafts, 1943), 5. This memorandum was produced at Robinson Boulton's request at the time of his father's death, a fact that must be borne in mind.

4. Rita McLean, "Introduction: Matthew Boulton, 1728–1809," in *Matthew Boulton*, ed. Shena Mason, 1–6.

5. Eric Hopkins, *Birmingham: The First Manufacturing Town in the World, 1760–1840* (London: Weidenfeld & Nicolson, 1989), chapter 5.

6. Peter M. Jones, *Industrial Enlightenment: Science, Technology and Culture in Birmingham and the West Midlands 1760–1820* (Manchester: Manchester University Press, 2008).

7. The classic study of the Lunar Society is Robert E. Schofield, *The Lunar Society of Birmingham: A Social History of Provincial Science and Industry in Eighteenth-Century England* (Oxford: Clarendon Press, 1963). An accessible recent study is Jenny Uglow, *The Lunar Men: The Friends Who Made the Future 1730–1810* (London: Faber & Faber, 2002).

8. Shena Mason, *The Hardware Man's Daughter: Matthew Boulton and His "Dear Girl"* (Chichester: Phillimore & Co., 2005), 2–3. This excellent book offers numerous insights into the Boulton family's financial and domestic circumstances.

9. McLean, "Introduction: Matthew Boulton," 3–4.

10. For an analysis of changing depictions of the partnership and their varied functions, see David Philip Miller, "Scales of Justice: Assaying Matthew Boulton's Reputation and the Partnership of Boulton and Watt," *Midland History* 34 (2009): 58–76.

11. Boulton's financial problems and recklessness were long ago recognized in a classic study by J. E. Cule, "Finance and Industry in the Eighteenth Century: The Firm of Boulton and Watt," *Economic History Supplement to Economic Journal* 4 (1940): 319–25, based on J. E. Cule, "The Financial History of Matthew Boulton" (MComm thesis, University of Birmingham, 1935). See also Jenni-

fer Tann, "Boulton and Watt's Organization of Steam Engine Production before the Opening of Soho Foundry," *Transactions of the Newcomen Society* 49 (1977–78): 51, note 1. On Boulton's scientific and technical capacities, see Jennifer Tann, "Matthew Boulton—Innovator," in *Matthew Boulton: Enterprising Industrialist of the Enlightenment*, ed. Kenneth Quickenden et al., 33–50; David Philip Miller, "Was Matthew Boulton a Scientist? Operating between the Abstract and the Entrepreneurial," in *Matthew Boulton: Enterprising Industrialist of the Enlightenment*, ed. Kenneth Quickenden et al., 51–65; Jim Andrew, "Was Matthew Boulton a Steam Engineer?," in *Matthew Boulton: A Revolutionary Player*, ed. Malcolm Dick (Studley: Brewin Books, 2009), 107–15.

12. Matthew Boulton to Thomas Wilson, 28 November 1792, Wilson Correspondence, Cornwall Record Office, volume 5, AD1583/5/69. The wager never went ahead. Watt probably talked Boulton out of it!

13. Watt, *Memoir of Matthew Boulton*, 5.

14. George Demidowicz, "Power at the Soho Manufactory and Mint," in *Matthew Boulton: A Revolutionary Player*, ed. Malcolm Dick, 118.

15. Richard L. Hills, *Power from Steam: A History of the Stationary Steam Engine* (Cambridge: Cambridge University Press, 1989), 31–33.

16. See Benjamin Franklin to Matthew Boulton, 19 March 1766, in response to Matthew Boulton to Benjamin Franklin, 22 February 1766, both accessible via the online Franklin Papers at http://www.franklinpapers.org.

17. Watt to Boulton, 20 October 1768 in Muirhead, *Origin and Progress*, I: 30.

18. Watt to Boulton, 20 October 1768 in Muirhead, *Origin and Progress*, I: 31.

19. There are clues that Roebuck did change his mind. Thus George Jardine, Roebuck's secretary, in a letter written about this time said of Roebuck that "the more he is convinced of the practicality of the scheme, the keener he is of carrying it to practice yourselves for your mutual advantage" (George Jardine to Watt, "Monday" 1768, MS3219/4/59). Watt told Small in April 1769 that "the nearer it approaches to certainty, he [Roebuck] grows the more tenacious of it" (Watt to Small, 28 April 1769, in Muirhead, *Origin and Progress*, I: 55).

20. Roebuck to Boulton, 12 December 1768, quoted in Hills, *James Watt, Volume 1*, 419.

21. Boulton to Watt, 7 February 1769, in Muirhead, *Origin and Progress*, I: 41–43.

22. For a full treatment of the wheel-engine, see Hills, *James Watt, Volume 1*, 424–32.

23. Henry Hamilton, "The Failure of the Ayr Bank, 1772," *The Economic History Review* 8 (1956): 405–17.

24. R. H. Campbell, "Roebuck, John," *Oxford Dictionary of National Biography*, 2004, online edition updated 2013: http://www.oxforddnb.com/view/article/23944, accessed 14 February 2017.

25. See Hills, *James Watt, Volume 1*, 446.

26. Journal Notebook of James Watt, MS3219/4/135.

27. Watt to William Small, no date, in Muirhead, *Origin and Progress*, II: 62.

28. William Small to Watt, 8 October 1773, in Muirhead, *Origin and Progress*, II: 63.

29. Journal Notebook of James Watt, 26 April 1773–May 1774, MS3219/4/135.

30. Watt to William Small, 11 December 1773, in Muirhead, *Origin and Progress*, II: 69.

31. Journal Notebook of James Watt, 26 April 1773–May 1774, MS3219/4/135.

32. The fullest study of the Lunar Society remains Schofield, *The Lunar Society of Birmingham*. However, an important, incisive examination of its activities and ethos is at the center of Jones, *Industrial Enlightenment*, 82–94. Jones argues for seeing the Lunar Society more as a local cultural manifestation of Enlightenment sensibility than as a harbinger and architect of industrial development. The latter interpretation was pushed by Schofield and has been favored by many historians who argue that there was a close relationship between the pursuit of science, technological change, and economic growth during this period.

33. See Roy S. Porter, "Science, Provincial Culture and Public Opinion in Enlightenment England," *Journal for Eighteenth-Century Studies* 3 (1980): 20–46. There is a proliferation of local enlightenments centered on such societies. See, for example, Jon Mee and Jennifer Wilkes, "Transpennine Enlightenment: The Literary and Philosophical Societies and Knowledge Networks in the North, 1781–1830," *Journal for Eighteenth-Century Studies* 38 (2015): 599–612.

34. See the classic study, Arnold Thackray, "Natural Knowledge in Cultural Context: The Manchester Model," *American Historical Review* 79 (1974): 672–709.

35. See Eric Robinson, "The Lunar Society: Its Membership and Organization," *Transactions of the Newcomen Society* 35 (1962–63): 153–77; Eric Robinson, "The Origins and Life-Span of the Lunar Society," *University of Birmingham Historical Journal* 11 (1967): 14–15.

36. Jones, *Industrial Enlightenment*, 90.

37. William Small to Watt, 23 February 1774, in Muirhead, *Origin and Progress*, II: 74.

38. On this logbook see Dickinson, *James Watt: Craftsman and Engineer*, 87, where he gives an idea of its contents: records of the number of strokes by the engines, the weight of coal burned and the weight of water evaporated by the boiler. It also indicates that the piston and its packing was the biggest problem and the focus of attention at this stage.

39. Watt to Watt Sr., 11 December 1774, in Muirhead, *Origin and Progress*, II: 79.

40. On Wilkinson see H. W. Dickinson, *John Wilkinson, Ironmaster 1728–1808* (Ulverston: Hume Kitchin, 1914); Frank C. Dawson, *John Wilkinson: King of the Ironmasters* (Stroud: The History Press, 2012). For insights into Wilkinson as part of an intellectual network, see Roger N. Bruton, "The Shropshire Enlightenment: A Regional Study of Intellectual Activity in the Late Eighteenth and Early Nineteenth Centuries" (PhD diss., University of Birmingham, 2015).

41. See Alexander Murdoch, "Wedderburn, Alexander, first earl of Rosslyn (1733–1805)," *Oxford Dictionary of National Biography*, https://doi.org/10.1093/ref:odnb/28954.

42. MS3219/6/222. The text is reproduced as appendix 2 in Hills, *James Watt, Volume 2*, 240–43, quotation at 241. The collaboration between Watt and Small ended with the latter's untimely death in February 1775 from malaria, contracted during his time in Virginia.

43. "Petition to Parliament for Engine Act," MS3219/4/227.

44. *Journals of the House of Commons 29 November 1774 . . . to 15 October 1776*. Reprinted 1803, vol. 35, 142.

45. *Journals of the House of Commons*, vol. 35, 168–69.

46. Alexander Cumming (1732?–1814) was an instrument maker who long worked for Archibald Campbell, Third Duke of Argyll, for whom he constructed models of scientific instruments and steam engines of various designs, becoming an acknowledged expert on the latter by virtue of that activity.

(see Hills, *James Watt, Volume 1*, 114, 321; Emerson, *An Enlightened Duke*, 135, 207, 337, 351). On the wide deployment of models during this period and their importance, see Fox, *The Arts of Industry*, 135–77.

47. *Journal of the House of Commons*, vol. 35, 169.

48. Eric Robinson, "II. Matthew Boulton and the Art of Parliamentary Lobbying," *The Historical Journal* 7 (1964): 209–29 gives a detailed guide to who was recruited to the cause and by whom.

49. For the most recent account of this episode, see Sally Baggott, "Hegemony and Hallmarking: Matthew Boulton and the Battle for the Birmingham Assay Office," in *Matthew Boulton: Enterprising Industrialist of the Enlightenment*, ed. Quickenden et al., 147–61.

50. Matthew Boulton to the Earl of Dartmouth, 22 February 1775, extracted in *English Historical Documents, 1714–1783*, ed. D. B. Horn and Mary Ransome (London: Eyre & Spottiswoode, 1957), X: 474–75.

51. See Robinson, "Boulton and the Art of Parliamentary Lobbying," 210–16; Brian D. Bargar, "Matthew Boulton and the Birmingham Petition of 1775," *The William and Mary Quarterly*, series 3, 13 (1956): 26–39.

52. *Journal of the House of Commons*, vol. 35, 191, 207.

53. "Comparison of Mr Blakey's and Mr Watt's Steam Engines," in MS3219/4/227.

54. The Committee's proceedings are recorded in "Copy Minutes on Recommitment of Mr Watt's Engine Bill," MS3219/4/221.

55. "Case of James Watt, Engineer, Inventor of the New Improvements upon the Steam Engine," Matthew Boulton Papers, Archives of Soho, Library of Birmingham, MS3782/12/76/193.

56. Robinson, "Boulton and the Art of Parliamentary Lobbying," 223.

57. Watt to Watt Sr., 8 May 1775, in Muirhead, *Origin and Progress*, II: 89.

58. See Hills, *James Watt, Volume 2*, 40; Erasmus Darwin to James Watt, 29 March 1775, in *The Collected Letters of Erasmus Darwin*, ed. Desmond King-Hele (Cambridge: Cambridge University Press, 2007), 133–34; John Marr to Watt, 19 December 1775, MS3219/4/77; Muirhead, *Origin and Progress*, I: clxvi. Watt had received a prior offer in 1771, again through Robison, of the post of "Master Founder of Iron Ordnance to Her Imperial Majesty." In that case Robison had expounded on Watt's knowledge of metallurgy and mechanics as well as his "intimate acquaintance with all the process as carry'd on at Carron." See Robison to Watt, 22 April 1771, in Robinson and MacKie, *Partners in Science*, 24.

59. Not Nancy Millar as Hills (*James Watt, Volume 2*, 79) records. Nancy, formally Agnes, was the wife of John Marr, Watt's old friend who had watched over his instrument making sojourn in London. Marr and Nancy left for Canada in 1774. Betty Millar, the sister of Nancy and of Watt's late wife Margaret, wrote a number of letters to Watt in 1775 recording the situation of the children. After John Marr died in October 1786, Nancy lived with Betty in Glasgow and was then involved with the care of Watt's daughter Peggy, who at the time was about twenty years old.

60. Gilbert Hamilton to Watt, Glasgow, 5 September 1774, MS3219/4/17. Also Betty Millar to Watt, 29 August 1774, MS3219/4/77.

61. Betty Millar to Watt, 16 November 1774, MS3219/4/77. Betty had previously been at "Croy," presumably Croy Leckie, which by 1774 was the home of Watt's cousin Robert Muirhead, his uncle John, the previous incumbent, having died in 1769. The reference to young Jamie's "briches" concerns

the so-called "breeching ceremony" that celebrated the passage of a young boy (usually age four to six) from petticoats to trousers.

62. Betty Millar to Watt, 22 February 1775, MS3219/4/77.

63. Betty Millar to Watt, 22 February 1775, MS3219/4/77. Watt had given similar concerns to his cousin Robert Muirhead, who inquired in a letter of 22 February, "Are you intending to settle in England, or come back to the Land of Cakes [Scotland]?" Muirhead's attitude seemed to be that if Watt was receiving the encouragement he deserved in England, then he should stay there (Robert Muirhead to Watt, 22 February 1775, MS3219/4/17).

64. Betty Millar to Watt, 22 February 1775, MS3219/4/77.

65. Robert Muirhead to Watt, 22 February 1775, MS3219/4/17; Hills, *James Watt, Volume 2*, 78.

66. Muirhead to Watt, 3 May 1775, MS3219/4/17.

67. Gilbert Hamilton to Watt, 5 April 1775, MS3219/4/17.

68. Gilbert Hamilton to Watt, 3 August 1774, MS3219/4/17. The Lord Kilmaurs referred to here was James Cunningham (1749–1791), who became Fourteenth Earl of Glencairn in 1775. He is best known as the patron of Robbie Burns, but the Cunningham family from the seventeenth century had made major investments in collieries on the Forth and on the Clyde. See John U. Nef, *The Rise of the British Coal Industry* (Abingdon: Frank Cass, 1966), 6. James Buchanan was provost of Glasgow from 1774 to 1776.

69. Gilbert Hamilton to Watt, 22 May 1775, MS3219/4/17. The "imperious" engineer was presumably John Smeaton.

70. Gilbert Hamilton to Watt, 14 June 1775, MS3219/4/17.

71. See Dickinson and Jenkins, *James Watt and the Steam Engine*, 43.

72. Matthew Boulton to Watt, [August/September] 1775, in Muirhead, *Origins and Progress*, II: 94.

73. Watt to Matthew Boulton, 3 July 1776, quoted in Hills, *James Watt, Volume 2*, 89.

74. Betty Millar to Watt, 30 October/4 November 1775, MS3219/4/77.

75. In Watt's will all his correspondence and papers were bequeathed to Watt Jr., with the exception of letters exchanged between Watt and Annie, which were left to her. However, some of their correspondence from later times does survive in the Watt Papers (Will of Doctor James Watt, Doctor of Laws, Handsworth, Staffordshire, Proved 13 October 1819, The National Archives, Kew PROB11/1621/182).

76. Gilbert Hamilton to Watt, 18 October 1775, MS3219/4/17.

77. Given that Betty Millar's slightly later letter presumed a relationship between Watt and Annie and portrayed her father as the only obstacle, it is possible that Watt had asked the question of Annie before he left Scotland but had not received the answer until it was conveyed by "the Jewel."

78. Gilbert Hamilton to Watt, 2 November 1775, MS3219/4/17.

79. Gilbert Hamilton to Watt, 2 November 1775, MS3219/4/17.

80. Gilbert Hamilton to Watt, 8 February 1776, MS3219/4/17.

81. Gilbert Hamilton to Watt, 8 March 1776, MS3219/4/17.

82. Gilbert Hamilton to Watt, 17 March 1776, MS3219/4/17.

83. Gilbert Hamilton to Watt, 19 April 1776, MS3219/4/17.

84. Watt to Matthew Boulton, 8 July 1776, quoted in Hills, *James Watt, Volume 2*, 83.

85. Watt to Boulton, 28 July 1776, quoted in Hills, *James Watt, Volume 2*, 83.

86. The entry is available at https://www.nrscotland.gov.uk/research/learning/hall-of-fame/hall -of-fame-a-z/watt-james. The original is at National Records of Scotland, OPR 500/2. It appears that the banns were not read some weeks before in the normal way, but rather a proclamation of intent was made, followed immediately by the marriage.

87. Watt to Boulton, 8 July 1776, in Hills, *James Watt, Volume 2*, 83.

88. *Jewellery Quarter Conservation Area, Character Appraisal and Management Plan* (Birmingham City Council, 2002), 11, https://www.birmingham.gov.uk/downloads/download/237/jewellery _quarter_conservation_area.

89. Hills, *James Watt, Volume 2*, 81; Dickinson, *James Watt, Craftsman and Engineer*, 94.

90. See Dickinson and Jenkins, *James Watt and the Steam Engine*, 43.

91. Watt to William Small, 28 May 1769, in Muirhead, *Origin and Progress*, I: 63–64.

92. Boulton and Watt outlined this process to various interested parties. See Boulton to Watt, 24 April 1775, in Muirhead, *Origin and Progress*, II: 85; Watt to J. D. H. van Liender, 18 July 1775, transcribed in Jan Adrianus Verbruggen, "The Correspondence of Jan Daniel Huichelbos Van Liender" (PhD diss., University of Twente, 2005), 103.

93. See Jim Andrew, "The Soho Steam-Engine Business," in *Matthew Boulton: Selling What All the World Desires*, ed. Shena Mason, 66.

94. On the origins of the foundry, see Laurence Ince, "The Soho Engine Works 1796–1895," *Stationary Power*, no. 16 (2000): 5–10.

95. The following section relies on Hills, *James Watt, Volume 2*, 57–58.

96. See John Kanefsky and John Robey, "Steam Engines in 18th-century Britain: A Quantitative Assessment," *Technology and Culture* 21 (1980): 161–86.

97. Quoted in Dickinson and Jenkins, *James Watt and the Steam Engine*, 113–14.

98. See Hills, *James Watt, Volume 2*, 64–65.

99. Bernard Deacon, "'The Hollow Jarring of the Distant Steam Engines': Images of Cornwall between West Barbary and the Delectable Duchy," in *Cornwall: The Cultural Construction of Place*, ed. Ella Westland (Newmill: Patten Press, 1997), 7–24.

100. Matthew Boulton to Watt, [September] 1775, quoted in Hills, *James Watt, Volume 2*, 93.

101. Dickinson and Jenkins, *James Watt and the Steam Engine*, 129; Hills, *James Watt, Volume 2*, 95–96.

102. Watt to Boulton, 13 September 1777, quoted in Hills, *James Watt, Volume 1*, 100.

103. Watt to Boulton, 20 September 1777, quoted in Hills, *James Watt, Volume 1*, 100. On Budge see Dickinson and Jenkins, *James Watt and the Steam Engine*, 308–09.

104. Extensive incoming correspondence to Wilson is preserved and held at the Cornwall Record Office. Digital transcriptions are at http://www.cornishmining.net/story/bwpapers.htm.

105. For a description of the house and its comforts, see Mason, *The Hardware Man's Daughter*, 50–51.

106. This document is reprinted in Dickinson and Jenkins, *James Watt and the Steam Engine*, 375–98.

107. See Hills, *James Watt, Volume 2*, 113–17.

108. Alessandro Nuvolari and Bart Verspagen, "*Lean's Engine Reporter* and the Development of the Cornish Engine: A Reappraisal," *Transactions of the Newcomen Society* 77 (2007): 167–89.

109. The following relies on: Richard L. Hills, "James Watt and His Copying Machine," in *Studies in British Paper History, Volume 1, The Oxford Papers*, ed. Peter Bowers (Kidlington: British Association of Paper Historians, 1996), 81–88; James H. Andrew, "The Copying of Engineering Drawings and Documents," *Transactions of the Newcomen Society* 53 (1981–82): 1–15; Hills, *James Watt, Volume 2*, 190–211; Barbara Rhodes and William Streeter, *Before Photocopying: The Art and History of Mechanical Copying, 1780–1938* (Delaware: Oak Knell Press, 1999).

110. Despite this, they did find some users, a notable one being Thomas Jefferson. See Silvio Bedini, *Thomas Jefferson and His Copying Machines* (Charlottesville: University of Virginia Press, 1984).

111. John Marr to Watt, 24 December 1779, MII/11/24.

112. John Marr to Watt, 19 April 1780, MII/11/25.

113. On Magellan's activities in this specific connection, see Watt to Magellan, 1 September 1780, 20 September 1780, 16 October 1780 and Magellan to Watt, 16 September 1780, 6 October 1780, in *For the Love of Science: The Correspondence of J. H. de Magellan (1722–1790)*, ed. R. W. Home et al., 2 vols. (Bern: Peter Lang, 2017), II: 1162–67. On Magellan see the biographical introduction to *For the Love of Science*, I: 1–48.

114. On longer-term developments see David Philip Miller, "'Men of Letters' and 'Men of Press Copies': The Cultures of James Watt's Copying Machine," in *The Romance of Science: Essays in Honour of Trevor H. Levere*, ed. Jed Buchwald and Larry Stewart (Cham: Springer, 2017), 65–79.

115. Just before Watt's third trip to Cornwall, Annie took the brave step of communicating her concerns about her husband to Boulton, who was away. She depicted Watt as suffering bad health and low spirits, and being in danger of "sinking under that fatal depression." She mentioned as preying on his mind "the bond he is engaged in to Vere's house," an old, unsettled account from the beginning of the business. Annie was pleading with Boulton to ensure that Watt had "no unsettled scores to look back to and brood over in his mind" (Annie Watt to Boulton, 15 April 1781, quoted in Smiles, *Lives of Boulton and Watt*, 217).

116. Watt to Boulton, 30 October 1779, quoted in Hills, *James Watt, Volume 1*, 164.

117. Quoted in Samuel Smiles, *Lives of the Engineers: The Steam Engine. Boulton and Watt* (London: J. Murray, 1778), 213.

118. See the long cri de coeur in Watt to Boulton, 31 October 1780, MS3219/4/122.

119. Eric Roll, *An Early Experiment in Industrial Organisation: Being a History of the Firm of Boulton & Watt, 1775–1805* (Abingdon: Frank Cass, 1930), 91–94. See also G. C. Allan, "An Eighteenth-Century Combination in the Copper Mining Industry," *Economic Journal* 33 (1923): 74–85 and J. R. Harris and R. O. Roberts, "Eighteenth Century Monopoly: The Cornish Metal Company Agreements of 1785," *Business History* 5 (1963): 69–82.

120. Roll, *An Early Experiment*, 94.

121. The patent trials will be dealt with in chapter 5.

122. For broad views of the longevity of waterpower, and of steam as adjunct, see Andreas Malm, *Fossil Capital: The Rise of Steam Power and the Roots of Global Warming* (London: Verso, 2016), 37–57; Robert C. Allen, *The British Industrial Revolution in Global Perspective* (Cambridge: Cambridge University Press, 2009), 169–72.

123. Quoted in Hills, *Power from Steam*, 62.

124. Hills, *Power from Steam*, 59–62.

125. Gilbert Hamilton to James Watt, 22 March 1782, MS3219/4/19.

126. Hills, *James Watt, Volume 3*, 53–54.

127. Watt to Gilbert Hamilton, 28 September 1783, MS3219/4/123; Hills, *James Watt, Volume 3*, 60.

128. Hills, *Power from Steam*, 85–86.

129. John Rennie was a Scottish millwright who had studied with Robison and Black at Edinburgh University. He was recruited by Boulton & Watt, trained at Soho, and assigned to Albion Mill. He became the country's leading civil engineer in business on his own account. He and Watt remained friends until Watt's death. See "Rennie, John FRS FRSE (1761–1821)," in *A Biographical Dictionary of Civil Engineers in Great Britain and Ireland, I: 1500–1830*, ed. A. W. Skempton et al. (London: Thomas Telford Publishing Ltd, 2002), 554–66.

130. Hills, *James Watt, Volume 3*, 79–80; Dickinson and Jenkins, *James Watt and the Steam Engine*, 220–24.

131. "Appendix by Mr Watt [to Robison's essay on the steam engine]," in John Robison, *A System of Mechanical Philosophy* (Edinburgh: John Murray, 1822), II: 155.

132. Compare Hills, *James Watt, Volume 3*, 63–64, and *The Selected Papers of Boulton & Watt, Volume 1: The Engine Partnership 1775–1825*, ed. Jennifer Tann (London: Diploma Press, 1981), 6–7.

133. The following draws on Hills, *James Watt, Volume 3*, 69–70.

134. Hills, *James Watt, Volume 3*, 80; Matthew Boulton to Watt, 26 June 1790, in Dickinson and Jenkins, *James Watt and the Steam Engine*, 197.

135. Watt to Matthew Boulton, 17 April 1786, quoted in Dickinson, *James Watt*, 149–150.

136. "Fire at Albion Mills," *Hampshire Chronicle*, 7 March 1791, 3.

137. Watt to Annie Watt, 26 May 1787, MS3219/4/267A.

138. "Royal Visit at Mr. Whitbread's Brewery," *Norfolk Chronicle*, 2 June 1787, 4.

139. Jennifer Tann, "Marketing Methods in the International Steam Engine Market: The Case of Boulton and Watt," *The Journal of Economic History* 38 (1978): 363–91.

140. Tann, "Marketing Methods," 363–64.

141. Jennifer Tann, "Steam and Sugar: The Diffusion of the Stationary Steam Engine to the Caribbean Sugar Industry 1779–1840," *History of Technology* 19 (1998): 63–84.

142. Watt to John Roebuck, 3 February 1787, in Muirhead, *Origin and Progress*, II: 214–15.

143. Paul Naegel and Pierre Teissier, "Obtaining a Royal Privilege in France for the Watt Engine, 1776–1786," *International Journal for the History of Engineering and Technology* 83 (2013): 96–118.

144. Verbruggen, "The Correspondence of Jan Daniel Huichelbos van Liender," 12–16.

145. Boulton to Watt, 3 July 1778, quoted in Tann, "Marketing Methods," 368.

146. Tann, "Marketing Methods," 368.

147. Tann, "Marketing Methods," 388, table 3.

148. The chief sources on Trevithick are: Francis Trevithick, *Life of Richard Trevithick, with an Account of His Inventions*, 2 vols. (London: F. & F. N. Spon, 1872); H. W. Dickinson and Arthur Tit-

ley, *Richard Trevithick: The Engineer and the Man* (Cambridge: Cambridge University Press, 1934); Anthony Burton, *Richard Trevithick: Giant of Steam* (London: Aurum Press, 2000). In many respects the most judicious and even-handed account of the relations of the Watt and Trevithick camps is given in William Pole, *A Treatise on the Cornish Pumping Engine; in Two Parts* (London: John Weale, 1844).

149. See Rhys Jenkins, "A Cornish Engineer: Arthur Woolf 1766–1837," *Transactions of the Newcomen Society* 13 (1932): 55–73.

150. Boulton, Watt & Co. was the name of the new partnership set up on the establishment of the Soho Foundry. It involved Boulton and his sons Watt Jr. and Gregory Watt, but not Watt himself.

151. Quoted in Dickinson and Titley, *Richard Trevithick*, 60, from Trevithick to Davies Giddy, 1 October, 1803.

152. "Dreadful Accident," *Philosophical Magazine* 16 (1803): 373.

153. For example: "The innumerable accidents which have occurred by the explosion of the boilers of high-pressure steam engines, or those on the plan of Trevithick, ought, long ago, to have induced the legislature of this country to have prohibited their use, since those on the construction of Bolton and Watt, are not only safe, but have been found to answer every useful purpose" ("Letter from a Correspondent to the Editor," *The Newcastle Courant*, 15 May 1819).

154. Alessandro Nuvolari, "Collective Invention during the British Industrial Revolution: The Case of the Cornish Pumping Engine," *Cambridge Journal of Economics* 28 (2004): 347–63.

155. Richard Hills, "The Development of the Steam Engine from Watt to Stephenson," *History of Technology* 25 (2004): 192–93. Another steam packet explosion at Norwich in 1822 elicited a public lament that high-pressure engines ("ungovernable machines") were in some cases being preferred "to the safe and truly manageable engines on the condensing principle." (*The Morning Post*, 9 April 1817). It is very likely that Watt Jr., who was engaged in trials of his own steamboat engines at this time, had something to do with this report.

156. Hills, "Development of the Steam Engine," 194. On the history of the railway see the insightful account in Ben Marsden and Crosbie Smith, *Engineering Empires: A Cultural History of Technology in Nineteenth-Century Britain* (Houndmills: Palgrave Macmillan, 2005), 129–177.

157. See Dickinson and Titley, *Richard Trevithick*, 4–5. The original publication of this statement was in Trevithick, *Life of Richard Trevithick*, II: 395, where the eminent scientific personage is identified as John Isaac Hawkins (1771–1855).

158. Dickinson and Titley, *Richard Trevithick*, 24.

159. For the range of arguments see Michele Boldrin and David K. Levine, "2003 Lawrence R. Klein Lecture: The Case against Intellectual Monopoly," *International Economic Review* 45 (2004): 348–49 and George Selgin and John Turner, "James Watt as Intellectual Monopolist: Comment on Boldrin and Levine," *International Economic Review* 47 (2006): 1341–48.

160. See chapter 7. On Watt's private fears see Ben Marsden, *Watt's Perfect Engine* (Cambridge: Icon Books, 2002), 142–43.

Chapter 4: Watt as Natural Philosopher

1. I use the term "natural philosophy" in a loose sense here to denote the range of inquiries into the natural world, thus the eighteenth-century analogue of "science" in our time. In earlier periods, systems of natural philosophy more clearly connoted a system of thought and inquiry that included strong

metaphysical and theological strands; as John Schuster puts it, they "purported to describe and explain the entire universe and the relationship of that universe to God, however conceived" (John Schuster, "The Scientific Revolution," in *Companion to the History of Modern Science*, ed. R. C. Olby et al. (London: Routledge, 1990), 224.

2. My initial take on this theme was David Philip Miller, "Puffing Jamie: The Commercial and Ideological Importance of Being a 'Philosopher' in the Case of the Reputation of James Watt (1736–1819)," *History of Science* 38 (2000): 1–21. The filial aspect of Watt Jr's efforts to secure and shape his father's reputation were a central concern of Miller, *Discovering Water*.

3. See especially MacLeod, *Heroes of Invention;* Christine MacLeod, "Concepts of Invention and the Patent Controversy in Britain," in *Technological Change: Methods and Themes in the History of Technology*, ed. Robert Fox (Amsterdam: Harwood Academic Publishers, 1996), 137–53; Christine MacLeod, "James Watt, Heroic Invention and the Idea of the Industrial Revolution," in *Technological Revolutions in Europe: Historical Perspectives*, ed. M. Berg and K. Bruland (Cheltenham: Edward Elgar, 1998), 96–116.

4. Christine MacLeod and Jennifer Tann, "From Engineer to Scientist: Re-inventing Invention in the Watt and Faraday Centenaries 1919–1931," *The British Journal for the History of Science* 40 (2007): 389–411.

5. See Jacob and Stewart, *Practical Matter;* Joel Mokyr, *The Gifts of Athena: Historical Origins of the Knowledge Economy* (Princeton and Oxford: Princeton University Press, 2002), 51–52.

6. For an influential articulation of the conventional view, see Herbert Butterfield, *The Origins of Modern Science, 1300–1800*, rev. ed. (New York: The Free Press, 1957), 203–21. For early challenges to that view, see J. R. R. Christie and J. V. Golinski, "The Spreading of the Word: New Directions in the Historiography of Chemistry 1600–1800," *History of Science* 20 (1982): 235–66; for the various interpretations of the chemical revolution up to recent times, see John G. McEvoy, *The Historiography of the Chemical Revolution* (London: Pickering & Chatto, 2010).

7. An important treatment of the chemical revolution in Britain that provides the context for Watt's efforts is in Jan V. Golinski, *Science as Public Culture: Chemistry and Enlightenment in Britain, 1760–1820* (Cambridge: Cambridge University Press, 1992), 129–52.

8. Jones, *Industrial Enlightenment*, 92–94.

9. Watt to J. A. De Luc, October 1786, MS3219/4/123.

10. P. A. Tunbridge, "Jean André De Luc, F.R.S.," *Notes and Records of the Royal Society of London* 26 (1971): 15–33; M. J. S. Rudwick, "Jean-André De Luc and Nature's Chronology," in *The Age of the Earth from 4004 BC to AD 2002*, ed. C. L. E. Lewis and S. J. Knell (London: The Geological Society, 2001), 51–60.

11. Those accounts were: François Arago, *Historical Eloge of James Watt*, trans. from the French with additional notes and an appendix by James Patrick Muirhead (London: John Murray, 1839); J. P. Muirhead, ed., *Correspondence of the Late James Watt on His Discovery of the Theory of the Composition of Water* (London: John Murray, 1846); J. P. Muirhead, ed., *The Origins and Progress of the Mechanical Inventions of James Watt*, 3 vols. (London: John Murray, 1854); Muirhead, *The Life of James Watt;* Smiles, *Lives of Boulton and Watt*.

12. The significant exception here was Wilson in his "biography" of Henry Cavendish: George Wilson, *The Life of the Honourable Henry Cavendish* (London: The Cavendish Society, 1851). Wilson

reached the conclusion that Watt should be considered a significant chemist by the standards of his own time.

13. For the history and surprises of Newton's papers, see Sarah Dry, *The Newton Papers: The Strange and True Odyssey of Isaac Newton's Manuscripts* (Oxford: Oxford University Press, 2014).

14. For Rankine's renaming of the indicator diagram as the "diagram of energy" as a crucial move in the thermodynamic conception of heat engines, see Ben Marsden, "Engineering Science in Glasgow: W. J. M. Rankine and the Motive Power of Air" (PhD diss., University of Kent at Canterbury, 1992), 161–62. Marsden describes Rankine as making this move with "an appropriate Whig-historical genuflection" to Watt. See also Miller, *James Watt, Chemist*, chapter 6; David Philip Miller, "The Mysterious Case of James Watt's ' "1785" Steam Indicator': Forgery or Folklore in the History of an Instrument?" *International Journal for the History of Engineering and Technology* 81 (2011): 129–50.

15. See Smith, *The Science of Energy*, 32–34.

16. M. Norton Wise and Crosbie Smith, "Work and Waste: Political Economy and Natural Philosophy in Nineteenth-Century Britain, Parts I, II and III," *History of Science* 27 (1989): 263–301, 391–449; 28 (1990): 221–61.

17. See Miller, *James Watt, Chemist*, 164–69.

18. See Hills, *James Watt, Volume 2*, chapter 9.

19. Watt to Joseph Black, 13 December 1782, in Robinson and McKie, *Partners in Science*, 117–19. Victor D. Boantza, "Collecting Airs and Ideas: Priestley's Style of Experimental Reasoning," *Studies in History and Philosophy of Science* 38 (2007): 506–22 provides an insightful counter to both the depreciative and sympathetic historical accounts of Priestley's approach to his work.

20. By the 1830s the notion of being engaged in a "train of research," rather than occupational or professional employment in research, was used in British Association circles as a key criterion in distinguishing "serious" natural philosophers from the rest. See Jack Morrell and Arnold Thackray, *Gentlemen of Science: Early Years of the British Association for the Advancement of Science* (Oxford: Clarendon Press, 1981).

21. Muirhead, *Life of James Watt*, 50, identifies the premises near the beef market off King Street. This is based on the testimony of Robert Hart. The Delftfield, Anderston Walk, location is nominated in "Inaugural Address as Rector of Glasgow University by Sir James Mackintosh," in *Inaugural Addresses by Lords Rectors of the University of Glasgow*, ed. John Barras Hay (Glasgow: David Robertson, 1839), 23–34, note XVII; 195–96. John Gardner himself recalled working on steam experiments with Watt at premises above the Cross in Glasgow (see Gregory Watt to James Watt, September 1796, MS3219/4/8).

22. See Cleland, *Enumeration of the Inhabitants of the City of* Glasgow, 145, which also nominates "an apartment in the Delph Work" as where Watt did his experiments with the aid of a single assistant, who is identified as Gardner. See also Hills, *James Watt, Volume 1*, 106–07.

23. The notebook is at MS3219/4/170. A full transcription of it, together with an explanatory introduction and notes, is provided in Robinson and McKie, *Partners in Science*, 425–90.

24. John Robison, *A System of Mechanical Philosophy with Notes by David Brewster*, 4 vols. (Edinburgh: John Murray, 1822). Watt's and John Southern's notes on Robison's articles "Steam" and "Steam-Engines" were published separately as well, and prior: *The Articles Steam and Steam-Engines, written for the Encyclopaedia Britannica, by the late John Robison . . . with Notes and*

Additions, by James Watt . . . and a Letter on Some Properties of Steam by the Late John Southern (London: John Murray, 1818).

25. Robinson and McKie, *Partners in Science*, 435.

26. See Erasmus Darwin to Matthew Boulton, 12 December 1765, in King-Hele, *The Collected Letters of Erasmus Darwin*, 65–66. See also Miller, "Was Matthew Boulton a Scientist?" 60–61.

27. "Copy of Mr Watt's Remarks on Mr Robison's Edition of Dr Black's Lectures, Communicated to Mr Playfair 4 Jany 1809," transcribed in Robinson and McKie, *Partners in Science*, 416–18.

28. The experiment is described in the 1765 notebook; see Robinson and McKenzie, *Partners in Science*, 436–37.

29. Notebook in Robinson and McKie, *Partners in Science*, 438.

30. Henry Guerlac, "Joseph Black's Work on Heat," in *Joseph Black 1728–1799: A Commemorative Symposium*, ed. A. D. C. Simpson (Edinburgh: The Royal Scottish Museum, 1982), 13–22.

31. Robinson and McKie, *Partners in Science*, 439.

32. Guerlac, "Black's Work on Heat," 19.

33. On "Watt's Law" see Miller, *James Watt, Chemist*, 144–45.

34. Watt to William Small, 17 August 1773, in Muirhead, *Origin and Progress*, II: 58–59.

35. Watt to William Small, 17 August 1773, in Muirhead, *Origin and Progress*, II: 59.

36. Watt to William Small, 3 March 1774, in Muirhead, *Origin and Progress*, II: 75.

37. Watt to Matthew Boulton, 10 December 1782, in Muirhead, *Origin and Progress*, II: 167–68.

38. Watt to William Small, 16 March 1770, in Muirhead, *Origin and Progress*, I: 99.

39. See Hills, *James Watt, Volume 2*, 220.

40. Watt to Boulton, 10 December 1782, in Muirhead, *Origin and Progress*, II: 167–68.

41. Watt to Joseph Black, 13 December 1782, in Anderson and Jones, *The Correspondence of Joseph Black*, I: 566–67.

42. See Miller, *Discovering Water*. I have previously reconstructed Watt's route to these ideas in *James Watt, Chemist*, and I rely heavily on that work in this section.

43. Robert E. Schofield, *The Enlightened Joseph Priestley: A Study of his Life and Work from 1773–1804* (University Park, PA: The Pennsylvania State University Press, 2004), 160–61.

44. Matthew Boulton to Watt, 21 July 1781, MS3147/3/5.

45. "Memoirs of Dr. Joseph Priestley (written by himself)," reproduced in *Autobiography of Joseph Priestley* (Bath: Adams & Dart, 1970), 120.

46. Schofield, *The Enlightened Joseph Priestley*, 161; Schofield, *Lunar Society*, 201–02, 241, 333; Watt to Priestley, "Thursday," Royal Society of London Library, Box PH 13, cited in Hills, *James Watt, Volume 2*, 218.

47. Joseph Priestley, *Experiments and Observations Relating to Various Branches of Natural Philosophy with a Continuation of the Observations on Air* (Birmingham: J. Johnson, 1781), II: 388.

48. The Common Place Book is a thick folio notebook, bound in vellum, with entries covering the period 1782 to 1812, MS3219/4/171. When Priestley discovered an air that he interpreted as being produced when common air was deprived of its phlogiston, he called it, quite logically, "dephlogisticated air."

49. James Watt, "Thoughts on the Constituent Parts of Water and of Dephlogisticated Air; with an Account of Some Experiments on That Subject. In a Letter from Mr James Watt, Engineer, to Mr De Luc, F.R.S.," *Philosophical Transactions of the Royal Society of London* 74 (1784): 329–53.

50. Watt, "Thoughts on the Constituent Parts of Water," 333.

51. Watt, "Thoughts on the Constituent Parts of Water," 334.

52. Watt, "Thoughts on the Constituent Parts of Water," 336.

53. Watt to Joseph Priestley, 2 May 1783, in Muirhead, ed., *The Correspondence of the Late James Watt*, 27.

54. James Watt, "Sequel to the Thoughts on the Constituent Parts of Water and Dephlogisticated Air," *Philosophical Transactions of the Royal Society of London* 74 (1784): 354–57.

55. On Cavendish see successive editions of Christa Jungnickel and Russell McCormmach, *Cavendish: The Experimental Life* (Lewisburg, PA: Bucknell, 1999); Russell McCormmach, *Speculative Truth: Henry Cavendish, Natural Philosophy, and the Rise of Modern Theoretical Science* (Oxford: Oxford University Press, 2004); Russell McCormmach, *The Personality of Henry Cavendish* (Cham: Springer, 2014).

56. Henry Cavendish, "Experiments on Air," *Philosophical Transactions of the Royal Society of London* 74 (1784): 128.

57. These complications are important but need not detain us here. See Miller, *Discovering Water*, 28–29 for details.

58. Joseph Priestley, "Experiments Relating to Phlogiston, and the Seeming Conversion of Water into Air," *Philosophical Transactions of the Royal Society of London* 73 (1783): 414. It subsequently emerged in the nineteenth-century water controversy that Priestley's supposed repetition of Cavendish's experiment was not exactly such and that it could not have produced the results that he claimed. This was taken to have consequences for Watt's claims to the extent that he relied on Priestley's experiment.

59. Watt to Joseph Black, 21 April 1783, in Robinson and McKie, *Partners in Science*, 126; Watt to Gilbert Hamilton, 22 April 1783, and James Watt to Jean André De Luc, 26 April 1783, extracted in Muirhead, ed., *The Correspondence of the Late James Watt*, 20–23.

60. Their experiments eventually culminated in the famous experiment, often seen as decisive evidence supporting Lavoisier's chemistry, in which water was decomposed into inflammable and dephlogisticated air and then synthesized from those products. See Henry Guerlac, "Chemistry as a Branch of Physics: Laplace's Collaboration with Lavoisier," *Historical Studies in the Physical Sciences* 7 (1976): 205–16; Robert J. Morris, "Lavoisier and the Caloric Theory," *The British Journal for the History of Science* 6 (1972): 1–38.

61. J. A. De Luc to Watt, 1 March 1784, in Muirhead, ed., *The Correspondence of the Late James Watt*, 43.

62. Watt to J. A. De Luc, 6 March 1784, in Muirhead, ed., *The Correspondence of the Late James Watt*, 47–49.

63. See Edward Seymour, *History of the Wars Resulting from the French Revolution: With a Relation of the Circumstances Which Led to That Important Event*, 2 vols. (London: Thomas Crabb, 1815), II: 212.

64. We can be confident that Watt would be among the cultural conservatives who found the antics of the duchess everything that they despised: her "inappropriate" assumption of a political role, her gambling, her conspicuous and fashionable consumption. See Anna Clark, *Scandal: The Sexual Politics of the British Constitution* (Princeton: Princeton University Press, 2013), 69–83.

65. Among many accounts of the dissensions see: Harold B. Carter, *Sir Joseph Banks, 1743–1820* (London: British Museum [Natural History], 1988), 194–202; John Heilbron, "A Mathematicians' Mutiny with Morals," in *World Changes: Thomas Kuhn and the Nature of Science*, ed. Paul Horwich (Cambridge, MA: MIT Press, 1993), 81–129; Benjamin Sutherland Wardhaugh, "Charles Hutton and the 'Dissensions' of 1783–84: Scientific Networking and Its Failures," *Notes and Records: The Royal Society Journal of the History of Science* 71 (2017): 41–60.

66. Russell McCormmach, "Henry Cavendish on the Proper Method of Rectifying Abuses," in *Beyond History of Science: Essays in Honor of Robert E. Schofield*, ed. Elizabeth Garber (Bethlehem, PA: Lehigh University Press, 1990), 35–51.

67. Watt to Joseph Banks, 12 April 1784, in Muirhead, ed., *Correspondence of the Late James Watt*, 53.

68. Watt to Joseph Banks, 21 May 1784, Charles Blagden to Watt, 25 May 1784, and Watt to Blagden, 27 May 1784, all extracted in Muirhead, ed., *Correspondence of the Late James Watt*, 62–65.

69. Watt to Mr. Fry, 15 May 1784, in Muirhead, ed., *Correspondence of the Late James Watt*, 61. A more accurate estimate of Cavendish's wealth would have been £1,000,000, which is probably what Watt intended to write. Joseph Fry (1728–1787) was a Quaker apothecary and doctor who founded the soon-to-be-famous family chocolate business (Hills, *James Watt, Volume 2*, 179; James Mosley, "Fry, Joseph [1728–1787]," *Oxford Dictionary of National Biography*, http://www.oxforddnb.com/view/article/10212, accessed 24 July 2017).

70. Henry Cavendish and Charles Blagden, "Computations and Observations in Journey 1785," Devonshire Collections, Chatsworth House, Cavendish Mss X(a) 4, 36–37. See Miller, *James Watt, Chemist*, 162–63.

71. Muirhead, *Life of James Watt*, 399.

72. Annie Watt to Watt, 25 December 1786, MS3219/4/5.

73. See Hills, *James Watt, Volume 3*, 147–50.

74. David F. Larder, "An Unpublished Chemical Essay of James Watt," *Notes and Records of the Royal Society of London* 25 (1970): 193–210. For part of the nomenclature discussions between Watt and Berthollet, see Watt to Berthollet, 18 [December] 1788, MS3219/4/123,268; Watt to Berthollet, 24 January 1789, MS3219/4/123,279; Berthollet to Watt, 28 December 1788, MS3219/4/102.

75. Larder, "An Unpublished Chemical Essay," 194.

76. Larder, "An Unpublished Chemical Essay," 202.

77. The full title of the first edition continues: *In Two Parts, Part 1 by Thomas Beddoes, M.D. Part II by James Watt Esq* (Bristol, 1794). Second and third editions were published in 1795 and 1796 with slightly variant titles and containing additional material, mainly in the form of medical cases. In what follows I use the third edition.

78. See Leslie Tomory, *Progressive Enlightenment: The Origins of the Gaslight Industry, 1780–1820* (Cambridge, MA: The MIT Press, 2012), 68.

79. See Mike Jay, *The Atmosphere of Heaven: The Unnatural Experiments of Dr Beddoes and His Sons of Genius* (New Haven: Yale University Press, 2009), 99–100.

80. Hills, *James Watt, Volume 3*, 152–58.

81. See Gregory Watt to James Watt, September 1796, MS3219/4/8; Annie Watt to Watt, 2 October 1796, MS3219/4/7.

82. See David Philip Miller and Trevor H. Levere, "'Inhale it and See?' The Collaboration between Thomas Beddoes and James Watt in Pneumatic Medicine," *Ambix* 55 (2008): 5–28, on which I draw here. An excellent treatment is Trevor Levere, "Chemistry, Consumption, and Reform," in Trevor Levere, Larry Stewart, and Hugh Torrens with Joseph Wachelder, *The Enlightenment of Thomas Beddoes: Science, Medicine and Reform* (London: Routledge, 2017), 10–78.

83. Trevor Levere, "Dr Thomas Beddoes at Oxford: Radical Politics in 1788–1793 and the Fate of the Regius Chair in Chemistry," *Ambix* 28 (1961): 61–69; Trevor Levere, "Dr Thomas Beddoes (1750–1808): Science and Medicine in Politics and Society," *The British Journal for the History of Science* 17 (1984): 187–204.

84. Watt to Sir Joseph Banks, 7 December 1794, MS3219/4/27; Banks to James Watt, 10 December 1794, extracted in *The Banks Letters*, ed. Warren R. Dawson (London: British Museum, 1958), 860. See Larry Stewart, "A Jacobin Cloven Foot," in Levere et al., *The Enlightenment of Thomas Beddoes*, 133; Jay, *Atmosphere of Heaven*, 106–07 is useful but mistakenly has Watt Jr. approaching Banks. Beddoes's efforts to raise funds for pneumatic medicine are dealt with at length in Frank A. J. L. James, "'The first example . . . of an extensive scheme of pure scientific medical investigation': Thomas Beddoes and the Medical Pneumatic Institution in Bristol, 1794 to 1799," The Eighth Wheeler Lecture, Royal Institution, 2015, *RSCHG Occasional Paper No. 8*, 2016, http://www.rsc.org/images/Final%20Online%20James%20Wheeler%20Lecture_tcm18-248985.pdf.

85. James Watt Jr. to John Ferriar, 19 December 1794, quoted in Trevor H. Levere, "Dr Thomas Beddoes and the Establishment of His Pneumatic Institution: A Tale of Three Presidents," *Notes and Records of the Royal Society of London* 32 (1977): 41–49.

86. James Lind to Watt, 20 February 1795, MS3219/4/27, as quoted in Levere, "Chemistry, Consumption and Reform," 48.

87. Watt to Thomas Beddoes, 25 October 1794, MS3219/4/124. Gregory had gone to Glasgow with his mother. The potatoes are revealing! Watt grew them and a variety of vegetables in the gardens at Heathfield. The fact that the potatoes had been harvested and his laboratory was full of them confirms that the laboratory in question was the one that plans show on the ground floor next to the brewhouse (see chapter 6). Watt's famous garret was not the site of the experiments because the potatoes from Watt's gardens would not have been hauled up there! (cf. Simon Werrett, "Household Oeconomy and Chemical Inquiry," in *Compound Histories: Materials, Governance and Production, 1760–1840*, ed. Lissa S. Roberts and Simon Werrett [Leiden/Boston: Brill, 2018], 41).

88. Among Beddoes's explicitly political effusions at about this time were *Essay on the Public Merits of Mr Pitt* (London: J. Johnson, 1796) and *A Word in Defence of the Bill of Rights against the Gagging Bill* (Bristol: N. Biggs, 1795).

89. Watt to Thomas Beddoes, 2 March 1795, MS3219/4/124. The Latin quotation is from Ovid, a contraction of "fas est et ab hoste doceri," meaning "it is right to be taught even by an enemy."

90. Watt to John Robison, 30 January 1798, quoted in Larry Stewart, "At the Medical Edge or, The Beddoes Effect," in *The Romance of Science: Essays in Honour of Trevor H. Levere*, ed. Jed Buchwald and Larry Stewart (Springer, 2017), 47–64.

91. *Considerations*, 3rd ed. (Bristol: J. Johnson, 1796), 181–219.

92. *Considerations*, 3rd ed., 107–18.

93. J. R. Partington, *History of Chemistry* (London: Macmillan, 1962–1964), III: 345–62.

94. *Considerations*, 3rd ed., 212–13.

95. *Considerations*, 3rd ed., 212–13.

96. *Considerations*, 3rd ed., 108.

97. *Considerations*, 3rd ed., 108 (my italics).

98. *Considerations*, 3rd ed., 212.

99. For example, Claude-Louis Berthollet, *Recherches sur les Lois de l'Affinité* (Paris: Badouin, 1801).

100. See Partington, *History of Chemistry*, III: 515. Pere Grapi and Mercè Izquierdo, "Berthollet's Conception of Chemical Change in Context," *Ambix* 44 (1997): 113–30, locates Berthollet's ideas about affinity in the years 1789 to 1793 in association with niter production, while they are traced to his work on dyeing by Barbara Whitney Keyser, "Between Science and Craft: The Case of Berthollet and Dyeing," *Annals of Science* 47 (1990): 213–60, esp. 234–35.

101. See, for example, Watt to Erasmus Darwin, July ? 1794, MS3219/4/124, and Watt to Dr. Ewart, 21 September 1794, MS3219/4/124.

102. Watt to Thomas Beddoes, 4 September 1794, MS3219/4/124.

103. Watt to Thomas Beddoes, 30 August 1794, MS3219/4/124.

104. Watt to Thomas Beddoes, 30 August 1794, MS3219/4/124.

105. Watt to Thomas Beddoes, 4 September 1794, JWP MS3219/4/124.

106. Joseph Priestley, *Experiments on the Generation of Air from Water* (London: J. Johnson, 1793), 12.

107. Joseph Priestley, *The Doctrine of Phlogiston Established, and That of the Composition of Water Refuted* (Northumberland, PA, 1800), 51–52, where Priestley cites Watt's contribution to the *Considerations*.

108. A. Rupert Hall, "What Did the Industrial Revolution in Britain Owe to Science?," in *Historical Perspectives: Studies in English Thought and Society in Honour of J. H. Plumb*, ed. Neil McKendrick (London: Europa, 1974), 129–51.

109. See A. E. Musson and Eric Robinson, *Science and Technology in the Industrial Revolution* (Manchester: Manchester University Press, 1969), 79–80.

110. See Mokyr, *The Gifts of Athena*, chapters 1 and 2; Joel Mokyr, *An Enlightened Economy: An Economic History of Britain 1700–1850* (New Haven: Yale University Press, 2009).

111. Jones, *Industrial Enlightenment*, 116–29.

112. See Larry Stewart, "'Ordinary' People and Philosophers in the Laboratories and Workshops of the Early Industrial Revolution," in *In Praise of Ordinary People: Early Modern Britain and the Dutch Republic*, ed. Margaret C. Jacob and Catherine Secretan (Palgrave Macmillan, 2013), 95–122.

113. Mokyr, *The Gifts of Athena*, chapter 1.

114. The case against any place for latent heat in Watt's improvements was early made by Donald Fleming, "Latent Heat and the Invention of the Watt Engine," *Isis* 43 (1952): 3–5.

115. Jim Andrew, "Boulton, Watt and Wilkinson: The Birth of the Improved Steam Engine," in *Matthew Boulton, Enterprising Industrialist of the Enlightenment*, ed. Kenneth Quickenden et al. (Farnham: Ashgate, 2013), 85–100.

116. Smith, *The Science of Energy*, and Marsden, "Engineering Science in Glasgow," are careful to avoid such easy equations between the way that Watt used the notion of "perfect engine" and the way that Rankine, Thomson, and other early thermodynamicists deployed it in contrast with "actual" engines. Hills was less concerned with such distinctions, though alert to them. Other historians have, in my view, been rather cavalier. See, for example, A. J. Pacey, "Some Early Heat Engine Concepts and the Conservation of Heat," *The British Journal for the History of Science* 7 (1974): 135–45; Miller, *James Watt, Chemist*, 164–67.

117. Matthew Boulton to Lord Lansdowne, 12 November 1794, MS3782/12/39, quoted in Jones, *Industrial Enlightenment*, 119.

118. For a fuller account see David Philip Miller, "A New Perspective on the Natural Philosophy of Steam in the Long Eighteenth Century and Its Relation to the Steam Engine," *Technology and Culture* (forthcoming).

119. See Miller, *James Watt, Chemist*, 126–35.

120. Watt to De Luc, 2 February 1784, MS3219/4/123.

121. Watt also tended to argue from medical cases involving the effects of pneumatic treatments on patients back to the nature of airs. See, for example, Watt to John Robison, 31 March 1797, MS3219/4/118, where he uses such reasoning to argue that oxygene made in the laboratory is different from the component of common air. We might, with danger of anachronism, see this as philosophical traffic from the clinic to the laboratory analogous to traffic from the workshop or manufactory to the laboratory.

122. For advocacy of this approach and examples of it, see *The Mindful Hand: Inquiry and Invention from the Late Renaissance to Early Industrialization*, ed. Lissa Roberts, Simon Schaffer, and Peter Dear (Amsterdam: Koninklijke Nederlandse Akademie van Wetenschappen, 2007). Also see Hjalmar Fors, *The Limits of Matter: Chemistry, Mining & Enlightenment* (Chicago: The University of Chicago Press, 2015).

123. Watt to Joseph Black, 3 February 1783, printed in Anderson and Jones, ed., *The Correspondence of Joseph Black*, I: 610–12.

124. Joseph Priestley, *Experiments and Observations on Different Kinds of Air . . . In Three Volumes* (Birmingham: Thomas Pearson, 1790), II: 409.

125. Hills, *James Watt, Volume 2*, 220.

126. For discussion of Priestley's experiments and the problems with them, see Partington, *A History of Chemistry*, III: 346. My own attempt to sort out what Watt meant by these observations is in Miller, *James Watt, Chemist*, 109–10 and notes.

127. Stewart, "'Ordinary' People," 115. See also Larry Stewart, "Assistants to Enlightenment: William Lewis, Alexander Chisholm, and Invisible Technicians in the Industrial Revolution," *Notes and Records of the Royal Society* 62 (2008): 17–29.

128. Russell, *James Watt: Making the World Anew*.

129. David Philip Miller, "The Usefulness of Natural Philosophy: The Royal Society of London and the Culture of Practical Utility in the Later Eighteenth Century," *The British Journal for the History of Science* 32 (1997): 192–93.

130. Miller, "Usefulness of Natural Philosophy," 194. Watt's "Thoughts upon Patents, or Exclusive Privileges for New Inventions" is reproduced in Robinson and Musson, ed., *James Watt and the Steam Revolution*, 214–28.

131. Miller, "Watt in Court."

132. Watt to William Small, 17 August 1773, in Muirhead, *Origin and Progress*, II: 58–59.

133. Matthew Boulton to Watt, 10 July 1776, quoted in Verbruggen, "The Correspondence of Jan Daniel Huichelbos van Liender," 107–08.

134. Matthew Boulton to Watt, 14 May 1780, quoted in Smiles, *Lives of Boulton and Watt*, 267.

135. J. H. Magellan to Watt, 22 February 1780 and 7 March 1780, in *For the Love of Science: The Correspondence of J. H. de Magellan*, ed. Home, Malaquias, and Thomaz, II: 1106–07, 1110–11. The work finally appeared as J. H. de Magellan, *Essai sur la Nouvelle Théorie du Feu élémentaire, et de la Chaleur des Corps* (London: W. Richardson, 1780), which includes remarks from Watt in "additions" and in a postscript.

136. Watt to J. H. Magellan, 9 March 1780, in Robinson and McKie, *Partners in Science*, 80–81. Watt's letters to Magellan and also to and from Black in this connection can be read at 76–89. See also the transcriptions of the relevant letters in Anderson and Jones, ed., *The Correspondence of Joseph Black*.

137. Watt to J. H. Magellan, 9 March 1780, in Robinson and McKie, *Partners in Science*, 81.

138. Joseph Black to Watt, 15 March 1780, in Anderson and Jones, ed., *The Correspondence of Joseph Black*, I: 414–15.

139. Watt to Joseph Black, 30 May 1780, in Anderson and Jones, ed., *The Correspondence of Joseph Black*, I: 421–22.

140. Watt to J. H. Magellan, 9 March 1780, in *For the Love of Science*, ed. Home, Malaquias, and Thomaz, II: 1113.

141. Watt to William Small, 28 May 1769, quoted in Muirhead, *Origin and Progress*, I: 63.

142. Leslie Tomory, "Fostering a New Industry in the Industrial Revolution: Boulton & Watt and Gaslight 1800–1812," *The British Journal for the History of Science* 46 (2013): 199–229. The story is also told in Tomory, *Progressive Enlightenment*, 148–51.

143. Watt to Ambrose Weston, 3 April 1809, MS3219/4/119.

144. Watt Jr. was consciously making a hero inventor out of Murdoch, attributing to him alone developments that had actually been a cooperative venture involving many in the firm as well as mill owners who had conducted trials in their factories. So this was an early example of the deliberate construction of a heroic inventor for commercial purposes, which was to become an important theme in the mid- to late nineteenth century. See MacLeod, *Heroes of Invention*; David Philip Miller, "Of Patents, Principles, and the Construction of Heroic Invention: The Case of Neilson's Hot Blast in Iron Production," *Proceedings of the American Philosophical Society* 160 (2016): 361–422. An important study of the creation of heroic inventor identities through patent disputes in the later nineteenth cen-

tury is Stathis Arapostathis and Graeme Gooday, *Patently Contestable: Electrical Technologies and Inventor Identities on Trial in Britain* (Cambridge, MA: The MIT Press, 2013).

145. James Watt Jr. to Watt, 26 February 1808, MS3219/4/33, quoted in Tomory, *Progressive Enlightenment*, 150.

146. The same purpose was served by the persistent attempts to portray Watt as a "philosopher." See Miller, "'Puffing Jamie.'"

147. See John Gascoigne, *Science in the Service of Empire: Joseph Banks, the British State and the Uses of Science in the Age of Revolutions* (Cambridge: Cambridge University Press, 1998), 121–22. For an overview of Boulton's relations with the Royal Mint, see also David Symons, "Matthew Boulton and the Royal Mint," in *Matthew Boulton: A Revolutionary Player*, ed. Malcolm Dick (Studley: Brewin Books, 2009), 170–84; Sue Tungate, "Matthew Boulton's Mints: Copper to Customer," in *Matthew Boulton: Selling What All the World Desires*, ed. Shena Mason, 80–88.

148. Watt to Annie Watt, 22 March 1792, MS3219/4/269.

149. See the account of visitors to Soho and the management of tensions between philosophical openness and commercial prudence in Jones, *Industrial Enlightenment*, 95–101.

150. Matthew Robinson Boulton to Matthew Boulton, [March 1800], quoted in J. Marc MacDonald, "Crossroads of Enlightenment 1685–1850: Exploring Education Science and Industry across the Delessert Network," (PhD diss., University of Saskatchewan, 2015), 383–90.

Chapter 5: "Team Watt"

1. On bringing such figures to the foreground, see Steven Shapin, "The Invisible Technician," *American Scientist* 77 (1989): 554–63. For this issue in Watt's circles, see Larry Stewart, "Assistants to Enlightenment: William Lewis, Alexander Chisholm and Invisible Technicians in the Industrial Revolution," *Notes and Records: The Royal Society Journal of the History of Science* 62 (2008): 17–30; Larry Stewart, "'Ordinary' People and Philosophers."

2. On Gardner's role alongside Watt, see Cleland, *Enumeration of the Inhabitants of the City of Glasgow*, 145. Gardner inherited Watt's instrument making business from circa 1773 and founded a firm, Gardner & Co., in 1799, well known throughout the nineteenth century.

3. See MacLeod, *Heroes of Invention;* Miller, "Watt in Court." An important insight into Boulton, Watt & Co. from "below stairs" through the eyes of William and Henry Creighton, engine erectors, is in Jennifer Tann, "Two Knights at Pandemonium: A Worm's-Eye View of Boulton, Watt & Co, c. 1800–1820," *History of Technology* 20 (1998): 47–72. Although this deals with the end of our period, one could reasonably assume that the working culture thus revealed was of long standing.

4. Joseph Melling, "Dark Satanic Millwrights? Forging Foremanship in the Industrial Revolution: Matthew Boulton and the Leading Hands of Boulton and Watt," in *Matthew Boulton: Enterprising Industrialist of the Enlightenment*, ed. Kenneth Quickenden et al., 163–77.

5. On Harrison see Dickinson and Jenkins, *James Watt and the Steam Engine*, 280.

6. Dickinson and Jenkins, *James Watt and the Steam Engine*, 281.

7. On espionage and the movement of skilled men, see Jones, *Industrial Enlightenment*, 149–60. The classic study is J. R. Harris, *Industrial Espionage and Technology Transfer: Britain and France in the Eighteenth Century* (Aldershot: Ashgate, 1998). For a different approach to recovering the Eu-

ropean networks of steam engineers at this time, see Lissa Roberts, "Full Steam Ahead: Entrepreneurial Engineers as Go-Betweens during the Late Eighteenth Century," in *The Brokered World: Go-Betweens and Global Intelligence, 1770–1820*, ed. Simon Schaffer et al. (Sagamore Beach: Science History Publications, 2009), 193–238.

8. Jones cites the case of the agreement between the Boulton & Watt partnership and the engine erector J. Varley in 1790 as prohibiting the disclosure of "any Secret, Art or Mystery whatsoever relative to the Business." (Jones, *Industrial Enlightenment*, 158, n. 116).

9. On the Hornblowers see Jennifer Tann, "Mr Hornblower and his Crew: Watt Engine Pirates at the End of the 18ᵗʰ Century," *Transactions of the Newcomen Society* 51 (1979–80): 95–107.

10. Melling, "Dark Satanic Millwrights?," 174.

11. Isaac Perrins to Boulton & Watt, 13 December 1794, quoted in Dickinson and Jenkins, *James Watt and the Steam Engine*, 283. This letter was docketed in the Boulton & Watt collection with the annotation "Impertinence." On Perrins's career see W. H. Chaloner, "Isaac Perrins, 1751–1801: Prize Fighter and Engineer," *History Today* 23 (October 1973): 740–43; Tony Gee, "Perrins, Isaac (1750–1801)," *Oxford Dictionary of National Biography*, http://www.oxforddnb.com/view/article /60190, accessed 15 May 2017.

12. Royal Society of London, election certificate, EC/1811/25, http://collections.royalsociety .org. On Lawson see Dickinson and Jenkins, *James Watt and the Steam Engine*, 286–88.

13. On the complexities of the production of drawings in the firm of Boulton & Watt, their function in production, in managerial control of the workforce, and the graphical cultures on which they drew, see Frances Robertson, "Ruling the Line: Learning to Draw in the First Age of Mechanical Reproduction," (PhD diss., Glasgow School of Art, 2011).

14. Watt to Matthew Boulton, 5 November 1785, quoted in H. W. Dickinson, "Some Unpublished Letters of James Watt," *Proceedings of the Institution of Mechanical Engineers* 89 (1915): 514–16.

15. Watt to Matthew Boulton, 5 November 1785, in Dickinson, "Some Unpublished Letters," 514–16.

16. John Griffiths, *The Third Man: The Life and Times of William Murdoch 1754–1839* (London: André Deutsch, 1992), 344–50. I have adopted the original spelling of Murdoch's name. He later changed it to "Murdock."

17. Dickinson and Jenkins, *James Watt and the Steam Engine*, 290.

18. We find in the case of the partnership of Boulton & Watt the prehistory of dealing with employee innovation in the nineteenth century. See Christine MacLeod, "Negotiating the Rewards of Invention: The Shop-Floor Inventor in Victorian Britain," *Business History* 41 (1999): 22–23. Melling, "Dark Satanic Millwrights?," makes the important point that careful management of employee invention at Soho did not begin with Watt Jr. and the Soho Foundry but had been a feature of the original Boulton & Watt partnership. More generally, the ownership of an employee's ideas was, of course, an age-old problem that cropped up prominently at the Royal Society in its early days and was treated there in light of the long history of craft practice. See Stephen Pumphrey, "Ideas above His Station: A Social Study of Hooke's Curatorship of Experiments," *History of Science* 29 (1991): 1–37; Steven Shapin, "Who Was Robert Hooke?" in *Robert Hooke: New Studies*, ed. Michael Hunter and Simon Schaffer (Woodbridge: The Boydell Press, 1990), 253–85.

19. Matthew Boulton to Watt, 6 August 1784, as quoted in Dickinson and Jenkins, *James Watt and the Steam Engine*, 293.

20. Matthew Boulton to Watt, 2 September 1786, transcribed in *The Selected Papers of Boulton & Watt*, ed. Jennifer Tann, 217–18.

21. Tomory, *Progressive Enlightenment*, 71–74.

22. Watt to Matthew Boulton, 3 January 1782, in Muirhead, *Origin and Progress*, II: 136. See Hills, *James Watt, Volume 3*, 19–24.

23. Described in Matthew Boulton to Watt, 13 January 1782 [misdated 1781], in *The Selected Papers of Boulton & Watt*, ed. Jennifer Tann, 50–53.

24. Dickinson and Jenkins, *James Watt and the Steam Engine*, 285–86.

25. Watt to Boulton, 1 October 1791, in Muirhead, *Origin and Progress*, II: 131.

26. Watt to Annie Watt, 25 May 1787; Watt to Annie Watt, 1 June 1787, MS3219/4/267.

27. Dickinson and Jenkins, *James Watt and the Steam Engine*, 212–13, 268–69. After 1790 the drawing office was established at Soho Manufactory.

28. Watt to John Southern, 21 April 1792, quoted in Dickinson and Jenkins, *James Watt and the Steam Engine*, 69.

29. Dickinson and Jenkins, *James Watt and the Steam Engine*, 229–30.

30. See Miller, *James Watt, Chemist*, 147–68; Miller, "The Mysterious Case of James Watt's '1785' Steam Indicator." It is important to recognize that historians such as Ben Marsden and Crosbie Smith, who have given us modern treatments of the origins of thermodynamics, and particularly the re-description of the indicator diagram as the "diagram of energy," recount these earlier appropriations of Watt but do not subscribe to them.

31. These issues are dealt with in considerable detail in Miller, *James Watt, Chemist*, 46–49.

32. Watt to Mr. Weston, 6 August 1815, MS3219/4/120.

33. Annie Watt to Watt, 26 September 1784, MS3219/4/5. Shortly thereafter when Annie was arranging to visit Scotland again, she observed: "The expence does not appear to me of any consequence as Mr B will continue to get it so engaged that if ever we will receive any benefit from the Fruits of your Labour it will be when we can not enjoy it" (Annie Watt to Watt, "June" 1785, MS3219/4/5). Annie clearly resented the impact of Boulton's business practices on Watt's income, implying that they would be in decrepit old age before they could enjoy it.

34. Annie Watt to Watt, 10 November 1786, MS3219/4/5. Pearson, a Scotsman, had been engaged in 1775 and remained at Soho as bookkeeper and cashier until 1817.

35. Annie Watt to Watt, 1 September 1786, MS3219/4/5.

36. Annie Watt to Watt, 7 November 1786, MS3219/4/5.

37. See Roll, *An Early Experiment*, 261–62, as quoted in MacLeod, "Negotiating the Rewards of Invention," 22.

38. See Melling, "Dark Satanic Millwright?," 167.

39. Tann, "Two Knights at Pandemonium," 63.

40. Though Henry Creighton's letter is undated, it is probably after 1806 since Watt received his honorary LLD from Glasgow College in that year. See chapter 7.

41. Betty Millar to Watt, [?] November 1775, MS3219/4/77.

42. Calton, previously a separate village, was being subsumed into Glasgow itself. It was an industrial area dominated by weavers but also had well-to-do residents, among them Watt's late father-in-law and Betty's father, Daniel Millar.

43. Eric Robinson, "Watt, James (1769–1848)," *Oxford Dictionary of National Biography*, http://www.oxforddnb.com/view/article/28881, accessed 26 May 2017.

44. Annie Watt to Watt, 4 October 1779, MS3219/4/4. Annie wrote this from Newcastle on her way to Glasgow.

45. Annie Watt to Watt, 17 January 1780, MS3219/4/4.

46. Annie Watt to Watt, 21 February 1785, MS3219/4/5.

47. Annie Watt to Watt, 1 September 1786, MS3219/4/6. When this letter was written, James Jr. was about seventeen years of age, Peggy nineteen, Gregory nine, and Janet (Jessy) seven.

48. Copy of letter, Watt to Peggy Watt, 28 August 1787, MS3219/4/8.

49. Watt to Peggy Watt, 28 August 1787, MS3219/4/8.

50. James Watt to Mrs. Marr, 21 December 1787, MII/11/29. Major Marr had died in October 1786.

51. Peter M. Jones, "Living the Enlightenment and the French Revolution: James Watt, Matthew Boulton, and Their Sons," *The Historical Journal* 42 (1999): 166. I rely heavily on this detailed and judicious account in what follows.

52. Annie Watt to Watt, June 1785, MS3219/4/5.

53. Annie Watt to James Watt Jr., 15 March 1786, MS3219/6/1.

54. Watt to Annie Watt, London, 26 October 1787, MS3219/4/267A.

55. Watt to Annie Watt, Birmingham, 16 November 1787, MS3219/4/267A.

56. Watt to Watt Jr., 16 September 1790, reproduced in Tann, ed., *The Selected Papers of Boulton & Watt*, 182–83.

57. Watt Jr. to Watt, 22 March 1792, MS3219/4/13. On Lavoisier's attachments to the *Ancien Régime*, see Arthur Donovan, *Antoine Lavoisier: Science, Administration, and Revolution* (Cambridge: Cambridge University Press, 1996).

58. Edmund Burke, *The Speeches of the Right Honourable Edmund Burke in the House of Commons and in Westminster Hall*, 4 vols. (London: Longman, Hurst, Rees, Orme and Brown, 1816), IV: 48. Burke communicated privately with Lord Grenville, then foreign secretary, in August 1792 about the continued danger from the "faction of English Jacobins," referring to the radical revolutionaries of France "and their Brethren the Priestleys, the Coopers and the Watts" (Edmund Burke to Lord Grenville, 18 August 1792, in *The Correspondence of Edmund Burke, Volume VII [January 1792–August 1794]*, ed. P. J. Marshall and John A. Woods [Cambridge: Cambridge University Press, 1968], 177). More generally, see Maurice Crosland, "The Image of Science as a Threat: Burke versus Priestley and the 'Philosophic Revolution,'" *The British Journal for the History of Science* 20 (1987): 277–307.

59. Annie Watt to Watt, 9 May 1792, MS3219/4/6. The reply was Thomas Cooper, *A Reply to Mr. Burke's Invective against Mr. Cooper and Mr. Watt, in the House of Commons on the 30th of April 1792* (London: J. Johnson, 1792). Burke might have welcomed, and aided, the confusion between father and son because of his earlier sympathy for rival engine interests.

60. Jones, "Living the Enlightenment," 175. Those biographies were financed by Watt Jr.

61. Robert Southey to Archibald Alison, 17 April 1833, in *The Life and Correspondence of Robert*

Southey, ed. C. C. Southey, 6 vols. (London: Longman, Brown, Green and Longmans, 1850), VI: 209. Southey was in revolutionary France at the same time as Watt Jr., who had told him the story. He remained a friend of the Watt sons.

62. This was recounted in Muirhead, *Origin and Progress*, I: cclxii–cclxiii.

63. Seymour Cohen, "Two Refugee Chemists in the United States, 1794: How We See Them," *Proceedings of the American Philosophical Society* 126 (1982): 305–07.

64. Watt to Watt Jr., 9 October 1793, MIV/15.

65. Watt Jr. to T. & R. Walker, 5 November 1793, MS3219/6/6.

66. Eric Robinson, "An English Jacobin," 355.

67. James Watt Jr. to Thomas Wilson, 22 February 1794, Wilson Correspondence, vol. 7, Cornwall Record Office, AD1583/7/13.

68. James Watt Jr. to Jan van Liender, 5 April 1794, MS3219/6/7.

69. Watt Jr. to Thomas Cooper, 7 April 1794, MS3219/6/7.

70. James Watt Jr to François Arago, 29 September 1834, MS3219/6/106

71. *The Selected Papers of Boulton & Watt*, ed. Jennifer Tann, 107.

72. See Miller, "Usefulness of Natural Philosophy," 192–95; Eric Robinson, "James Watt and the Law of Patents," *Technology and Culture* 13 (1972): 115–39.

73. My account is slanted against the Hornblowers in that I am telling this story from Watt's perspective. For an important corrective historiography, see Hugh Torrens, "Jonathan Hornblower (1753–1815) and the Steam Engine: A Historiographic Analysis," in *Perceptions of Great Engineers: Fact and Fantasy*, ed. Denis Smith (London/Liverpool: Science Museum/University of Liverpool, 1994), 23–34.

74. Jonathan Hornblower, *An Address to the Lords, Adventurers and Others Concern'd in the Mines of Cornwall* (Penryn: n.p., 1788), 4, 5–6. Hornblower did *not* assert that the separate condenser was fifty years old (as A. C. Todd, *Beyond the Blaze: A Biography of Davies Gilbert* [Truro: Bradford Barton, 1967], 59, has it). Rather, his claim was that his engine did not use a "separate condenser" since the steam was condensed in the second *cylinder*.

75. Watt to Thomas Wilson, 28 January 1790, Wilson Correspondence, vol. 4, Cornwall Record Office, AD1583/4/4.

76. Thomas Wilson, *A Comparative Statement of the Effects of Messrs. Boulton and Watt's Steam Engines with Newcommen's and Mr. Hornblower's* (Truro: W. Harry, 1792), 11. See also Hills, *James Watt, Volume 3*, 175–179; John Farey, *A Treatise on the Steam Engine: Historical , Practical and Descriptive*, (London: Longman, Rees, Orme, Brown and Green, 1827), II: 384–93. A useful perspective on the contests comes from Davies Gilbert, who was a friend to the Hornblower cause but not an uncritical one. See Todd, *Beyond the Blaze*, 57–78.

77. Modern technology studies suggest that such arguments are objectively unresolvable, requiring to be closed by social and political means. See Miller, "Watt in Court," esp. 69–70, drawing particularly on Harry Collins, *Changing Order: Replication and Induction in Scientific Practice* (Chicago: The University of Chicago Press, 1992).

78. *Observations on the Part of Messrs Boulton and Watt Concerning Mr Hornblower's Steam-Engine Bill, 17 April 1792*, MS3219/4/227.

79. Watt to J. A. De Luc, 14 March 1792, MS3219/4/124.

80. John Rennie had, as we saw earlier, been a key figure in the Albion Mill project.

81. Watt to Gilbert Hamilton, 30 May 1792, MS3219/4/124.

82. Annie Watt to Watt, 6 April 1792, MS3219/4/6.

83. Annie Watt to Watt, 6 April 1792, MS3219/4/6.

84. Annie Watt to Watt, 3 May 1792, MS3219/4/6.

85. Watt to Thomas Wilson, 18 July 1792, Wilson Correspondence, vol. 5, Cornwall Record Office, AD1583/5/32.

86. Watt to Thomas Wilson, 26 November 1788, Wilson Correspondence, vol. 3, Cornwall Record Office, AD1583/3/57.

87. Watt to Thomas Wilson, 22 September 1789, Wilson Correspondence, vol. 3, Cornwall Record Office, AD1583/3/95.

88. With the Bull trial postponed from November 1792 to the following year, Boulton entertained an alternative approach, once again directed at the Hornblowers. This was sparked by an anonymous article signed "A Disinterested Cornish Miner," which appeared in the *Sherborne Mercury* and which incensed both Boulton and Watt not only by its invidious comparison of their engines with Hornblower's but also by exaggerating their profits drawn from Cornwall. Boulton sent to Thomas Wilson an answer to that paper, which, when subsequently rewritten by Watt, appeared anonymously as *An Address to the Mining Interest of Cornwall on the Subject of Messrs Boulton and Watt's and Mr Hornblower's Engines* (Truro, n.p.: 1793). But Boulton had a further idea for a direct challenge to the Hornblowers by building a new engine at Tin Croft mine at his and Watt's expense (or, if Watt did not like the gamble, at his own expense) that would be demonstrably superior. This was not proceeded with. See Matthew Boulton to Thomas Wilson, 28 November 1792, Wilson Correspondence, vol. 5, Cornwall Record Office, AD1583/5/69.

89. The clouds are courtesy of the meteorological journal kept for this period at the Royal Society's apartments and appended to its *Philosophical Transactions* 84 (1794). Green Lettice Lane, now Laurence Pountney Hill in the city, was about two and a half miles from Westminster.

90. The witnesses were listed thus in Watt Jr. to Arago, 13 October 1834, MS3219/6/106. On Mitchell and the coaching of him and Murdoch as witnesses by Thomas Wilson, see A. and R. Weston to Thomas Wilson, 10 June 1793, Wilson Correspondence, vol. 6, Cornwall Record Office, AD1583/6/32.

91. Ambrose Weston to Thomas Wilson, 29 June 1795, Wilson Correspondence, vol. 8, Cornwall Record Office, AD1583/8/37.

92. Watt to Thomas Wilson, 23 July 1795, Wilson Correspondence, vol. 8, Cornwall Record Office, AD1583/8/65.

93. Watt to Joseph Black, 24 May 1795, in Anderson and Jones, *Correspondence of Joseph Black*, vol. 2, 1263–64.

94. Watt to Joseph Black, 7 January 1796, in Anderson and Jones, *Correspondence of Joseph Black*, vol. 2, 1278.

95. Maberley was a shadowy figure who has been variously named as John, David, or Stephen. John Maberley was a London currier who had purchased the steam engine patent of Isaac Mainwaring and employed Jabez Hornblower to improve it. They became business partners in engines. See Susan W. Howard, "Jabez Carter Hornblower, Engineer and Inventor (1744–1814)," http://www.penwood

.famroots.org/jabez_carter_hornblower.htm; A. P. Woolrich, "Hornblower and Maberley Engines in London, 1805," *Transactions of the Newcomen Society* 56 (1984–85): 159–68.

96. This document is reproduced in Robinson and McKie, eds., *Partners in Science*, 253–56.

97. Robison's prepared evidence ("Memorial by Dr. John Robison relative to his first acquaintance with Mr. Watt and the improvements of the latter in the Steam Engine") is reproduced in Robinson and McKie, eds., *Partners in Science*, 256–60.

98. John Robison to Watt, 3 February 1797, in Robinson and McKie, eds., *Partners in Science*, 266. Mrs. Sarah Siddons (1755–1831) was the undisputed queen of Drury Lane for whom theater audiences went wild in the 1780s and 1790s.

99. Watt to Matthew Boulton, 25 January 1799, in Muirhead, *Origin and Progress*, II: 260.

100. See Annie Watt to Watt, 1 September and 3 September 1796, MS3219/4/7. Annie's father had been very ill since August.

101. Watt to Agnes Marr, 12 June 1796, MII/11/37.

102. Matthew Boulton to Thomas Wilson, 13 January 1795, Wilson Correspondence, vol. 8, Cornwall Record Office, AD1583/8/4.

103. See, for example, "Considerations upon the measure most proper to be adopted in the present state of affairs with Maberley," in Watt Jr.'s hand, reproduced in Tann, ed., *The Selected Papers of Boulton & Watt*, 136–41.

104. Gregory also played his part. In September 1796 he was in Glasgow and visited John Gardner, Watt's assistant or apprentice during the early steam and engine experiments, with a view to recruiting him and possibly another early employee, Mr. McMillan, to testify. See Gregory Watt to Watt, "Sepr" 1796, MS3219/4/8.

105. James Watt Jr. to Thomas Wilson, 31 January 1795, Wilson Correspondence, vol. 8, Cornwall Record Office, AD1583/8/9.

106. See James Watt Jr. to Watt, 11 November 1797, MS3219/4/14.

107. Hills, *James Watt, Volume 3*, 215.

Chapter 6: The Fruits of Success

1. Betty Millar to Watt, 22 February 1775, MS3219/4/77. Black's observation is reported in Annie Watt to James Watt, 8 October 1779, MS3219/4/4. Annie was in Glasgow visiting family, and Watt was in Cornwall on engine business at the time.

2. Annie Watt to Watt, 21 November 1779 and 1 November 1779, MS3219/4/4.

3. Dickinson and Jenkins, *James Watt and the Steam Engine*, 19.

4. Watt to James Watt Jr., 13 April 1791, MS3219/4/124.

5. Annie Watt to Watt, 6 April 1792, MS3219/4/6.

6. James Watt Jr. to Stephen Delessert, 6 September 1794, MS3219/6/7.

7. Annie Watt to Watt, 30 August 1796, MS3219/4/7.

8. Will of Doctor James Watt, Doctor of Laws, Handsworth, Staffordshire, Proved 13 October 1819, The National Archives, Kew, PROB 11/1621/182.

9. If this sum was provided by secure government stocks earning 3 percent per year, it would require a capital investment of about £48,000.

10. Annie's will (of 1832) tells us that before she died she had made various gifts of money and

property (the nature and extent of which is unspecified) to twelve of the children of her late nephews and nieces Robert Hamilton, Archibald Hamilton, and Ann Wallace. The will itself left £3,000 to be invested in the interest of her sister Janet McGrigor for her lifetime. At Janet's death that sum was to be divided six ways between various nephews and nieces. There were also a number of small legacies (see Probate Copy of Will of Ann Watt, 1 November 1832, MS3219/6/123). James Watt Jr.'s will (1848) left the estates at Heathfield and Doldowlod, and various other property including coal estates, as well as many tens of thousands of pounds, in trust to provide income for a number of nephews and nieces. The main beneficiary was James Gibson, son of Watt Jr.'s niece Agnes Gibson (she being the granddaughter of Watt, daughter of Margaret Miller, and wife of James Gibson), who succeeded to the estates on attaining his majority, duly changing his name to include "Watt" as the will required. He became James Gibson Watt.

11. Dickinson, *James Watt, Craftsman and Engineer*, 199.

12. Jennifer Tann, "Riches from Copper: The Adoption of the Boulton & Watt Engine by Cornish Mine Adventurers," *Transactions of the Newcomen Society* 67 (1995): 34, table 3.

13. This compares with an estimate of total premiums payable from both types of engine between 1775 and 1799 of £471,265. This estimate is made in Jennifer Tann, "Fixed Capital Formation in Steam Power, 1775–1825: A Case Study of the Boulton and Watt Engine," in *Studies in Capital Formation in the United Kingdom 1750–1920*, ed. Charles H. Feinstein and Sidney Pollard (Oxford: Clarendon Press, 1988), 179, table 7.10.

14. Thus figures for premium income from 1788 were £5,247 from fifteen Cornish engines, which is approximately £350 per engine; £2,188 from twenty-seven non-Cornish reciprocating engines, which is an average of £81 per engine; and £1,792 from thirty-seven rotative engines, an average of £48 per engine. See Hills, *James Watt, Volume 3*, 72.

15. There was little precedent for what this markup should be, but Boulton & Watt appears to have converged on the figure of 50 percent, before dropping it to 33 percent or even 25 percent at the end of the century. See Robert Williams, "Management Accounting Practice and Price Calculation at Boulton and Watt's Soho Foundry: A Late 18th Century Example," *The Accounting Historians Journal* 26 (1999): 65–88.

16. Cule, "Finance and Industry in the Eighteenth Century."

17. The matter is complicated because on rotative engines the split between Boulton and Watt was 50/50, not 66.6:33.3 as was the arrangement regarding the reciprocating engines.

18. See H. W. Dickinson, *Matthew Boulton* (Cambridge: Cambridge University Press/Babcock and Wilcox, 1936), 175–76.

19. See note 14. The figures there allow us to calculate average premium income from various types of engines, specifically: on average, Cornish reciprocating engines earned £350 per year; non-Cornish reciprocating engines £81 per year; rotative engines £48 per year. If we make some assumptions about the average paying life of engines between 1775 and 1800, then we can calculate what they would have earned: 49 reciprocating engines in Cornwall at £350 per year for, say, 10 years yields £171,500; 134 reciprocating engines outside Cornwall at £81 per year for, say, 10 years yields £108,540; 266 rotative engines at £48 per year for, say, 7 years yields £89,376.

20. Watt to Joseph Banks, 15 November 1803, MS3219/4/119.

21. See Martin Daunton, *Trusting Leviathan: The Politics of Taxation in Britain, 1799–1914* (Cambridge: Cambridge University Press, 2001), 44–45.

22. Quoted in Hills, *James Watt, Volume 3*, 216. Note that his wife's income was being treated as his own for the purpose of tax liability, as was that of Gregory, then in his early twenties.

23. This is an enclosure in Watt to Mr. Woodward, 21 November 1803, MS3219/4/119.

24. This declaration is in Watt to Gilbert Hamilton, 26 November 1803, MS3219/4/119.

25. Watt to Gilbert Hamilton, 15 February 1804, MS3219/4/119.

26. See Watt to Ambrose Weston, 24 April 1808, MS3219/4/119, and the following correspondence to Weston and Mr. Woodward.

27. Watt subsequently wrote a detailed document describing the progress of the negotiations with Birch for the land and cottages adjacent to Heathfield House (See MS3219/4/229). See also Annie Watt to James Watt, 7 March 1787, MS3219/4/5.

28. James Watt to Annie Watt, 16 October 1787, MS3219/4/267A; Annie Watt to James Watt, 23 October 1787, MS3219/4/5.

29. Watt to Samuel Wyatt, 19 May 1789, MS3219/4/124.

30. Charles Pye, *A Description of Modern Birmingham; Whereunto Are Annexed, Observations Made during an Excursion around the Town, in the Summer of 1818* (Birmingham: John Lowe, 1820), 113–14.

31. A ground-floor plan of Heathfield showing the "Elaboratory" can be seen at https://www.search.birminghamimages.org.uk/Details.aspx?&ResourceID=1025&PageIndex=6&SearchType=2&ThemeID=559. I have been unable to locate the original of this plan. It was reproduced in G. W. Beard, "Works by Samuel Wyatt" in *Country Life* magazine in 1930.

32. Watt to Mrs. C. Matthews, 15 July 1795; Watt to Bedford (his lawyer), 24 August 1795, quoted in W. K. V. Gale, "Soho Foundry: Some Facts and Fallacies," *Transactions of the Newcomen Society* 34 (1961–62): 76–77. Watt was in fact repaid a few years later.

33. On Boulton's mint ventures see Sue Tungate, "Matthew Boulton's Mints: Copper to Customer," in *Matthew Boulton: Selling What All the World Desires*, ed. Shena Mason, 80–88. More broadly see Richard G. Doty, *Soho Mint and the Industrialization of Money* (London: Spink & Son Ltd., 1998).

34. Larry Stewart, "James Watt's Paine: Mob Rules, Democrats and Demons," in *James Watt (1736–1819): Culture, Innovation and Enlightenment*, ed. Malcolm Dick and Caroline Archer (Liverpool: Liverpool University Press, forthcoming 2019).

35. Watt to Mrs. Charlotte Matthews, 5 March 1797, MS3219/4/118. The Bank of England had suspended specie payments on 25 February 1797.

36. Watt to Ambrose Weston, 12 March 1797, MS3219/4/118.

37. The following account of the acquisition of Watt's Welsh estates draws on: J. B. Sinclair and R. W. D. Fenn, "Mr Watt Buys Some Farms, Part 1," *Transactions of the Radnorshire Society* 64 (1994): 79–98; J. B. Sinclair and R. W. D. Fenn, "Mr Watt Buys Some Farms, Part II," *Transactions of the Radnorshire Society* 65 (1995): 85–108; J. B. Sinclair and R. W. D. Fenn, "James Watt, Father & Son," *Transactions of the Radnorshire Society* 62 (1992): 51–65.

38. Francis Garbett (1743–1800) was the son of Samuel Garbett (1716–1803), a minor member

of the Lunar Society group in Birmingham and erstwhile partner of John Roebuck in various ventures. Francis had been high sheriff of Radnorshire in 1790.

39. Sinclair and Fenn, "Mr Watt Buys Some Farms, Part II," 104.

40. See Virginia Glenn, "George Bullock, Richard Bridgens and James Watt's Regency Furnishing Schemes," *Furniture History* 15 (1979): 54–67.

41. Annie Watt to Watt, 8 September 1787, MS3219/4/5. The "weeping philosopher" refers to Heraclitus of Ephesus, the pre-Socratic philosopher known for his melancholy disposition and his philosophy that saw strife and change as the natural state of things. We must conclude that Annie saw her husband, with considerable justice, as the "weeping philosopher."

42. Peter Borsay, "From Port to Resort: Tenby and Narratives of Transition, 1760–1914," in *Resorts and Ports: European Seaside Towns since 1700*, ed. Peter Borsay and John K. Walton (Bristol: Channel View Publications, 2011), 88–91.

43. *The Cambrian Directory, or Cursory Sketches of the Welsh Territories* (Salisbury, 1800) and *the Cambrian Tourist: Or, Post-Chaise Companion through Wales*, 5th ed. (London, 1814). Comparison of the accounts of Tenby in these works marks its transformation during that period.

44. The trip was described in Watt to Thomas Beddoes, 20 August 1797, MS3219/4/118.

45. Watt to Robert Muirhead, 16 December 1808, MS3219/4/119.

46. Watt to Mr. Hamilton, 23 October 1808, MS3219/4/119.

47. Watt to Robison, 26 April 1803, in Robinson and McKie, *Partners in Science*, 378. On Lunar group networks with France at this time, see MacDonald, "Crossroads of Enlightenment 1685–1850," 383–90.

48. Henry Cockburn, *Life of Lord Jeffrey: With a Selection from His Correspondence*, 2 vols. (Edinburgh: Adam & Charles Black, 1852), II: 20–21.

49. Watt to Gilbert Hamilton, 26 May 1793, MS3219/4/124.

50. Watt to Gilbert Hamilton, 26 May 1793, MS3219/4/124.

51. Ince, "The Soho Engine Works, 1796–1895," 11.

52. Hills, *James Watt, Volume 3*, 124–25.

53. On Davy see Jan Golinski, *The Experimental Self: Humphry Davy and the Making of a Man of Science* (Chicago: University of Chicago Press, 2016); David Knight, *Humphry Davy: Science and Power* (Cambridge: Cambridge University Press, 1998).

54. See Jay, *The Atmosphere of Heaven*, 154–55.

55. Gregory Watt to Watt Jr., 1 June 1804, MS3219/7/24.

56. Watt to Hon. Charles Francis Greville, 1 November 1804, MS3219/4/119.

57. Watt to Gilbert Hamilton, 1 November 1804, MS3219/4/119.

58. Watt to Robert Muirhead, [?] December 1804, MS3219/4/119.

59. Watt to Robert Muirhead, [?] December 1804, MS3219/4/119.

60. Watt to Gilbert Hamilton, 4 January 1805, MS3219/4/119.

61. A list of recipients of rings and lockets in Glasgow is in Watt to [?], 13 December 1804, MS3219/4/119.

62. Watt to Watt Jr., 13 December 1804, MS3219/4/119.

63. Gregory Watt, "Observations on Basalt, and on the Transition from the Vitreous to the Stony Texture, Which Occurs in the Gradual Refrigeration of Melted Basalt; With Some Geological Re-

marks," *Philosophical Transactions of the Royal Society of London* 94 (1804): 279–314. On Gregory's "geological" work see Hugh S. Torrens, "The Geological Work of Gregory Watt, His Travels with William Maclure in Italy (1801–1802), and Watt's 'proto-geological' map of Italy," in *The Origins of Geology in Italy*, ed. Gian Battista Vai and W. Glen E. Caldwell (Boulder, CO: Geological Society of America, 2006), 179–97.

64. Watt to C. F. Greville, 19 November 1804, MS3219/4/119. The things in print but not under Gregory's name were probably the reviews that he undertook for the *Edinburgh Review*. See Torrens, "The Geological Work of Gregory Watt," 188.

65. See Lynn Hunt and Margaret Jacob, "The Affective Revolution in 1790s Britain," *Eighteenth-Century Studies* 34 (2001): 491–521. The authors see a transgressive "affective revolution" as conjoined to religious dissent and political radicalism among a number of Lunar Society figures and especially the younger generation. Their intimation that Watt would have participated in similar sentiments had he not been diverted by his opposition to the French Revolution and reform is not, I think, plausible. Watt almost always kept emotional matters under tight control even among close friends. This was in line with his cultural Presbyterianism, which melded into his personality. On William Creighton as employee see Tann, "Two Knights at Pandemonium."

66. On the trunk's contents, which were mainly childhood memorabilia, see Russell, *James Watt: Making the World Anew*, 178–79.

67. Watt to Robert Muirhead, 9 March 1815, MS3219/4/120.

68. See Watt to Dr. Christie (Cheltenham), 18 January 1815, MS3219/4/120; Watt to Joseph Banks, 1 March 1815, MS3219/4/120; Banks to Watt, 31 March 1815, advised the Cape (in Muirhead, *Origin and Progress*, II: 363–65.

69. Watt to John Robison Jr., 5 February 1805, MS3219/4/119.

70. See Malcolm Dick, "The Death of Matthew Boulton 1809: Ceremony, Controversy and Commemoration," in *Matthew Boulton: Enterprising Industrialist of the Enlightenment*, ed. Kenneth Quickenden et al., 247–66.

71. Watt Jr. to Watt, 25 August 1809, MS3219/4/33.

72. The memoir as Watt sent it to Robinson Boulton is to be found as "Biographical sketch from James Watt," 17 September 1809, MS3782/13/37. Dickinson, *Matthew Boulton*, 203–08.

73. See Miller, "Scales of Justice."

74. In what follows I rely on the excellent account of this long-neglected episode in Mason, *The Hardware Man's Daughter*, 142–48.

75. Annie Watt to Watt Jr., 13 June 1810, MS3219/6/1/355, quoted in Mason, *Hardware Man's Daughter*, 142.

76. Watt to Watt Jr., 25 June 1810, MS3219/6/1/357. Watt Jr.'s annotation of this letter from his father ("Miss B's difficulty with her brother. Thinks I should write to him") removes the thin cloak of anonymity thrown over the matter by Watt.

77. Watt Jr. to Robinson Boulton, 1 July 1810, MS3782/13/40/24.

78. Watt Jr. to Watt, 27 August 1810, MS3219/4/34/16.

79. Annie Watt to Watt Jr., 14 February 1811, MS3219/6/1/378.

80. See H. W. Dickinson, *The Garret Workshop of James Watt*, Science Museum Technical Pamphlet No. 1 (London: HMSO, 1929); Russell, *James Watt, Making the World Anew*, 205–10.

81. Sangamesh R. Deepak and G. K. Ananthasuresh, "James Watt and His Linkages," *Resonance* (June 2009): 530–43; Jane Insley, "James Watt and the Reproduction of Sculpture," *Sculpture Journal* 22 (2013): 37–65.

82. See Barbara Fogarty, "The Mechanical Paintings of Matthew Boulton and Francis Eginton," in *Matthew Boulton: Enterprising Industrialist of the Enlightenment*, ed. Kenneth Quickenden et al., 111–26. On copying machines see chapter 3.

83. Insley, "Watt and the Reproduction of Sculpture," 43; Hills, *James Watt, Volume 3*, 234.

84. Watt to Ambrose Weston, 3 April 1809, MS3219/4/119.

85. Watt to Robert Hamilton, 8 August 1814; Watt to Francis Chantrey, 17 November 1814, MS3219/4/120.

86. Watt to J. C. De Boffe, 5 February 1810, MS3219/4/120.

87. Watt to Sir Joseph Banks, 1 March 1815, in Robinson and McKie, *Partners in Science*, 420. Biot was probably working on his *Traité de physique expérimentale et mathématique*, published in 1816, which does briefly mention Watt's results for the latent heat of steam.

88. Watt to Pierre Lévêque, 14 March 1810, extracted in Muirhead, *Origin and Progress*, II: 313–314. He communicated much the same to Berthollet (Watt to Count Berthollet, 26 December 1810, MS3219/4/120).

89. Annie Watt to Watt, 16 October 1796, MS3219/4/7.

90. Ambrose Weston to Watt, 29 May 1798, MS3219/4/40.

91. Henry Brougham, *Lives of Men of Letters and Science, Who Flourished in the Time of George III* (London: Charles Knight, 1845), 383; Walter Scott, *The Monastery: A Romance* (Edinburgh: Longman, Hurst, Rees, Orme and Brown, 1820), 63. See Dickinson and Jenkins, *James Watt and the Steam Engine*, 74–75.

Chapter 7: The Living Legend

1. Annie Watt to Watt, 10 November 1792, MS3219/4/6.

2. Annie Watt to Watt, 15 November 1792, MS3219/4/6.

3. I do not disagree with Christine MacLeod's view that Watt "received little public acclaim during his lifetime" in terms of public marks of distinction. See MacLeod, *Heroes of Invention*, 84. This is certainly true when compared with such marks in the period after his death. But I do contend that in many other respects Watt had the mark of legendary status before his demise.

4. The process continues; see Kathleen Bell, "A Place for Ingenuity: Re-Imagining the Life of James Watt in a Sequence of Poems," *Life Writing* 14 (2017): 267–75.

5. Erasmus Darwin, *The Botanic Garden: A Poem in Two Parts. Part I: Containing the Economy of Vegetation. Part II: The Loves of the Plants. With Philosophical Notes* (London: J. Johnson, 1791), part 1, canto 1, lines 253–62.

6. Watt to Erasmus Darwin, 31 January 1790, MS3219/4/287. "Seraffo" is the rendition in Italian of the name of an old Persian coin, representing in Watt's meaning here, perhaps, "ancient treasure."

7. Erasmus Darwin to Watt, 19 January 1790, in *The Collected Letters of Erasmus Darwin*, ed. Desmond King-Hele, 358–59.

8. William Hutton, *A Brief History of Birmingham*, 2nd ed. (Birmingham: Grafton & Reddell,

1802), xxiii. The barrister was J. Morfitt, Esq., and his poem was presented as a prospect of Birming-ham seen initially from afar without the sounds of its "whirling lathes" and "faint hammers." But having quickly focused on Soho House, the poem celebrates Boulton, Watt, and Francis Eginton. Watt is helpfully identified in a footnote as "Mr. James Watt, the celebrated mechanic and chemist, grown grey in works of genius and utility," which in itself suggests that at this point his general celebrity had its limits.

9. [John Robison], "Steam-Engine," in *Encyclopaedia Britannica*, 3rd ed., 18 vols. (Edinburgh: A. Bell and C. Macfarquhar, 1797), 36, XVII, 769–70.

10. [Robison], "Steam-Engine," 772.

11. Watt to John Robison, 24 October 1796, in Robinson and McKie, *Partners in Science*, 235.

12. John Robison, "Steam Engine," in *Supplement to the Third Edition of the Encyclopaedia Britannica* (Edinburgh: Bonar, 1801), II: 522–24.

13. See W. A. Smeaton, "Some Comments on James Watt's Published Account of His Work on Steam and Steam Engines," *Notes & Records of the Royal Society of London* 26 (1971): 35–42. John Playfair, who wrote a biography of Robison, had begun the edition of Robison's works but got no further before his death. Brewster took over the task. The work finally appeared as John Robison, *A System of Mechanical Philosophy*, 4 vols. (Edinburgh: John Murray, 1822).

14. Watt to Sir Joseph Banks, 1 March 1815, MS3219/4/120, in Robinson and McKie, *Partners in Science*, 419–21.

15. Robison, *A System of Mechanical Philosophy*, II: 149n.

16. Torrens, "Jonathan Hornblower (1753–1815) and the Steam Engine," 27. A similar strategy had been adopted earlier in an article by Playfair in the *Edinburgh Review*, with Watt Jr. behind the scenes. See below.

17. Anon., "James Watt," in *Public Characters of 1802–1803* (London: Richard Phillips, 1803), 501–18. This work, published between 1798 and 1806, was mainly edited by the prolific writer, stationer, and publisher Richard Phillips. It leaned very much, like Phillips himself, toward sympa-thy with radical dissenting views. See MacLeod, *Heroes of Invention*, 72. On Phillips, see Thomas Secombe, "Phillips, Sir Richard (1767–1840)," rev. M. Clare Loughlin-Chow, *Oxford Dictionary of National Biography*, http://www.oxforddnb.com/view/article/22167.

18. *Public Characters of 1802–1803*, 501–02.

19. *Public Characters of 1802–1803*, 515–17.

20. See the "Introductory Epistle from Captain Clutterbuck," and "Answer by the Author of 'Wa-verley,'" in Walter Scott, *The Monastery: A Romance* (Edinburgh: Longman, Hurst, Rees, Orme and Brown, 1820), 1–60, at 48–51; Christiana C. Hankin, ed., *Life of Mary Anne Schimmelpenninck*, 2 vols. (London: Longman, Green, 1858), 287, gives a recollection from as early as the 1780s of Watt regularly spending his evenings in "some light amusing reading." Schimmelpenninck (1778–1856) was the daughter of Samuel Galton of Birmingham and spent time with the Watt family when they lived in Regent's Place, Harper's Hill.

21. Amid all the books on philosophy, engineering, history, and the like that came down through the Watt family, we find, for example (all inscribed "Heathfield" and acquired during Watt's lifetime): the works of the Scottish and English poets; Byron's *Childe Harold's Pilgrimage*, 8th ed. (1814), and his *Lara, a Tale. Jacqueline, a Tale* (1814); James Hogg, *The Queen's Wake* (1813); Walter Scott, *Har-*

old the Dauntless (1817); Humphrey Hedgehog [John Agg], *The General-Post Bag; or, News! Foreign and Domestic. To Which Is Added La Bagatelle* (1814); J. F. Davis, translator, *Laou-Seng-Urh, or an Heir in His Old Age: A Chinese Drama* (1817); [John Roby], *Jokeby, a Burlesque on Rokeby: A Poem* (1817). See Sotheby's, *The James Watt Sale*, 112–14; Smiles, *The Lives of Boulton and Watt*, 504.

22. Maria Edgeworth and Richard Lovell Edgeworth, *Practical Education*, 2 vols. (London: J. Johnson, 1798). On Beddoes's program of educational reform, see Hugh Torrens and Joseph Wachelder, "Models, Toys, and Beddoes' Struggle for Educational Reform, 1790–1800," in Levere, Stewart, and Torrens, with Wachelder, *The Enlightenment of Thomas Beddoes*, 206–37.

23. Maria Edgeworth, *Harry and Lucy Concluded, Being the Last Part of Early Lessons* (London: R. Hunter, 1825).

24. Edgeworth and Edgeworth, *Practical Education*, 592–94.

25. Christine MacLeod shows that Maria Edgeworth's treatment of Watt is perhaps assimilable more to a democratic than to a heroic tradition. See MacLeod, *Heroes of Invention*, 171–73.

26. *Life of Mary Anne Schimmelpenninck*, 286–87.

27. Maria Edgeworth to Mrs. Edgeworth, 4 March 1819, in *The Life and Letters of Maria Edgeworth*, 2 vols., ed. Augustus J. Hare (New York: Houghton, Mifflin and Company, 1895), I: 276–77.

28. See Nicholas Philipson, "Dalrymple, Sir John, of Cousland, fourth baronet (1726–1810)," *Oxford Dictionary of National Biography*, http://www.oxforddnb.com/view/article/7055. Dalrymple had corresponded with Matthew Boulton many years before and was clearly known to Watt.

29. Watt to Sir John Dalrymple, 2 March and 14 March 1797, MS3219/4/119.

30. Watt to Sir John Dalrymple, 1 June 1797, MS3219/4/119. As early as 1769 Watt had been introduced to waterproofing by a Swiss dyer, Chaillet. See Clow and Clow, *The Chemical Revolution*, 216.

31. Watt to Sir John Dalrymple, 1 June 1797, MS3219/4/119.

32. Joseph Huddart (the younger), *Memoir of the Late Captain Joseph Huddart F.R.S. &c* (London: W. Phillips, 1821), 100.

33. Huddart, *Memoir of the Late Captain Joseph Huddart*, vi.

34. Discussed in Bottomley, *The British Patent System during the Industrial Revolution 1700–1852*, 152–53.

35. Watt to Ambrose Weston, 15 February 1804, MS3219/4/119.

36. Watt to Ambrose Weston, 17 February 1804, MS3219/4/119. That Watt did attend is reported in Watt to Archibald Hamilton, 1 March 1804, MS3219/4/119.

37. See Hills, *James Watt, Volume 3*, 232–33. Ambrose Weston to Watt, Dec 1806–Jan 1807, MS3219/4/48, 91–99. The specification is contained in *Repertory of Arts, Manufactures, and Agriculture* 12 (1808): 174–78.

38. Speer enrolled a patent "for an improvement in the construction of hydrometers" on 2 August 1802 and was author of *An Inquiry into the Causes of the Errors and Irregularities Which Take Place in Ascertaining the Strengths of Spirituous Liquors by the Hydrometer* (London: Payne and MacKenlay, 1802). He was also a member of the Royal Irish Academy. See William J. Ashworth, *Customs and Excise: Trade, Production, and Consumption in England, 1640–1845* (Oxford: Oxford University Press, 2003), 273–74; Peter Mathias, *The Brewing Industry in England 1700–1830* (Cambridge: Cambridge University Press, 1959), 69.

39. William Speer to Watt, 19 February 1810, MS3219/4/50.

40. Watt to Gilbert Hamilton, 8 April and 15 April 1806, MS3219/4/119.

41. James D. Marwick, *Glasgow, the Water Supply of the City* (Glasgow: R. Anderson, 1901), 79–80.

42. John Robison Jr., "Account of the Flexible Water Main, Contrived by the Late Mr Watt for the Glasgow Water-Work Company," *The Edinburgh Philosophical Journal* 3 (1820): 60–62.

43. This tureen remained in the family and is pictured in Sotheby's *James Watt Sale*, 142–43. On Watt as Glasgow benefactor, see MacLeod, *Heroes of Invention*, 113.

44. Stephen Cowley, *Rational Piety and Social Reform in Glasgow: The Life, Philosophy, and, Political Economy of James Mylne (1757–1839)* (Eugene, OR: Wipf & Stock, 2015).

45. In this account I draw in part on Hills, *James Watt, Volume 3*, 233–34, and also on Samuel Smiles, *Lives of the Engineers: Harbours–Lighthouses–Bridges. Smeaton and Rennie*, rev. ed. (London: John Murray, 1874), 329–42.

46. Joseph Banks to Watt, 29 June 1812, MS3219/4/51.

47. The universalist theme in Gavin de Beer, *The Sciences Were Never at War* (London: Thomas Nelson & Sons, 1960), was gently queried by A. Hunter Dupree, "Nationalism in Science—Sir Joseph Banks and the Wars with France," in *A Festschrift for Frederick B. Artz*, ed. D. H. Pinkney and T. Ropp (Durham, NC: Duke University Press, 1964), 37–51. An excellent recent treatment is Elise Lipkowitz, "'The Sciences are Never at War?' The Scientific Republic of Letters in the Era of the French Revolution, 1789–1815" (PhD diss., Northwestern University, 2009).

48. Jones, *Industrial Enlightenment*, 210–19.

49. Lipkowitz, "'The Sciences Are Never at War?'" 61–73, charts the collapse of the networks of a number of major intelligencers, including Banks, Blagden, Cuvier, and Laplace. Watt's own international correspondence noticeably slims during this period due to the combined effects of this general breakdown and his declining interest and health. My interpretation of this affair owes a debt to the treatment in Jones, *Industrial Enlightenment*, 224–25.

50. The documents are in the James Watt Papers at MS3219/4/284.

51. News of the arrival of Delambre's letter was conveyed in Watt Jr. to Watt, 27 July 1808, addressed to Watt at Kington, Herefordshire, near the Welsh border, MS3219/4/33.

52. "Copy of letter to Monsr Delambre," 8 August 1808, MS3219/5/1.

53. Watt to M. Lévêque, 14 March 1810, extracted in Muirhead, *Origin and Progress*, II: 313–14.

54. See John Gascoigne, *Joseph Banks and the English Enlightenment: Useful Knowledge and Polite Culture* (Cambridge: Cambridge University Press, 1994), 244–45; Lipkowitz, "'The Sciences Are Never at War?'" 243–47.

55. Count Berthollet to Watt, 18 February 1815, in Muirhead, *Origin and Progress*, II: 361–62.

56. Watt to James Millar Jr., 19 March 1815, MS3219/4/120. It had taken a month for Berthollet's letter to arrive at Heathfield.

57. W. Taylor to Watt, 27 May 1806, MS3219/4/48.

58. See *Inaugural Addresses Delivered by Lords Rectors of the University of Glasgow* (Edinburgh: University of Glasgow, 1848), where he is invoked by Sir James Mackintosh, the Marquis of Lansdowne, Lord Stanley, Henry Brougham, and Robert Peel.

59. Hart, "Reminiscences of James Watt," 3–4.

60. Watt to Professor William Taylor, 3 June 1808; Watt to J[ohn]? Young, 4 June 1808, MS3219/4/119. See also W. Taylor to Watt, 27 May 1806, MS3219/4/48.

61. Watt to Gilbert Hamilton, 10 August 1803, and Watt to Robert Muirhead, 4 August 1803, MS3219/4/119.

62. James Watt to Rt. Hnble. Lord Chief Baron, 18 November 1803, MS3219/4/114. The holder of the office at this time was Sir Archibald Macdonald (1747–1826).

63. Ambrose Weston to Watt, 12 January 1804, MS3219/4/114.

64. Dickinson and Jenkins, *James Watt and the Steam Engine*, 78.

65. Watt to Ambrose Weston, 15 January 1804, MS3219/4/119. Strictly, Scottish Presbyterianism, recognized as the official church of Scotland in the Act of Union, was not a dissenting religion. The grounds on which Weston and Watt (and the mob) assimilated it to dissent probably involved the doctrinal commitment of many English-based Presbyterians to Unitarianism.

66. Watt to Ambrose Weston, 16 January 1804, MS3219/4/119.

67. Watt to Robert Muirhead, 23 November 1816, quoted in Muirhead, *Life of Watt*, 489–90. See Hills, *James Watt, Volume 3*, 247.

68. Muirhead, *Life of Watt*, 400.

69. H. T. Wade, "The James Watt Centenary," *Scientific American* 121 (30 August 1919): 219.

70. Watt's dislike of aristocracy as an institution came out on various occasions. One was his anger and defiance toward the "illustrious House of Cavendish" during his contretemps with Henry of that ilk over the discovery of the composition of water, when De Luc implied that Watt might need to soft-pedal in the dispute because of the possible business consequences of resisting a Cavendish. A second was during his involvement with Boulton and Wedgwood in pursuing the trade policy of the manufacturing interest in the mid-1780s. He was outraged when the Board of Trade was incorporated into the Privy Council, so that the latter could exert aristocratic dominance over the board. At the time he observed that landed gentlemen saw the manufacturers as "poor mechanics and slaves to be looked down on with contempt" (see Witt Bowden, *Industrial Society in England towards the End of the Eighteenth Century*, 2nd ed. [London: Cass, 1965], 155). In the 1780s during the struggle against Cornish mine owners for payment of engine dues, Watt had attacked Sir Francis Bassett in particular. Remarking on the claimed "inconvenience" to the mine owners of payment of engine dues, Watt said: "It is also very inconvenient for the man who wishes to get a slice of the squire's land, that there should be a law tying it up by an entail. Yet the squire's land has not been of his own making, as the condensing engine has been of mine. He has only passively inherited his property, while this invention has been the product of my own labour, and of God knows how much anguish of mind and body" (Watt to Matthew Boulton, 31 October 1780, quoted in Smiles, *Lives of Boulton and Watt*, 280).

71. James Watt Jr. to J. P. Muirhead, 3 November 1839, Muirhead Papers, Special Collections, University of Glasgow, MS GEN 1354/405. The review of the eloge was in *The Civil Engineer and Architect's Journal* 2 (1839): 399–419.

72. James Watt Jr. to J. P. Muirhead, 3 November 1839, Muirhead Papers, Special Collections, University of Glasgow, MS GEN 1354/405.

73. In what follows I am indebted much to Hugh Torrens, "Jonathan Hornblower (1753–1815) and the Steam Engine."

74. Olinthus Gregory, *A Treatise of Mechanics, Theoretical, Practical and Descriptive*, 3 vols. (London: George Kearsley, 1806), II: 355.

75. Gregory, *A Treatise of Mechanics*, 360–61.

76. Gregory, *A Treatise of Mechanics*, 361–62.

77. Gregory, *A Treatise of Mechanics*, 362.

78. Gregory, *A Treatise of Mechanics*, 365.

79. [John Playfair], "Account of Steam Engines—from A Treatise on Mechanics, Theoretical, Practical and Descriptive. By Olinthus Gregory, A.M. Second Edition. London 1807," *Edinburgh Review* 13 (Oct 1808–Jan 1809): 311–33.

80. [Playfair], "Account of Steam Engines," 313, 319.

81. [Playfair], "Account of Steam Engines," 325.

82. [Playfair], "Account of Steam Engines," 332–33.

83. See Olinthus Gregory, "Misrepresentation in the Edinburgh Reviewer's account of Mr Ol. Gregory's 'Treatise of Mechanics,'" *Monthly Magazine* (1 August 1809): 29–32; "Appendix—to Olinthus Gregory," *Edinburgh Review* 15 (1809): 245–54; Olinthus Gregory, "Dr Olinthus Gregory's second answer to the Edinburgh Reviewers," *Literary Panorama* 7 (1810): 694–96.

84. Olinthus Gregory, *A Treatise of Mechanics*, 3rd ed. (London, 1815), 381n.

85. Watt to Professor Playfair, 27 February 1809, MS3219/4/119.

86. Watt to Watt Jr., 11 November 1808, MS3219/6/1.

87. Watt to Professor Playfair, 27 February 1809, MS3219/4/119. The first Latin quotation is short for "magna est veritas et praevalet"—"Great is truth, and it prevails." The second, "judicet qua posteritas," refers to the appeal to the judgment of posterity.

88. A prominent example is Joseph Bramah, *A Letter to the Rt. Hon. Sir James Eyre, Lord Chief Justice of the Common Pleas: on the Subject of the Cause Boulton and Watt v Hornblower and Maberley, for Infringement of Mr Watt's Patent for an Improvement on the Steam Engine* (London: J. Stockdale, 1797).

89. A.P. Woolrich, "John Farey and His *Treatise on the Steam Engine* of 1827," *History of Technology* 22 (2000): 63–106; John Farey, *A Treatise on the Steam Engine, Historical, Practical and Descriptive* (London: Longman, Rees, Orme, Brown and Green, 1827); Thomas Tredgold, *The Steam Engine* (London: J. Taylor, 1827); Elijah Galloway, *History and Progress of the Steam Engine* (London: Thomas Kelly, 1836), 94–95 ("not one half of his schemes answered"; "he, like all men, was liable to misconception and error"; "[he had a] disposition to monopoly"); Robert Stuart, *A Descriptive History of the Steam* Engine (London: John Knight and Henry Lacey, 1824), 14, 21, 84, 193 (the "illustrious," "venerable," and "celebrated" Mr. Watt; "no man ever before bestowed such a gift on his kind"); Charles F. Partington, *An Historical and Descriptive Account of the Steam Engine* (London: J. Taylor, 1826), 185 (the "great and comprehensive genius of the late Mr. Watt").

90. Boulton held feasts for the Soho workmen on various occasions, such as Robinson Boulton's coming of age in 1791, when seven hundred men processed to the feast to be thanked in Boulton's speech for the fact that not one of them was absent during the late riots (See *Derby Mercury*, 18 August 1791). The opening of Soho Foundry was another major, widely reported, occasion on which the workers were fed, watered, and lectured to. The new young master, Robinson Boulton, presid-

ed, but Boulton Sr. was the high priest amid much religious imagery. In his "benediction" Boulton pronounced: "As the smith cannot do without the striker, so neither can the master do without his workmen. Let each perform his part well, and do their duty in that state which it has pleased God to call them, and this they will find to be the true rational ground of equality." The affair concluded with a full chorus of "God Save the King" and the discharge of six cannons (see *Leeds Intelligencer*, 8 February 1796, 3). Watt and Watt Jr. were nowhere mentioned.

91. The following draws in part on G. F. Tyas, "Matthew Murray: A Centenary Appreciation," *Transactions of the Newcomen Society* 6 (1925): 111–43.

92. "The King against Murray," *Leeds Intelligencer*, 18 July 1803, 3.

93. *Leeds Intelligencer*, 25 July 1803, 3.

94. Tyas, "Matthew Murray," reproduces this correspondence as appendix B, 133–43.

95. Watt Jr. to Robinson Boulton, 14 June 1802, in Tyas, "Matthew Murray," 137.

96. Margaret Maria Gordon, *The Home Life of Sir David Brewster* (Edinburgh: David Douglas, 1881), 69. The quotation remained (see Maria Edgeworth, *Harry and Lucy Concluded* 4 vols., 2nd ed. [London: R. Hunter, 1827], II: 335–36), describing Watt as "the man whose genius discovered the means of multiplying our national resources," as "giving the feeble arm of man the momentum of Afrite—commanding manufactures to arise, as the rod of the prophet produced water in the desert—affording the means of dispersing with that time and tide which wait for no man; and of sailing without that wind. . . . This magician, whose cloudy machinery has produced a change in the world, the effects of which, extraordinary as they are, perhaps are only now beginning to be felt."

97. Gordon, *The Home Life of Sir David Brewster*, 70.

98. Gordon, *The Home Life of Sir David Brewster*, 70.

99. These portraits are reproduced in Dickinson and Jenkins, *James Watt and the Steam Engine*, between 10–11 and 14–15.

100. Annie Watt to James Watt, 26 October 1785, MS3219/4/5. A "picture of charity" as a genre in Christian art usually depicted a mother and her children. The uncle would have been John Watt or John Muirhead.

101. "Inventory of Furniture in each room of James Watt's House [Heathfield] 1791," MS3219/4/238.

102. Watt to Annie Watt, 4 February 1787, and 8 March 1787, MS3219/4/267A. On Mrs Wray see http://www.profilesofthepast.org.uk/mckechnie/wray-mary-mrs-mckechnie-section-1.

103. Sue McKechnie, *British Silhouette Artists and Their Work (1760–1860)* (London: Sotheby, Parke Bernet, 1978).

104. Important discussions of Watt portraiture include: Muirhead, *Origin and Progress*, I: cclxxiv–cclxxxiii; Dickinson and Jenkins, *James Watt*, 81–89; Jane Insley, "Picturing James Watt," *The British Art Journal* 11 (2011): 37–47, which is perhaps the most comprehensive account, though it sees the subject through Science Museum holdings and exhibitions. On von Breda see Karl Asplund, "Carl Fredrik von Breda," *The Burlington Magazine for Connoisseurs* 83 (1943): 296–301; Olga Baird, "The Lunar Society in Portraits by Carl-Fredrik von Breda," *Birmingham Historian*, no. 30 (Summer 2007): 27–32.

105. In the following I rely on Val Loggie, "Portraits of Matthew Boulton," in *Matthew Boulton: A Revolutionary Player*, ed. Malcolm Dick (Studley: Brewin Books, 2009), 63–76. See also Val Log-

gie, "Soho Depicted: Prints, Drawings and Watercolours of Matthew Boulton, His Manufactory and Estate, 1760–1809" (PhD diss., University of Birmingham, 2011), which convincingly develops as a major theme Boulton's use of images in marketing the Soho business and its products.

106. On the organization and functions of portraiture during this period, see Shearer West, *Portraiture* (Oxford: Oxford University Press, 2004), esp. 43–69 and 81–97; Desmond Shawe-Taylor, *The Georgians: Eighteenth-Century Portraiture and Society* (London: Barrue & Jenkins, 1990). On portraits of scientific and medical figures see, Ludmilla Jordanova, *Defining Features: Scientific and Medical Portraits 1660–2000* (London: Reaktion Books, 2000). An important specific study full of insights into portraiture, its dissemination, and its role in generating heroic status and "genius" is Patricia Fara, *Newton: The Making of Genius* (London: Macmillan, 2002).

107. My reading of this portrait is perhaps at odds with Jordanova, *Defining Features*, 42–43. Professor Jordanova notes that a number of portraits of Watt exhibited his melancholia. The von Breda portrait could certainly be seen in that way, but could also, as Jordanova notes, be understood in more positive terms as "modest or unassuming," or as I claim here, "philosophical." See also MacLeod, *Heroes of Invention*, 75–76.

108. Frances Robertson, "Ruling the Line: Learning to Draw in the First Age of Mechanical Reproduction" (PhD diss., Glasgow School of Art, 2011), 34.

109. On how the system of print production and distribution worked, and on its personal and business value, see Val Loggie, "Creating an Image: Portrait Prints of Matthew Boulton," in *Matthew Boulton: Enterprising Industrialist of the Enlightenment*, ed. Kenneth Quickenden, 231–45.

110. Watt wrote to Rennie on 9 June 1792 that he wished to talk with him about the proposed portrait "to which I am rather averse as I think it an honour I do not merit and that my countenance cannot be worth procuring" (Dickinson and Jenkins, *James Watt and the Steam Engine*, 82; Loggie, "Creating an Image," 69).

111. Dickinson and Jenkins, *James Watt and the Steam Engine*, 85.

112. Muirhead, *Life of James Watt*, 531.

113. Muirhead, *Life of James Watt*, 533. The quotation is from Watt to James Weston, 23 November 1815, MS3219/4/120.

114. Watt to Watt Jr., 23 June 1810, MS3219/6/1/356.

115. Rennie to Watt, 29 July 1816, and Watt to Rennie, 31 July 1816, in Muirhead, *Origin and Progress*, II: 366–67. The artist in question was Archibald Skirving (1749–1819) known best for his pastel portrait of Robert Burns.

116. For example, Watt told a friend in 1809 that he had sat again for a copy of the Beechey portrait that was "more like than the original." That copy was then in the collection of Mr. John Tuffen. See Muirhead, *Origin and Progress*, I: cclxxv.

117. Watt to James Watt Jr., 15 June 1813, MS3219/6/1.

118. Watt to James Weston, 23 November 1815, MS3219/4/120. Client-artist relations were further complicated when more than one artist was involved, in this case portraitist and engraver. Engravers were also often clients of the painters.

119. Muirhead, *Origins and Progress*, I: cclxxxiii.

120. Watt to Francis Chantrey, 14 April 1815, MS3219/4/120.

121. Watt to Francis Chantrey, 31 July 1815, MS3219/4/120. Watt's exchanges with Chantrey

show that he regarded himself as engaged in collaboration with the sculptor. Not only did Watt cri-tique the image, but he also advised Chantrey on making casts and molds, and on finding a better-quality plaster. Watt discussed engraving methods with Charles Turner and suggested the process of engraving in mezzotint on steel to him in 1812. See Charles Turner, "On the Invention, Progress and Advantages of the Art of Engraving in Mezzotint upon Steel," *Transactions of the Society Instituted at London for the Encouragement of Arts, Manufactures and Commerce* 42 (1823): 55.

122. Muirhead, *Life of James Watt*, 520–21.

123. Watt Jr. to John Rennie, 16 August 1819, National Library of Scotland, Rennie MSS 19824, ff. 132–33.

124. See letters in Rennie MSS 19823, ff. 134–52.

125. Watt Jr. to John Rennie, 26 August 1819, National Library of Scotland, Rennie MSS 19824, ff. 153–54.

126. Maria Edgeworth to Miss Honora Edgeworth, April 1820, in *The Life and Letters of Maria Edgeworth*, ed. Augustus Hare, I: 289–91. The "gay trellis" refers to a trellis covered with climbing plants that had long adorned the main entrance to Heathfield. See figure 6.1.

127. *Life and Letters of Maria Edgeworth*, ed. Augustus Hare, 290–91. Characteristically, Watt Jr. was in the process of furnishing Aston Hall in a very lavish style. He had moved in there a few months before his father's death. See Virginia Glenn, "George Bullock, Richard Bridgens and James Watt's Regency Furnishing Schemes," *Furniture History* 15 (1979): 54–67.

Chapter 8: Afterlife

1. See Miller, *James Watt, Chemist*, 21–23, 172–173.

2. Watt Jr. to John Rennie, 26 August 1819, National Library of Scotland, Rennie MSS 19824, ff. 153–54.

3. Dickinson, *James Watt: Craftsman and Engineer*, 199. See also Dickinson and Jenkins, *James Watt and the Steam Engine*, 79.

4. *Birmingham Chronicle*, 9 September 1819, 3.

5. Watt to Watt Jr., 30 August 1809, MS3219/6/1/339.

6. Watt Jr. funded the creation of the chapel in which a Chantrey statue of Watt sits as well as memorials to Boulton and William Murdoch. This monument later gained St. Mary's the sobriquet "Cathedral of the Industrial Revolution."

7. *Birmingham Post*, 9 September 1819, 3. Thomas Lane Freer (1777–1835) was rector of St. Mary's Handsworth from 1803 until his death.

8. Watt Jr. to John Rennie, 10 September 1819, National Library of Scotland, Rennie MSS 19824, ff. 155–56; Francis Jeffrey, "The Late Mr James Watt," *The Scotsman*, 4 September 1819, 5. This was reproduced as "Notice and Character of James Watt" in Francis Jeffrey, *Contributions to the Ed-inburgh Review*, 2nd ed., 3 vols. (Edinburgh: Longman, Brown, Green and Longmans, 1846), III: 693–98.

9. Presumably, *The British Gallery of Contemporary Portraits, Being a Series of Engravings of the Most Eminent Persons Now Living or Lately Deceased in Great Britain and Ireland: From Drawings Accurately Made from Life or from the Most Approved Original Pictures, Accompanied by Short Biographical Notices*, 2 vols. (London: T. Cadell & Davies, 1810–1822).

10. Watt Jr. to John Rennie, 10 September 1819, National Library of Scotland, Rennie MSS 19824, ff. 155–56.

11. On Playfair I draw on: Ian Spence and Howard Wainer, "Who Was Playfair?" *Chance* 10 (1997): 35–37; Ian Spence and Howard Wainer, "William Playfair: A Daring Worthless Fellow," *Chance* 10 (1997): 31–34; and especially Jean-François Dunyach, "William Playfair (1759–1823), Scottish Enlightenment from Below?" in *Jacobitism, Enlightenment and Empire, 1680–1820,* ed. Douglas J. Hamilton (London: Routledge, 2015), 159–72.

12. Dunyach, "William Playfair," 164. The modern estimate of the work is much more favorable, being credited as a landmark in the graphical representation of information. William Playfair, *Statistical Breviary* (London: T. Bensley, 1801) developed the use of the pie chart.

13. This copy, which is annotated "Copy of a letter recevd Oct 9 1819 W Playfair" was addressed to Playfair, signed "J. Smith," and dated from London, 7 October 1819. Rennie MSS, NLS 19824, f. 168.

14. Watt Jr. to John Rennie, 21 October 1819, Rennie MSS, NLS 19824, ff. 160–61.

15. *New Monthly Magazine,* 12 December 1819, 576.

16. See Messrs. Longman & Co. to Watt Jr., 21 December 1819 and 4 January 1820; Watt Jr. to Messrs. Longman & Co, 1 January 1819 [1820], MS3219/6/105.

17. Watt Jr. to John Rennie, 14 October 1819, MS3219/6/105.

18. The *Britannica* had included biographical articles from its second edition, but Napier was working hard to add articles on recent lives. See Richard Yeo, *Encyclopaedic Visions: Scientific Dictionaries and Enlightenment Culture* (Cambridge: Cambridge University Press, 2001), 260–64.

19. Macvey Napier to Watt Jr., 9 May 1823, MS3219/6/105.

20. Watt Jr. to Napier, 22 December and 31 December 1823, MS3219/6/105. For a more detailed account of their negotiations, see Miller, *Discovering Water,* 94–96.

21. John Barrow to George Rennie, 27 November 1823, National Library of Scotland, MSS 19938, ff. 38–41; Barrow to George Rennie, 7 January 1824, Macvey Napier Correspondence, British Library, Add. MSS 34.611, ff. 217–18.

22. On Arago see John Cawood, "François Arago, homme de science et homme politique," *la Recherche* 16 (1985): 1464–71. On his eloges, see Maurice Crosland, *Science under Control: The French Academy of Sciences 1795–1914* (New York: Cambridge University Press, 1992), 360–361; Dorinda Outram "The Language of Natural Power: The Éloges of Georges Cuvier," *History of Science* 16 (1978): 153–78.

23. J. B. Pentland to William Buckland, 15 August 1834, MS3219/6/106.

24. Watt Jr. to Francis Chantrey, 22 August 1834; Watt Jr. to William Buckland, 26 August 1834, MS3219/6/106.

25. Watt Jr. to George Rennie, 24 October 1834, MS3219/6/106.

26. Watt Jr. to François Arago, 22 January 1835, MS3219/6/106.

27. Watt Jr. to François Arago, 28 May 1839, MS3219/6/106.

28. Miller, *Discovering Water.* Among the key documents of the controversy were: François Arago, *Historical Eloge of James Watt,* trans. from the French with additional notes and an appendix by James Patrick Muirhead (London: John Murray, 1839); William Vernon Harcourt, "Address," *Report of the Ninth Meeting of the British Association for the Advancement of Science Held at Birmingham*

in August 1839 (1840), 3–69; James Patrick Muirhead, ed. *The Correspondence of the Late James Watt on His Discovery of the Theory of the Composition of Water* (London: John Murray, 1846); Francis Jeffrey, "The Discoverer of the Composition of Water; Watt or Cavendish?" *Edinburgh Review* 87 (1848): 67–137; Wilson, *The Life of the Honourable Henry Cavendish*.

29. See Ben Marsden, "Engineering Science in Glasgow: Economy, Efficiency and Measurement as Prime Movers in the Differentiation of an Academic Discipline," *The British Journal for the History of Science* 25 (1992): 319–46; Crosbie Smith, *The Science of Energy: A Cultural History of Energy Physics in Victorian Britain* (Chicago: The University of Chicago Press, 1998).

30. A point first made in Ben Marsden, "Engineering Science in Glasgow: W. J. M. Rankine and the Motive Power of Air," 162–63.

31. On Muirhead see B. M. Sturt, "Muirhead, James Patrick (1813–1898)," Rev. Richard L. Hills, *Oxford Dictionary of National Biography*, http://www.oxforddnb.com/view/article/19507. Lockhart Muirhead's parents were both Muirheads, and George Muirhead was his uncle.

32. Muirhead, ed., *The Correspondence of the Late James Watt.*

33. James Gibson Watt (1831–1891), who had been educated at Rugby and Magdalen College, Cambridge, based himself at Doldowlod Hall (the Elizabethan-style mansion erected by Watt Jr. in 1845) in Radnorshire, where he took a leading role in county affairs. He had added the name of Watt to his own by letters patent in 1856 when he turned twenty-five and inherited the estates left by Watt Jr. (Obituary, *Montgomeryshire Echo*, 27 June 1891, 5).

34. "Last Will and Testament of James Watt of Aston Hall," The National Archives, Kew, London, Prob 11/2078, 362.

35. "Last Will and Testament of James Watt of Aston Hall," 369.

36. "Last Will and Testament of James Watt of Aston Hall," 369.

37. See report in *Aris's Birmingham Gazette*, 24 August 1835, 3.

38. See Dickinson, *The Garret Workshop.*

39. T. Edgar Pemberton, *James Watt of Soho and Heathfield: Annals of Industry and Genius* (Birmingham: Cornish Brothers, 1905), 34.

40. Sotheby's, *The James Watt Sale.*

41. Muirhead, *Origin and Progress*, I: cclv.

42. Pemberton, *James Watt of Soho*, 62. The dust remained a matter of fascination. See Ben Russell, "Preserving the Dust: The Role of Machines in Commemorating the Industrial Revolution," *History & Memory* 26 (2014): 106–32.

43. Smiles, *Lives of Boulton and Watt*, 512–14.

44. The story of their exploits is told in Russell, *James Watt*, 224–30. I have adopted the date that Russell gives for Woodcroft's visit, rather than 4 May as specified by Dickinson, *The Garret Workshop*, 8.

45. Russell, *James Watt*, 228–29.

46. Samuel Timmins (1826–1902) was a Birmingham hardware manufacturer and bibliophile who was intensely proud of his city. Following a visit of the British Association to Birmingham, he edited a collection of accounts of its industries: Samuel Timmins, ed., *The Resources, Products and Industrial History of Birmingham and the Midland Hardware District* (London: Robert Hardwicke,

1866). See Stephen Roberts, "Timmins, Samuel (1826–1902)," *Oxford Dictionary of National Biography*, https://doi.org/10.1093/ref:odnb/104869.

47. Dickinson, *The Garret Workshop*, 11.

48. See Russell, *James Watt*, 231. Sir George Tangye (1835–1920) eventually headed an engineering business that had involved a number of the Tangye brothers. Their Cornwall works was located in Soho. Among the brothers' numerous philanthropic acts was the preservation of Boulton and Watt artifacts and materials that were presented to Birmingham Central Library. See J. F. Parker, "Some Notes on the Tangye Family," *Transactions of the Newcomen Society* 45 (1972–73): 191–204.

49. Subsequently published as E. A. Cowper, "On the Inventions of James Watt and His Models Preserved at Handsworth and South Kensington," *Proceedings of the Institution of Mechanical Engineers* 34 (1883): 599–631 and plates 55–87.

50. "Institution of Mechanical Engineers Meeting in Birmingham," *Birmingham Daily Post*, 2 November 1883, 8.

51. *Birmingham Daily Post*, 8 June 1885, 4, and 7 January 1886, 4. *The Graphic*, 8 May 1886, 8, emphasized the importance of keeping the relics at Heathfield as "they will not appeal to our imagination so powerfully or excite the same emotions, as they do when we see them in the room where the inventor worked."

52. J. W. Gibson Watt had petitioned Chancery in 1873 "praying that an order may be made vesting in the trustees of the will of James Watt, formerly of Aston Hall . . . powers of granting building leases for terms of years not exceeding 99 years." (*London Evening Standard*, 27 August 1873, 1).

53. On this and many other insights into the history of the workshop, see Ben Russell, "Watt's Workshop: Craft and Philosophy in the Science Museum," *Science Museum Group Journal*, no. 1 (Spring 2014), http://journal.sciencemuseum.ac.uk/browse/2014/Watts-workshop/.

54. Arthur Lawrence, "Romance in Hard Metal. An Interview with Mr. George Tangye," *Cornish Magazine* 2 (Jan–May 1899): 341–56.

55. *Birmingham Daily Gazette*, 18 July 1910, 6, and 20 August 1912, 6.

56. James Miller Gibson Watt (1875–1929) had inherited the estates on the death of his father in 1891. He was a leader in county affairs in Radnorshire, like his father, and he served in the South African campaign and the Great War.

57. "The Historic Home of James Watt," *Birmingham Daily Gazette*, 19 March 1920, 3.

58. "Heathfield Hall to Be Lost to the City?" *Birmingham Daily Gazette*, 25 February 1921, 3.

59. "James Watt's Home. Future of Heathfield Hall. Garden City Plan," *Birmingham Daily Gazette*, 21 November 1923, 7.

60. That heirloom clause tied the relics to the house. It had been open to any of the line of heirs to sell the relics, but their passing on by inheritance required that they remain with the house. The various heirs had been ultimately, but not without some wavering, unwilling to sell. The impending demise of Heathfield Hall itself forced their hand, and the Science Museum was the solution so far as the workshop was concerned. Other relics presumably went to Doldowlod Hall.

61. Dickinson, *James Watt: Craftsman and Engineer*.

62. H. Hopkins, "James Watt Plus Birmingham," *Birmingham Daily Gazette*, 17 January 1936, 6.

63. Gale, "Soho Foundry, Some Facts and Fallacies," 73–87; Griffiths, *The Third Man*, 333.

64. Ince, "Soho Foundry, 1796–1895," 11–14.

65. Watt Jr. to François Arago, 8 November 1835, MS3219/6/106.

66. Ince, "Soho Foundry, 1796–1895," 35–45.

67. See Miller, "The Mysterious Case."

68. On the negotiations regarding this inscription between Brougham and Watt Jr. see Miller, *Discovering Water*, 96–97. Watt Jr. balked at Brougham's original use of the phrase "trained in philosophic research" since it suggested that Watt was Black's pupil.

69. MacLeod, *Heroes of Invention*, 91–108.

70. *Morning Post*, 17 September 1832, 3; *Monmouthshire Merlin*, 22 September 1832, 4.

71. *The Athenaeum*, 3 December 1842, 1035.

72. Arthur Penrhyn Stanley, *Historical Memorials of Westminster Abbey*, 3rd rev. ed. (London: John Murray, 1886), 349–50.

73. MacLeod, *Heroes of Invention*, 98–99.

74. MacLeod, *Heroes of Invention*, 99 and fn36. See also Charles Hampden Turner to Josiah Wedgwood, 12 June [1824], and George Augustus Lee to Wedgwood, 5 June 1824, Wedgwood Manuscripts, E28-19922A and E33-25318, Wedgwood Museum, Barlaston, Stoke on Trent.

75. Watt Jr. to Josiah Wedgwood, 12 June 1824, Wedgwood Manuscripts, E4-3340, Wedgwood Museum.

76. Francis Jeffrey, "The Late Mr James Watt," *The Scotsman*, 4 September 1819, 5.

77. MacLeod, "James Watt, Heroic Invention, and the Idea of the Industrial Revolution."

78. MacLeod, *Heroes of Invention*, 98–101.

79. See A. Rupert Hall, *The Abbey Scientists* (London: R. & R. Nicholson, 1966).

80. On Don, a Scot, see http://www.westminster-abbey.org/our-history/people/alan-campbell-don

81. See Eric S. Abbott to General Sir Brian Robertson, 22 February 1960 and 12 October 1960. These and all subsequent items of correspondence referred to about the statue can be found in the National Archives, British Railways Board File AN111/638, "The James Watt Statue," and File AN111/639, "James Watt Statue. Clapham." On the removal see *The Times*, 22 December 1960, 10.

82. R. E. Trevithick to J. H. Scholes, 26 October 1962, and J. H. Scholes to R. E. Trevithick, 1 November 1962. Richard Ewart Trevithick (1891–1973) was a great-grandson of the steam engineer, himself an engineer like many other of the man's descendants, a captain in the Royal Engineers, and clearly a keeper of the flame.

83. David Gibson-Watt to J. H. Scholes, 1 August 1968.

84. Sir Iain Maxwell Stewart (1916–1985) was managing director and later chairman of Thermotank, a heating and ventilating company in Glasgow. He was very active in trying to rejuvenate Clydeside, especially through his Fairfields Shipyard Plan. After Sinclair met Sean Connery on the golf course and discovered their shared Scottish nationalism, Connery made a documentary film, *The Bowler and the Bunnet* (1967), about the Fairfield experiment, which was shown on Scottish TV. See Fred M. Walker, *Ships and Shipbuilders: Pioneers of Design and Construction* (Barnsley: Seaforth Publishing, 2010); Michael Feeney Callan, *Sean Connery* (New York: Random House, 2012).

85. Iain M. Stewart to Dr. Richard Beeching, 27 July 1962, and Stewart to Stephen Garratt, 3

August 1962. Beeching (1913–1985) became chairman of BTC in 1961. As this correspondence occurred he was hatching the plans that became the famous, or infamous, Beeching Report of March 1963, which recommended widespread closures in the British rail network. Garratt was the public relations adviser to BTC.

86. I owe this point to Ben Marsden.

87. Copy of Consultative Panel to Stephen Garratt, 19 September 1962.

88. In a number of follow up letters to the BTC, Stewart had sought to apply pressure in various ways: see Iain M. Stewart to S. Garratt, 10 August 1962 and 14 September 1962.

89. Consultative Panel to Stephen Garratt, 19 September 1962.

90. "James Watt Broods in a Bus Depot," *Sunday Times*, 17 June 1962, 16. Iain Stewart, who had visited the Clapham Museum unannounced and incognito on 23 May and quizzed staff about how the statue came to be there, probably supplied the story and photograph. The rather stunned staff on duty that day found him a "charming gentleman" and clearly "a man of education and some standing."

91. See Mike Horne, "Britain's National Railway Museum, Parts 1–4," especially part 4, "The Science Museum and the Battle of York," https://machorne.wordpress.com.

92. Eric Merrill to Chief Secretary & Legal Adviser, 3 January 1974; R. H. Lascelles to Eric Merrill, 7 January 1974.

93. Dr. Bernard Feilden to Eric Merrill, 3 February 1975. Feilden was surveyor of the fabric at St. Paul's.

94. "A Great Scot Comes Home after 150 Years in Exile," *The Sunday Herald*, 23 August 1996.

95. MacLeod, *Heroes of Invention*, 112–15.

96. MacLeod, *Heroes of Invention*, 347; Ray McKenzie, *Public Sculpture of Glasgow* (Liverpool: Liverpool University Press, 2001), 393–94, 404; Ray McKenzie, "Hard Lessons: Public Sculpture and the Education System in Nineteenth-Century Glasgow," in *Art, Community and Environment: Educational Perspectives*, ed. G. Coutts and T. Jokela (Bristol: Intellect Books, 2008), 241–62.

97. *Glasgow Herald*, 3 May 1870, 4. The irony was false, of course, given Watt's tenuous connection with railways.

98. Crosbie Smith, "'Nowhere but in a Great Town': William Thomson's Spiral of Classroom Credibility," in *Making Space for Science: Territorial Themes in the Shaping of Knowledge*, ed. Crosbie Smith and Jon Agar (Basingstoke: Macmillan, 1998), 140–41.

99. "Report from the Greenock Advertiser, of the Proceedings of a Meeting, held in the Assembly Rooms, Greenock, for the purpose of deliberating on the erection of a Monument to the Memory of the late James Watt, Esquire," reproduced in George Williamson, *Letters Respecting the Watt Family*, 50–67.

100. "Report from the Greenock Advertiser," 59.

101. See Robert Murray Smith, *The History of Greenock* (Greenock: Orr, Pollock & Co., 1921), 151. A lithograph of the proposed tower is in Williamson, *Memorials*, 256. The original plan in 1835 for the construction of the Washington Monument included the idea that marble and granite should be brought to Washington from each state.

102. "Proposed National Monument to James Watt. Meeting at Greenock," *Glasgow Herald*, 28 December 1886, 3; "Greenock Memorials of James Watt," *Glasgow Herald*, 12 March 1889, 4.

103. See Smith, *History of Greenock*, 152; Miller, *James Watt, Chemist*, 12–13, 28–31. This view of Watt, emphasizing his descent from the Celts and the Covenanters, pervaded the biography that Carnegie wrote of Watt: *James Watt* (New York: Doubleday, Page & Co., 1905).

104. MacLeod, *Heroes of Invention*, 347, 108–12.

105. MacLeod, *Heroes of Invention*, 348–49.

106. See "James Watt," in Terry Wyke, *Public Sculpture of Greater Manchester* (Liverpool: Liverpool University Press, 2005), 117–19.

107. *Art Journal*, 1 October 1857, 308–09, quoted in Wyke, *Public Sculpture of Greater Manchester*, 118.

108. George Noszlopy, *Public Sculpture of Birmingham* (Liverpool: Liverpool University Press, 1998), 32–33.

109. Noszlopy, *Public Sculpture of Birmingham*, 15–17.

110. Miller, *James Watt, Chemist*, chapter 1.

111. For an overview see Moureen Coulter, *Property in Ideas: The Patent Question in Mid-Victorian Britain* (Kirksville, MO: Thomas Jefferson Press, 1992).

112. Christine MacLeod, "Concepts of Invention and the Patent Controversy."

113. See Harold Dutton, *The Patent System and Inventive Activity in the Industrial Revolution, 1750–1852* (Manchester: Manchester University Press, 1984), 77–79.

114. John Timbs, *Stories of Inventors and Discoverers in Science and the Useful Arts* (London: Kent and Co., 1860); Henry Dircks, *Inventions and Inventors* (London: E. and F. N. Spon, 1867).

115. C. William Siemens, "Presidential Address, the Mechanical Section," *Report of the Thirty-Ninth Meeting of the British Association for the Advancement of Science* (London: John Murray, 1870), 201–02.

116. Robert Andrew McFie, *The Patent Question: A Solution of Difficulties by Abolishing or Shortening the Inventor's Monopoly, and Instituting National Recompenses* (London: W. J. Johnson, 1863).

117. See David Philip Miller, "Principle, Practice and Persona in Isambard Kingdom Brunel's Patent Abolitionism," *The British Journal for the History of Science* 41 (2008): 43–72.

118. *Report and Minutes of Evidence taken before the Select Committee of the House of Lords appointed to consider the Bill intituled 'An Act further to Amend the Law Touching Letters Patent for Inventions'* . . . (London: House of Commons, 1851), 214–15. Among other frequent invocations of Watt, by both sides, were the evidence of Thomas Webster (17–18) and of William Carpmael (38).

119. Adrian Johns, *Piracy: The Intellectual Property Wars from Gutenberg to Gates* (Chicago: The University of Chicago Press, 2009), 271.

120. MacLeod, "James Watt, Heroic Invention, and the Idea of the industrial Revolution," 98.

121. Arnold Toynbee, *Lectures on the Industrial Revolution in England* (London: Rivingtons, 1884).

122. Becky Conekin, *The Autobiography of a Nation: The 1951 Exhibition of Britain, Representing Britain in the Post-War World* (Manchester: Manchester University Press, 2003), 67–68, 140–141. The design of the section of the exhibition dealing with Watt can be seen in "Power for Industry-Steam-Drawings by Albert Smith: Work by James Watt on Steam," 1948–1952, The National Archives, Kew, London, WORK 25/183/C2/KH-E/2P.

123. Christine MacLeod and Jennifer Tann, "From Engineer to Scientist: Reinventing Invention in the Watt and Faraday Centenaries, 1919–31," *The British Journal for the History of Science* 40 (2007): 389–411.

124. *History and Regulations of the Watt Club of Greenock* (Greenock: The Advertiser Office, 1827).

125. Miller, *James Watt, Chemist*, 148–51. Among the more celebrated Watt Anniversary lecturers were Sir William Thomson (1869 and 1876), James Clerk Maxwell (1871), John Scott Russell (1867 and 1877), P. G. Tait (1873 and 1878), Sir Frederick Bramwell (1893), Sir Oliver Lodge (1900), and Sir William Ramsay (1903). The lecturers are listed in Greenock Philosophical Society, *Jubilee Celebrations, Friday 13 January 1911* (Greenock: The Society, 1911), 17–18.

126. *Greenock Telegraph*, 14 January 1891, 2; *Motherwell Times*, 28 January 1898, 2.

127. *Glasgow Herald*, 21 January 1895, 6. John Inglis (1842–1919) was educated at Glasgow University under Professors Thomson and Rankine, among others, and he served a full engineering apprenticeship in, and became manager of, the family shipyard, A. & J. Inglis. See "Mr John Inglis, LL.D., Pointhouse," in William S. Murphy, *Captains of Industry* (Glasgow: W. S. Murphy, 1901), 91–95.

128. *Glasgow Herald*, 25 January 1897, 6; "Is James Watt Forgotten?" *Evening Telegraph* (Dundee), 28 January 1896, 2.

129. "James Watt Anniversary Dinner in Glasgow," *The Scotsman*, 20 January 1902, 9. William Foulis (1838–1903) was gas engineer to the Corporation of Glasgow and known for his many improvements to the processes of gas manufacture. See https://www.gracesguijde.co.uk/William_Foulis.

130. "James Watt Anniversary, Glasgow Celebration," *The Scotsman*, 27 January 1911, 8.

131. See Eric Hobsbawm, "Mass-Producing Traditions: Europe, 1870–1914," in *The Invention of Tradition*, ed. Eric Hobsbawm and Terence Ranger (Cambridge: Cambridge University Press, 1983), 263–307; Roland Quinault, "The Cult of the Centenary, c. 1784–1914," *Historical Research* 71 (1998): 303–23.

132. "James Watt Centenary," *Aberdeen Press and Journal*, 6 September 1919, 4.

133. MacLeod and Tann, "From Engineer to Scientist."

134. William Mills (1856–1932) was an English engineer, knighted in 1922 mainly for his development and manufacture in Birmingham of the "Mills bomb," the hand grenade widely used by British and Imperial forces in World War I. He had many other inventions to his credit, including lifeboat improvements and aluminum golf clubs, and he was active in the Birmingham Chamber of Commerce.

135. A useful account of the program is in Dickinson and Jenkins, *James Watt and the Steam Engine*, 401–04.

136. J.H. Andrew, "The Smethwick Engine," *Industrial Archaeology Review* 8 (1985): 1–21.

137. This was, of course, Dickinson and Jenkins, *James Watt and the Steam Engine*.

138. See Robert Bud, "Framed in the Public Sphere: Tools for the Conceptual History of 'Applied Science,'" *History of Science* 51 (2013): 413–33; Robert Bud, "'Applied Science': A Phrase in Search of a Meaning," *Isis* 103 (2012): 537–45; Sabine Clarke, "Pure Science with a Practical Aim: The Meanings of Fundamental Research in Britain, circa 1916–1950," *Isis* 101 (2010): 285–311.

139. See "Bicentenary of James Watt," *The Times*, 20 January 1936, 19.

140. "Exhibition at the Science Museum," *The Times*, 20 January 1936, 19; "Watt Centenary Exhibition," *Aberdeen Press and Journal*, 21 January 1936, 8; "Wireless Programmes—Watt Bicente-

nary," *The Scotsman*, 20 January 1936, 13. Dunnett and Gough's play featured a discussion between Watt and Sir Walter Scott held during their meeting in Edinburgh about the effect of steam in the future, with flashbacks to Watt's early days.

141. "Proctor's Threat," *Sunderland Daily Echo and Shipping Gazette*, 28 January 1936, 4. See also "Jetsam Jottings," *The Stage*, 30 January 1936, 9.

142. "Scenes from Town Hall Pageant: James Watt's Memory Honoured in Song and Story," *The Greenock Telegraph*, 16 January 1936, 4.

143. "From Steam Engine to the Atom," *The Greenock Telegraph*, 18 January 1936, 7.

144. "The Children's Corner," *Greenock Telegraph*, 14 January 1936, 4.

145. "Order of Service. Bi-Centenary of James Watt. Commemoration Service Held in Handsworth Parish Church on Sunday, 1ˢᵗ March, 1936 at 11 a.m.," Additional Parish Records of St. Mary, Handsworth, Birmingham Archives and Heritage, Library of Birmingham, EP86/Box 1.

146. See Miller, "True Myths"; MacLeod, *Heroes of Invention*, 348.

147. On the yellow kettle see Miller, *James Watt, Chemist*, 18–19. For the advertisements see, in order, *The Sunday Times*, 4 September 1949, 1; *Illustrated London News*, 1 October 1949, 4; *Illustrated London News*, 19 January 1924, 114; *The Financial Times*, 16 March 1994, 7; *The Graphic*, 15 July 1911, 32; *Shields Daily Gazette*, 29 January 1904, 2; *Glasgow Herald*, 30 October 1900, 7.

148. "James Watt Figured It Out," *The Times*, 15 October 1985, 16.

149. This observation is based on large numbers of "What do you do?" dinner party conversations of the author in Australia, Britain, and the United States. To blame for this is the fact that in 1882 it was decided to name the electrical unit of power (strictly speaking, all forms of power) the "watt," and that most people's regular encounters with "watts" involve changing lightbulbs!

150. Paul J. Crutzen, "Geology of Mankind," *Nature* 415 (3 January 2002): 23.

151. Simon L. Lewis and Mark A. Maslin, "Defining the Anthropocene," *Nature* 519 (12 March 2015): 171–80; Colin N. Waters, et al., "The Anthropocene Is Functionally and Stratigraphically Distinct from the Holocene," *Science* 351 (8 January 2016): 137.

152. See for example, Clive Hamilton, "Getting the Anthropocene So Wrong," *The Anthropocene Review* 2 (2015): 102–07.

153. Andreas Malm and Alf Hornborg, "The Geology of Mankind? A Critique of the Anthropocene Narrative," *The Anthropocene Review* 1 (2014): 62–69. But for further critique of this position, see Daniel Cunha, "The Geology of the Ruling Class?" *The Anthropocene Review* 2 (2015): 262–66. Naomi Klein, however, comes close to blaming Watt when she states that "while making Europe richer, he also helped make many other parts of the world poorer, carbon-fueled inequalities that persist to this day" (Naomi Klein, *This Changes Everything: Capitalism vs. the Climate* [New York: Simon & Schuster, 2014], 175).

Chapter 9: Concluding Reflections on the "Great Steamer"

1. See Roderick Murchison to William Harcourt, 28 December 1839, in *Gentlemen of Science: Early Correspondence of the British Association for the Advancement of Science*, ed. Jack Morrell and Arnold Thackray (London: The Royal Historical Society, 1984), 328–29.

2. A. E. Musson and Eric Robinson, *Science and Technology in the Industrial Revolution* (Manchester: University of Manchester Press, 1969).

3. Margaret C. Jacob, *Scientific Culture and the Making of the Industrial West* (New York/Oxford: Oxford University Press, 1997), 116–30 and passim.

4. See, for example, Miller, "Was Matthew Boulton a Scientist?"; Andrew, "Was Matthew Boulton a Steam Engineer?"; Tann, "Matthew Boulton—Innovator."

5. Jones, *Industrial Enlightenment*, 16–18, 185–86; Ursula Klein, "Hybrid Experts."

6. Quoted in Robert Anderson, "Relations between Industry and Academe in Scotland, and the Case of Dyeing: 1760–1840," in *Compound Histories: Materials, Governance and Production, 1760–1840*, ed. Lissa Roberts and Simon Werrett (Leiden: Brill, 2018), 352–53.

Bibliography

Major Manuscript Sources

James Watt Papers, Matthew Boulton Papers, Boulton & Watt Archive

Archives of Soho, Wolfson Centre for Archival Research, Library of Birmingham

Muirhead and Watt Papers

Special Collections, University of Glasgow Library

Thomas Wilson Correspondence

Cornwall Record Office, Truro, Cornwall

John Rennie Papers

National Library of Scotland, Edinburgh

Macvey Napier Correspondence

British Library, Manuscript Collections, London

Wedgwood Manuscripts

Wedgwood Museum, Barlaston, Stoke on Trent

British Railways Board Files

The National Archives, Kew, London

Primary and Secondary Sources

Ahnert, Thomas. *The Moral Cultures of the Scottish Enlightenment, 1690–1805*. New Haven: Yale University Press, 2014.

Allan, G. C. "An Eighteenth-Century Combination in the Copper Mining Industry." *Economic Journal* 33 (1923): 74–85.

Allen, Robert C. *The British Industrial Revolution in Global Perspective*. Cambridge: Cambridge University Press, 2009.

Anderson, R. G. W., and Jean Jones. "Joseph Black: Life and Work." In *The Correspondence of Joseph Black*, edited by R. G. W. Anderson and Jean Jones, I: 25–57. 2 vols. Farnham: Ashgate, 2012.

Anderson, R. G. W., and Jean Jones, eds. *The Correspondence of Joseph Black*, 2 vols. Farnham: Ashgate, 2012.

Anderson, Robert. "Relations between Industry and Academe in Scotland, and the Case of Dyeing: 1760–1840." In *Compound Histories: Materials, Governance and Production, 1760–1840*, edited by Lissa L. Roberts and Simon Werrett, 333–53. Leiden: Brill, 2018.

Andrew, James H. "The Copying of Engineering Drawings and Documents." *Transactions of the Newcomen Society* 53 (1981–82): 1–15.

Andrew, Jim. "Boulton, Watt and Wilkinson: The Birth of the Improved Steam Engine." In *Matthew Boulton, Enterprising Industrialist of the Enlightenment*, edited by Kenneth Quickenden, Sally Baggott, and Malcolm Dick, 85–100. Farnham: Ashgate, 2013.

Andrew, Jim. "The Soho Steam-Engine Business." In *Matthew Boulton: Selling What All the World Desires*, edited by Shena Mason, 63–70. New Haven: Yale University Press, 2009.

Andrew, Jim. "Was Matthew Boulton a Steam Engineer?" In *Matthew Boulton: A Revolutionary Player*, edited by Malcolm Dick, 107–15. Studley: Brewin Books, 2009.

Andrew, Jim. "The Smethwick Engine." *Industrial Archaeology Review* 8 (1985): 1–21.

Arago, François. *Historical Eloge of James Watt*. Translated from the French with additional notes and an appendix by James Patrick Muirhead. London: John Murray, 1839.

Arapostathis, Stathis, and Graeme Gooday. *Patently Contestable: Electrical Technologies and Inventor Identities on Trial in Britain*. Cambridge, MA: The MIT Press, 2013.

Ashworth, William J. *Customs and Excise: Trade, Production, and Consumption in England, 1640–1845*. Oxford: Oxford University Press, 2003.

Asplund, Karl. "Carl Fredrik von Breda." *The Burlington Magazine for Connoisseurs* 83 (1943): 296–301.

Baggott, Sally. "Hegemony and Hallmarking: Matthew Boulton and the Battle for the Birmingham Assay Office." In *Matthew Boulton: Enterprising Industrialist of the Enlightenment*, edited by Kenneth Quickenden, Sally Baggott, and Malcolm Dick, 147–61. Farnham: Ashgate, 2013.

Baird, Olga. "The Lunar Society in Portraits by Carl-Fredrik von Breda." *Birmingham Historian*, no. 30 (Summer 2007): 27–32.

Bargar, Brian D. "Matthew Boulton and the Birmingham Petition of 1775." *The Wil-liam and Mary Quarterly* 3, no. 13 (1956): 26–39.

Beddoes, Thomas. *Essay on the Public Merits of Mr Pitt*. London: J. Johnson, 1796.

Beddoes, Thomas. *A Word in Defence of the Bill of Rights against the Gagging Bill*. Bristol: N. Biggs, 1795.

Beddoes, Thomas, and James Watt. *Considerations on the Medicinal Use of Facti-tious Airs and the Manner of Obtaining Them in Large Quantities in Two Parts: Part 1 by Thomas Beddoes, M.D. Part II by James Watt Esq*. Bristol: Bulgin and Rosser, 1794.

Bedini, Silvio. *Thomas Jefferson and His Copying Machines*. Charlottesville: Uni-versity of Virginia Press, 1984.

Bell, Kathleen. "A Place for Ingenuity: Re-Imagining the Life of James Watt in a Se-quence of Poems." *Life Writing* 14 (2017): 267–75.

Berg, Maxine. *The Age of Manufactures 1700–1820: Industry, Innovation and Work in Britain*. 4th ed. London: Routledge, 2005.

Berthollet, Claude-Louis. *Recherches sur les Lois de l'Affinité*. Paris: Badouin, 1801.

Black, Joseph. "Experiments on Magnesia Alba, Quicklime and Other Alcaline Sub-stances." *Essays and Observations, Physical and Literary* 2 (1756): 157–225.

Black, Joseph. *De Humore Acido a Cibis Orto, et Magnesis Alba*. Edinburgh: G. Hamilton and J. Balfour, 1754.

Boantza, Victor D. "Collecting Airs and Ideas: Priestley's Style of Experimental Rea-soning." *Studies in History and Philosophy of Science* 38 (2007): 506–22.

Boldrin, Michele, and David K. Levine. "2003 Lawrence R. Klein Lecture: The Case against Intellectual Monopoly." *International Economic Review* 45 (2004): 327–50.

Borsay, Peter. "From Port to Resort: Tenby and Narratives of Transition, 1760–1914." In *Resorts and Ports: European Seaside Towns since 1700*, edited by Peter Borsay and John K. Walton, 86–112. Bristol: Channel View Publications, 2011.

Bottomley, Sean. *The British Patent System during the Industrial Revolution 1700–1852: From Privilege to Property*. Cambridge: Cambridge University Press, 2014.

Bowden, Witt. *Industrial Society in England towards the End of the Eighteenth Century*. 2nd ed. London: Cass, 1965.

Bramah, Joseph. *A Letter to the Rt. Hon. Sir James Eyre, Lord Chief Justice of the Common Pleas: on the Subject of the Cause Boulton and Watt v Hornblower and Maberley, for Infringement of Mr Watt's Patent for an Improvement on the Steam Engine.* London: J. Stockdale, 1797.

The British Gallery of Contemporary Portraits, Being a Series of Engravings of the Most Eminent Persons Now Living or Lately Deceased in Great Britain and Ireland: From Drawings Accurately Made from Life or from the Most Approved Original Pictures, Accompanied by Short Biographical Notices, 2 vols. London: T. Cadell & Davies, 1810–1822.

Brock, William R. *Scotus Americanus: A Survey of the Sources for Links between Scotland and America in the Eighteenth Century.* Edinburgh: Edinburgh University Press, 1982.

Brougham, Henry. *Lives of Men of Letters and Science, Who Flourished in the Time of George III.* London: Charles Knight, 1845.

Bruton, Roger N. "The Shropshire Enlightenment: A Regional Study of Intellectual Activity in the Late Eighteenth and Early Nineteenth Centuries." PhD diss., University of Birmingham, 2015.

Bryden, David J. "James Watt, Merchant: The Glasgow Years, 1754–1774." In *Perceptions of Great Engineers: Fact and Fantasy,* edited by Denis Smith, 9–22. London: The Science Museum, 1994.

Bryden, David J. "The Jamaican Observatories of Colin Campbell F.R.S. and Alexander Macfarlane F.R.S." *Notes and Records of the Royal Society of London* 24 (1970): 261–72.

Bryden, David J. *James Short and His Telescopes.* Edinburgh: Royal Scottish Museum, 1968.

Buchanan, R. A. *The Engineers: A History of the Engineering Profession in Britain, 1750–1914.* London: Jessica Langley, 1989.

Bud, Robert. "'Applied Science': A Phrase in Search of a Meaning." *Isis* 103 (2012): 537–45.

Bud, Robert. "Framed in the Public Sphere: Tools for the Conceptual History of 'Applied Science.'" *History of Science* 51 (2013): 413–33.

Burke, Edmund. *The Speeches of the Right Honourable Edmund Burke in the House of Commons and in Westminster Hall,* 4 vols. London: Longman, Hurst, Rees, Orme and Brown, 1816.

Burton, Anthony. *Richard Trevithick: Giant of Steam.* London: Aurum Press, 2000.

Butterfield, Herbert. *The Origins of Modern Science, 1300–1800.* Rev ed. New York: The Free Press, 1957.

Cable, John A. "Early Scottish Science: The Vocational Provision." *Annals of Science* 30 (1973): 179–99.

Campbell, R. H. *Carron Company.* Edinburgh: Oliver and Boyd, 1961.

Carter, Harold B. *Sir Joseph Banks, 1743–1820.* London: British Museum (Natural History), 1988.

Cavendish, Henry. "Experiments on Air." *Philosophical Transactions of the Royal Society of London* 74 (1784): 119–53.

Cawood, John. "François Arago, homme de science et homme politique." *La Recherche* 16 (1985): 1464–71.

Chaloner, W. H. "Isaac Perrins, 1751–1801: Prize Fighter and Engineer." *History Today* 23 (October 1973): 740–43.

Christie, John R. R. "The Origins and Development of the Scottish Scientific Community, 1680–1760." *History of Science* 12 (1974): 122–41.

Christie, J. R. R., and J. V. Golinski. "The Spreading of the Word: New Directions in the Historiography of Chemistry 1600–1800." *History of Science* 20 (1982): 235–66.

Clark, Anna. *Scandal: The Sexual Politics of the British Constitution.* Princeton: Princeton University Press, 2013.

Clarke, Sabine. "Pure Science with a Practical Aim: The Meanings of Fundamental Research in Britain, circa 1916–1950." *Isis* 101 (2010): 285–311.

Cleland, John. *Enumeration of the Inhabitants of the City of Glasgow and County of Lanark.* 2nd ed. Glasgow: John Smith & Son, 1832.

Clow, Archibald, and Nan L. Clow. *The Chemical Revolution: A Contribution to Social Technology.* London: The Batchworth Press, 1952.

Cockburn, Henry. *Life of Lord Jeffrey: With a Selection from His Correspondence,* 2 vols. Edinburgh: Adam & Charles Black, 1852.

Cohen, Seymour. "Two Refugee Chemists in the United States, 1794: How We See Them." *Proceedings of the American Philosophical Society* 126 (1982): 301–15.

Collins, Harry. *Tacit and Explicit Knowledge.* Chicago: The University of Chicago Press, 2010.

Collins, Harry M. *Changing Order: Replication and Induction in Scientific Practice.* Chicago: The University of Chicago Press, 1992. First published 1985.

Conekin, Becky. *The Autobiography of a Nation: The 1951 Exhibition of Britain, Representing Britain in the Post-War World.* Manchester: Manchester University Press, 2003.

Cooper, Thomas. *A Reply to Mr. Burke's Invective against Mr. Cooper and Mr. Watt, in the House of Commons on the 30th of April 1792*. London: J. Johnson, 1792.

Corfield, Brian. "Thomas Newcomen the Man." *The International Journal for the History of Engineering and Technology* 83 (2013): 209–21.

Coulter, Moureen. *Property in Ideas: The Patent Question in Mid-Victorian Britain*. Kirksville, MO: Thomas Jefferson Press, 1992.

Cowley, Stephen. *Rational Piety and Social Reform in Glasgow: The Life, Philosophy, and, Political Economy of James Mylne (1757–1839)*. Eugene, OR: Wipf & Stock, 2015.

Cowper, E. A. "On the Inventions of James Watt and His Models Preserved at Handsworth and South Kensington." *Proceedings of the Institution of Mechanical Engineers* 34 (1883): 599–631.

Crosland, Maurice. *Science under Control: The French Academy of Sciences 1795–1914*. New York: Cambridge University Press, 1992.

Crosland, Maurice. "The Image of Science as a Threat: Burke versus Priestley and the 'Philosophic Revolution.'" *The British Journal for the History of Science* 20 (1987): 277–307.

Crutzen, Paul J. "Geology of Mankind." *Nature* 415 (3 January 2002): 23.

Cule, J. E. "Finance and Industry in the Eighteenth Century: The Firm of Boulton and Watt." *Economic History Supplement to Economic Journal* 4 (1940): 319–25.

Cule, J. E. "The Financial History of Matthew Boulton." MComm thesis, University of Birmingham, 1935.

Cunha, Daniel. "The Geology of the Ruling Class?" *The Anthropocene Review* 2 (2015): 262–66.

Cuthbertson, Kenneth L. *The Last Presbyterian? Remembering the Faith of My Forebears*. Eugene, OR: Resource Publications, 2013.

Darwin, Erasmus. *The Botanic Garden: A Poem in Two Parts. Part I: Containing the Economy of Vegetation. Part II: The Loves of the Plants. With Philosophical Notes*. London: J. Johnson, 1791.

Daunton, Martin. *Trusting Leviathan: The Politics of Taxation in Britain, 1799–1914*. Cambridge: Cambridge University Press, 2001.

David, Paul A. "Path Dependence, Its Critics and the Quest for 'Historical Economics.'" In *Evolution and Path Dependence in Economic Ideas: Past and Present*, edited by Pierre Garrouste and Stavros Ioannides, 15–40. Cheltenham: Edward Elgar, 2001.

Dawson, Frank C. *John Wilkinson: King of the Ironmasters*. Stroud: The History Press, 2012.

Dawson, Warren R., ed. *The Banks Letters*. London: British Museum, 1958.

de Beer, Gavin. *The Sciences Were Never at War*. London: Thomas Nelson & Sons, 1960.

Deacon, Bernard. "'The Hollow Jarring of the Distant Steam Engines': Images of Cornwall between West Barbary and the Delectable Duchy." In *Cornwall: The Cultural Construction of Place*, edited by Ella Westland, 7–24. Newmill: Patten Press, 1997.

Deepak, Sangamesh R., and G. K. Ananthasuresh. "James Watt and His Linkages." *Resonance* 14 (2009): 530–43.

Demidowicz, George. "The Origins of the Soho Manufactory and Its Layout." In *Matthew Boulton: Enterprising Industrialist of the Enlightenment*, edited by Kenneth Quickenden, Sally Baggott, and Malcolm Dick, 67–84. Farnham: Ashgate, 2013.

Demidowicz, George. "Power at the Soho Manufactory and Mint." In *Matthew Boulton: A Revolutionary Player*, edited by Malcolm Dick, 116–31. Studley: Brewin Books, 2009.

Demidowicz, George. "A Walking Tour of the Three Sohos." In *Matthew Boulton: Selling What All the World Desires*, edited by Shena Mason, 99–107. New Haven: Yale University Press, 2009.

Desaguliers, J. T. *A Course of Experimental Philosophy*, 2 vols. London, W. Innys, 1744.

Devine, T. M. "Scotland." In *The Cambridge Economic History of Modern Britain, Volume 1: Industrialization 1700–1860*, edited by Roderick Floud and Paul Johnson, 388–416. Cambridge: Cambridge University Press, 2004.

Devine, T. M. *The Tobacco Lords: A Study of the Tobacco Merchants of Glasgow and Their Trading Activities c. 1740–90*. Edinburgh: John Donald, 1975.

Dick, Malcolm. "The Death of Matthew Boulton 1809: Ceremony, Controversy and Commemoration." In *Matthew Boulton: Enterprising Industrialist of the Enlightenment*, edited by Kenneth Quickenden, Sally Baggott, and Malcolm Dick, 247–66. Farnham: Ashgate, 2013.

Dick, Malcolm, ed. *Matthew Boulton: A Revolutionary Player*. Studley: Brewin Books, 2009.

Dickinson, H. W. *A Short History of the Steam Engine*. Cambridge: Cambridge University Press, 1938.

Dickinson, H. W. *James Watt: Craftsman and Engineer.* Cambridge: Cambridge University Press, 1936.

Dickinson, H. W. *Matthew Boulton.* Cambridge: Cambridge University Press / Babcock and Wilcox, 1936.

Dickinson, H. W. *The Garret Workshop of James Watt.* Science Museum Technical Pamphlet No. 1. London: HMSO, 1929; reprinted 1958.

Dickinson, H. W. "Some Unpublished Letters of James Watt." *Proceedings of the Institution of Mechanical Engineers* 89 (1915): 487–534.

Dickinson, H. W. *John Wilkinson, Ironmaster 1728–1808.* Ulverston: Hume Kitchin, 1914.

Dickinson, H. W., and Arthur Titley. *Richard Trevithick: The Engineer and the Man.* Cambridge: Cambridge University Press, 1934.

Dickinson, H. W., and Rhys Jenkins. *James Watt and the Steam Engine.* Oxford: The Clarendon Press, 1927.

Dircks, Henry. *Inventions and Inventors.* London: E. and F. N. Spon, 1867.

Donovan, Arthur. *Antoine Lavoisier: Science, Administration, and Revolution.* Cambridge: Cambridge University Press, 1996.

Donovan, Arthur. *Philosophical Chemistry in the Scottish Enlightenment: The Doctrines and Discoveries of William Cullen and Joseph Black.* Edinburgh: Edinburgh University Press, 1975.

Doty, Richard G. *Soho Mint and the Industrialization of Money.* London: Spink & Son Ltd., 1998.

Dry, Sarah. *The Newton Papers: The Strange and True Odyssey of Isaac Newton's Manuscripts.* Oxford: Oxford University Press, 2014.

Dunyach, Jean-François. "William Playfair (1759–1823), Scottish Enlightenment from Below?" In *Jacobitism, Enlightenment and Empire, 1680–1820,* edited by Douglas J. Hamilton, 159–72. London: Routledge, 2015.

Dupree, A. Hunter. "Nationalism in Science—Sir Joseph Banks and the Wars with France." In *A Festschrift for Frederick B. Artz,* edited by D. H. Pinkney and T. Ropp, 37–51. Durham, NC: Duke University Press, 1964.

Dutton, Harold. *The Patent System and Inventive Activity in the Industrial Revolution, 1750–1852.* Manchester: Manchester University Press, 1984.

Edgeworth, Maria. *Harry and Lucy Concluded, Being the Last Part of Early Lessons.* London: R. Hunter, 1825.

Edgeworth, Maria, and Richard Lovell Edgeworth. *Practical Education,* 2 vols. London: J. Johnson, 1798.

Emerson, Roger L. *An Enlightened Duke: The Life of Archibald Campbell (1682–1761), Earl of Ilay, 3rd Duke of Argyll.* Edinburgh: Humming Earth, 2013.

Emerson, Roger L. *Academic Patronage in the Scottish Enlightenment: Glasgow, Edinburgh and St Andrews Universities.* Edinburgh: Edinburgh University Press, 2008.

Emerson, Roger L. "The Philosophical Society of Edinburgh, 1748–1768." *The British Journal for the History of Science* 14 (1981): 133–76.

Emerson, Roger L. "The Philosophical Society of Edinburgh, 1737–1747." *The British Journal for the History of Science* 12 (1979): 154–91.

Emerson, Roger L. "The Social Composition of Enlightened Scotland: The Select Society of Edinburgh, 1754–1764." *Studies on Voltaire and the Eighteenth Century* 114 (1973): 291–329.

Fara, Patricia. *Newton: The Making of Genius.* London: Macmillan, 2002.

Farey, John. *A Treatise on the Steam Engine: Historical, Practical and Descriptive.* London: Longman, Rees, Orme, Brown and Green, 1827.

Fenby, David V. "The Lectureship in Chemistry and the Chemical Laboratory, University of Glasgow, 1747–1818." In *The Development of the Laboratory: Essays on the Place of Experiments in Industrial Civilization*, edited by Frank A. J. L. James, 22–36. Houndmills: Macmillan Press, 1989.

Fleming, Donald. "Latent Heat and the Invention of the Watt Engine." *Isis* 43 (1952): 3–5.

Fogarty, Barbara. "The Mechanical Paintings of Matthew Boulton and Francis Eginton." In *Matthew Boulton, Enterprising Industrialist of the Enlightenment*, edited by Kenneth Quickenden, Sally Baggott, and Malcolm Dick, 111–26. Farnham: Ashgate, 2013.

Fors, Hjalmar. *The Limits of Matter: Chemistry, Mining & Enlightenment.* Chicago: The University of Chicago Press, 2015.

Fox, Celina. *The Arts of Industry in the Age of Enlightenment.* New Haven: Yale University Press, 2009.

Gale, W. K. V. "Soho Foundry: Some Facts and Fallacies." *Transactions of the Newcomen Society* 34 (1961–62): 73–87.

Galloway, Elijah. *History and Progress of the Steam Engine.* London: Thomas Kelly, 1836.

Gascoigne, John. *Science in the Service of Empire: Joseph Banks, the British State and the Uses of Science in the Age of Revolutions.* Cambridge: Cambridge University Press, 1998.

Gascoigne, John. *Joseph Banks and the English Enlightenment: Useful Knowledge and Polite Culture*. Cambridge: Cambridge University Press, 1994.

Glenn, Virginia. "George Bullock, Richard Bridgens and James Watt's Regency Furnishing Schemes." *Furniture History* 15 (1979): 54–67.

Golinski, Jan. *The Experimental Self: Humphry Davy and the Making of a Man of Science*. Chicago: The University of Chicago Press, 2016.

Golinski, Jan. *Science as Public Culture: Chemistry and Enlightenment in Britain, 1760–1820*. Cambridge: Cambridge University Press, 1992.

Gordon, Margaret Maria. *The Home Life of Sir David Brewster*. Edinburgh: David Douglas, 1881.

Gourvish, T. R. *British Railways 1948–1973: A Business History*. Cambridge: Cambridge University Press, 1986.

Grapi, Pere, and Mercè Izquierdo. "Berthollet's Conception of Chemical Change in Context." *Ambix* 44 (1997): 113–30.

Greenock Philosophical Society. *Jubilee Celebrations, Friday 13 January 1911*. Greenock: The Society, 1911.

Gregory, Olinthus. "Dr Olinthus Gregory's Second Answer to the Edinburgh Reviewers." *Literary Panorama* 7 (1810): 694–96.

Gregory, Olinthus. "Misrepresentation in the Edinburgh Reviewer's Account of Mr Ol. Gregory's 'Treatise of Mechanics.'" *Monthly Magazine* (1 August 1809): 29–32.

Gregory, Olinthus. *A Treatise of Mechanics, Theoretical, Practical and Descriptive*, 3 vols. London: George Kearsley, 1806.

Griffiths, John. *The Third Man: The Life and Times of William Murdoch 1754–1839*. London: André Deutsch, 1992.

Guerlac, Henry. "Joseph Black's Work on Heat." In *Joseph Black 1728–1799: A Commemorative Symposium*, edited by A. D. C. Simpson, 13–22. Edinburgh: The Royal Scottish Museum, 1982.

Guerlac, Henry. "Chemistry as a Branch of Physics: Laplace's Collaboration with Lavoisier." *Historical Studies in the Physical Sciences* 7 (1976): 205–16.

Hall, A. Rupert. "What Did the Industrial Revolution in Britain Owe to Science?" In *Historical Perspectives: Studies in English Thought and Society in Honour of J. H. Plumb*, edited by Neil McKendrick, 129–51. London: Europa, 1974.

Hall, A. Rupert. *The Abbey Scientists*. London: R. & R. Nicholson, 1966.

Hamilton, Henry. "The Failure of the Ayr Bank, 1772." *The Economic History Review* 8 (1956): 405–17.

Hankin, Christiana C., ed. *Life of Mary Anne Schimmelpenninck*, 2 vols. London: Longman, Green, 1858.

Harcourt, William Vernon. "Address." *Report of the Ninth Meeting of the British Association for the Advancement of Science Held at Birmingham in August 1839* (1840): 3–69.

Hare, Augustus J., ed. *The Life and Letters of Maria Edgeworth*, 2 vols. New York: Houghton, Mifflin and Company, 1895.

Harris, J.R. *Industrial Espionage and Technology Transfer: Britain and France in the Eighteenth Century*. Aldershot: Ashgate, 1998.

Harris, J. R., and R. O. Roberts. "Eighteenth Century Monopoly: The Cornish Metal Company Agreements of 1785." *Business History* 5 (1963): 69–82.

Hart, Robert. "Reminiscences of James Watt." *Transactions of the Glasgow Archaeological Society* 1 (1859): 1–7.

Heilbron, John. "A Mathematicians' Mutiny with Morals." In *World Changes: Thomas Kuhn and the Nature of Science*, edited by Paul Horwich, 81–129. Cambridge, MA: The MIT Press, 1993.

Hills, Richard L. *James Watt, Volume 3: Triumph through Adversity, 1785–1819*. Ashbourne: Landmark Publishing Ltd., 2006.

Hills, Richard L. *James Watt, Volume 2: The Years of Toil, 1775–1785*. Ashbourne: Landmark Publishing Ltd., 2005.

Hills, Richard L. "The Development of the Steam Engine from Watt to Stephenson." *History of Technology* 25 (2004): 181–98.

Hills, Richard L. *James Watt, Volume 1: His Time in Scotland, 1736–1774*. Ashbourne: Landmark Publishing Ltd., 2002.

Hills, Richard L. "James Watt and the Delftfield Pottery, Glasgow." *Proceedings of the Society of Antiquaries of Scotland* 131 (2001): 375–420.

Hills, Richard L. "James Watt and His Copying Machine." In *Studies in British Paper History, Volume 1: The Oxford Papers*, edited by Peter Bowers, 81–88. Kidlington: British Association of Paper Historians, 1996.

Hills, Richard L. *Power from Steam: A History of the Stationary Steam Engine*. Cambridge: Cambridge University Press, 1989.

History and Regulations of the Watt Club of Greenock. Greenock: The Advertiser Office, 1827.

Hobsbawm, Eric. "Mass-Producing Traditions: Europe, 1870–1914." In *The Invention of Tradition*, edited by Eric Hobsbawm and Terence Ranger, 263–307. Cambridge: Cambridge University Press, 1983.

Home, R. W., I. M. Malaquias, and M. F. Thomaz, eds. *For the Love of Science: The Correspondence of J. H. de Magellan (1722–1790)*, 2 vols. Bern: Peter Lang, 2017.

Hopkins, Eric. *Birmingham: The First Manufacturing Town in the World, 1760–1840*. London: Weidenfeld & Nicolson, 1989.

Horn, D. B., and Mary Ransome, eds. *English Historical Documents, 1714–1783*. London: Eyre & Spottiswoode, 1957.

Hornblower, Jonathan. *An Address to the Lords, Adventurers and Others Concern'd in the Mines of Cornwall* (Penryn: n.p., 1788).

Huddart, Joseph (the younger). *Memoir of the Late Captain Joseph Huddart F.R.S. &c*. London: W. Phillips, 1821.

Hunt, Lynn, and Margaret Jacob. "The Affective Revolution in 1790s Britain." *Eighteenth-Century Studies* 34 (2001): 491–521.

Hutton, William. *A Brief History of Birmingham*. 2nd ed. Birmingham: Grafton & Reddell, 1802.

Ince, Laurence. "The Soho Engine Works 1796–1895." *Stationary Power*, no. 16 (2000): 5–10.

Insley, Jane. "James Watt and the Reproduction of Sculpture." *Sculpture Journal* 22 (2013): 37–65.

Insley, Jane. "Picturing James Watt." *The British Art Journal* 11 (2011): 37–47.

Jacob, Margaret C. *Scientific Culture and the Making of the Industrial West*. New York/Oxford: Oxford University Press, 1997.

Jacob, Margaret C., and Larry Stewart. *Practical Matter: Newton's Science in the Service of Industry and Empire 1687–1851*. Cambridge, MA: Harvard University Press, 2004.

James, Frank A. J. L. "'The first example . . . of an extensive scheme of pure scientific medical investigation': Thomas Beddoes and the Medical Pneumatic Institution in Bristol, 1794 to 1799." The Eighth Wheeler Lecture, Royal Institution, 2015, *RSCHG Occasional Paper No. 8*, 2016.

"James Watt." In *Public Characters of 1802–1803*, 501–18. London: Richard Phillips, 1803.

Jay, Mike. *The Atmosphere of Heaven: The Unnatural Experiments of Dr Beddoes and His Sons of Genius*. New Haven: Yale University Press, 2009.

Jeffrey, Francis. "The Discoverer of the Composition of Water; Watt or Cavendish?" *Edinburgh Review* 87 (1848): 67–137.

Jeffrey, Francis. *Contributions to the Edinburgh Review*, 3 vols. 2nd ed. Edinburgh: Longman, Brown, Green and Longmans, 1846.

Jeffrey, Francis. "The Late Mr James Watt," *The Scotsman*, 4 September 1819, 5.

Jenkins, Rhys. "A Cornish Engineer: Arthur Woolf 1766–1837." *Transactions of the Newcomen Society* 13 (1932): 55–73.

Johns, Adrian. *Piracy: The Intellectual Property Wars from Gutenberg to Gates.* Chicago: The University of Chicago Press, 2009.

Jones, Peter M. "Becoming an Engineer in Industrializing Great Britain circa 1760–1820." *Engineering Studies* 3 (2011): 215–32.

Jones, Peter M. *Industrial Enlightenment: Science, Technology and Culture in Birmingham and the West Midlands 1760–1820.* Manchester: Manchester University Press, 2008.

Jones, Peter M. "Living the Enlightenment and the French Revolution: James Watt, Matthew Boulton, and Their Sons." *The Historical Journal* 42 (1999): 157–82.

Jordanova, Ludmilla. *Defining Features: Scientific and Medical Portraits 1660–2000.* London: Reaktion Books, 2000.

Jungnickel, Christa, and Russell McCormmach. *Cavendish: The Experimental Life.* Lewisburg, PA: Bucknell, 1999.

Kanefsky, John, and John Robey. "Steam Engines in 18th-Century Britain: A Quantitative Assessment." *Technology and Culture* 21 (1980): 161–86.

Keyser, Barbara Whitney. "Between Science and Craft: The Case of Berthollet and Dyeing." *Annals of Science* 47 (1990): 213–60.

King-Hele, Desmond, ed. *The Collected Letters of Erasmus Darwin.* Cambridge: Cambridge University Press, 2007.

Klein, Naomi. *This Changes Everything: Capitalism vs. the Climate.* New York: Simon & Schuster, 2014.

Klein, Ursula. "Hybrid Experts." In *The Structures of Practical Knowledge*, edited by M. Valleriani, 287–306. Cham: Springer, 2017.

Knight, David. *Humphry Davy: Science and Power.* Cambridge: Cambridge University Press, 1998.

Larder, David F. "An Unpublished Chemical Essay of James Watt." *Notes and Records of the Royal Society of London* 25 (1970): 193–210.

Lawrence, Arthur. "Romance in Hard Metal. An Interview with Mr. George Tangye." *Cornish Magazine* 2 (Jan-May 1899): 341–56.

Lehmann, William C. *John Millar of Glasgow 1735–1801*. Cambridge: Cambridge University Press, 1960.

Levere, Trevor. "Chemistry, Consumption, and Reform." In Trevor Levere, Larry Stewart, Hugh Torrens with Joseph Wachelder, *The Enlightenment of Thomas Beddoes: Science, Medicine and Reform*, 10–78. London: Routledge, 2017.

Levere, Trevor. "Dr Thomas Beddoes (1750–1808): Science and Medicine in Politics and Society." *The British Journal for the History of Science* 17 (1984): 187–204.

Levere, Trevor. "Dr Thomas Beddoes at Oxford: Radical Politics in 1788–1793 and the Fate of the Regius Chair in Chemistry." *Ambix* 28 (1961): 61–9.

Levere, Trevor H. "Dr Thomas Beddoes and the Establishment of his Pneumatic Institution: A Tale of Three Presidents." *Notes and Records of the Royal Society of London* 32 (1977): 41–9.

Levere, Trevor, Larry Stewart, and Hugh Torrens with Joseph Wachelder, *The Enlightenment of Thomas Beddoes: Science, Medicine and Reform*. London: Routledge, 2017.

Lewis, Simon L., and Mark A. Maslin. "Defining the Anthropocene." *Nature* 519 (12 March 2015): 171–80.

Lipkowitz, Elise. "'The Sciences Are Never at War?' The Scientific Republic of Letters in the Era of the French Revolution, 1789–1815." PhD diss., Northwestern University, 2009.

Loggie, Val. "Creating an Image: Portrait Prints of Matthew Boulton." In *Matthew Boulton: Enterprising Industrialist of the Enlightenment*, edited by Kenneth Quickenden, Sally Baggott, and Malcolm Dick, 231–45. Farnham: Ashgate, 2013.

Loggie, Val. "Soho Depicted: Prints, Drawings and Watercolours of Matthew Boulton, His Manufactory and Estate, 1760–1809." PhD diss., University of Birmingham, 2011.

Loggie, Val. "Portraits of Matthew Boulton." In *Matthew Boulton: A Revolutionary Player*, edited by Malcolm Dick, 63–76. Studley: Brewin Books, 2009.

MacDonald, J. Marc. "Crossroads of Enlightenment 1685–1850: Exploring Education Science and Industry across the Delessert Network." PhD diss., University of Saskatchewan, 2015.

Mackintosh, James. "Inaugural Address as Rector of Glasgow University by Sir James Mackintosh." In *Inaugural Addresses by Lords Rectors of the University of Glasgow*, edited by John Barras Hay, 23–34. Glasgow: David Robertson, 1839.

MacLeod, Christine. *Heroes of Invention: Technology, Liberalism and British Identity 1750–1914*. Cambridge: Cambridge University Press, 2007.

MacLeod, Christine. "Negotiating the Rewards of Invention: The Shop-Floor Inventor in Victorian Britain." *Business History* 41 (1999): 17–36.

MacLeod, Christine. "James Watt, Heroic Invention and the Idea of the Industrial Revolution." In *Technological Revolutions in Europe: Historical Perspectives*, edited by M. Berg and K. Bruland, 96–116. Cheltenham: Edward Elgar, 1998.

MacLeod, Christine. "Concepts of Invention and the Patent Controversy in Britain." In *Technological Change: Methods and Themes in the History of Technology*, edited by Robert Fox, 137–53. Amsterdam: Harwood Academic Publishers, 1996.

MacLeod, Christine. *Inventing the Industrial Revolution: The English Patent System, 1660–1800*. Cambridge: Cambridge University Press, 1988.

MacLeod, Christine, and Jennifer Tann. "From Engineer to Scientist: Re-inventing Invention in the Watt and Faraday Centenaries 1919–1931." *The British Journal for the History of Science* 40 (2007): 389–411.

Magellan, J. H. de. *Essai sur la nouvelle Théorie du Feu élémentaire, et de la Chaleur des Corps*. London: W. Richardson, 1780.

Malm, Andreas. *Fossil Capital: The Rise of Steam Power and the Roots of Global Warming*. London: Verso, 2016.

Malm, Andreas. "The Origins of Fossil Capital: From Water to Steam in the British Cotton Industry." *Historical Materialism* 21 (2013): 15–68.

Marsden, Ben. *Watt's Perfect Engine: Steam and the Age of Invention*. Cambridge: Icon Books, 2002.

Marsden, Ben. "Engineering Science in Glasgow: Economy, Efficiency and Measurement as Prime Movers in the Differentiation of an Academic Discipline." *The British Journal for the History of Science* 25 (1992): 319–46.

Marsden, Ben. "Engineering Science in Glasgow: W. J. M. Rankine and the Motive Power of Air." PhD diss., University of Kent at Canterbury, 1992.

Marsden, Ben, and Crosbie Smith. *Engineering Empires: A Cultural History of Technology in Nineteenth-Century Britain*. Houndmills: Palgrave Macmillan, 2005.

Marwick, James. *Glasgow, the Water Supply of the City*. Glasgow: R. Anderson, 1901.

Mason, Shena. *The Hardware Man's Daughter: Matthew Boulton and His "Dear Girl."* Chichester: Phillimore & Co., 2005.

Mason, Shena, ed. *Matthew Boulton: Selling What All the World Desires.* New Haven: Yale University Press, 2009.

Mathias, Peter. *The Brewing Industry in England 1700–1830.* Cambridge: Cambridge University Press, 1959.

McConnell, Anita. *Jesse Ramsden (1735–1800): London's Leading Instrument Maker.* Aldershot: Ashgate, 2007.

McCormmach, Russell. *The Personality of Henry Cavendish.* Cham: Springer, 2014.

McCormmach, Russell. *Speculative Truth: Henry Cavendish, Natural Philosophy, and the Rise of Modern Theoretical Science.* Oxford: Oxford University Press, 2004.

McCormmach, Russell. "Henry Cavendish on the Proper Method of Rectifying Abuses." In *Beyond History of Science: Essays in Honor of Robert E. Schofield*, edited by Elizabeth Garber, 35–51. Bethlehem, PA: Lehigh University Press, 1990.

McEvoy, John G. *The Historiography of the Chemical Revolution.* London: Pickering & Chatto, 2010.

McFie, Robert Andrew. *The Patent Question: A Solution of Difficulties by Abolishing or Shortening the Inventor's Monopoly, and Instituting National Recompenses.* London: W. J. Johnson, 1863.

McKechnie, Sue. *British Silhouette Artists and Their Work (1760–1860).* London: Sotheby, Parke Bernet, 1978.

McKenzie, Ray. "Hard Lessons: Public Sculpture and the Education System in Nineteenth-Century Glasgow." In *Art, Community and Environment: Educational Perspectives*, edited by G. Coutts and T. Jokela, 241–62. Bristol: Intellect Books, 2008.

McKenzie, Ray. *Public Sculpture of Glasgow.* Liverpool: Liverpool University Press, 2001.

McKinlay, Alan, and Alistair Mutch. "'Accountable Creatures': Scottish Presbyterianism, Accountability and Managerial Capitalism." *Business History* 57 (2015): 241–56.

McLean, Rita. "Introduction: Matthew Boulton, 1728–1809." In *Matthew Boulton: Selling What All the World Desires*, edited by Shena Mason, 1–6. New Haven: Yale University Press, 2009.

Mee, Jon, and Jennifer Wilkes. "Transpennine Enlightenment: The Literary and

Philosophical Societies and Knowledge Networks in the North, 1781–1830." *Journal for Eighteenth-Century Studies* 38 (2015): 599–612.

Melling, Joseph. "Dark Satanic Millwrights? Forging Foremanship in the Industrial Revolution: Matthew Boulton and the Leading Hands of Boulton and Watt." In *Matthew Boulton: Enterprising Industrialist of the Enlightenment*, edited by Kenneth Quickenden, Sally Baggott, and Malcolm Dick, 163–77. Farnham: Ashgate, 2013.

Milburn, J. R. *Benjamin Martin: Author, Instrument-Maker and "Country Showman."* Leyden: Noordhoff International Publishing, 1976.

Millar, John. *The Origin of the Distinction of Ranks*. Edited by John Craig. Edinburgh: W. Blackwood, 1806.

Miller, David Philip. "A New Perspective on the Natural Philosophy of Steam in the Long Eighteenth Century and Its Relation to the Steam Engine." *Technology & Culture* (forthcoming).

Miller, David Philip. "'Men of Letters' and 'Men of Press Copies': The Cultures of James Watt's Copying Machine." In *The Romance of Science: Essays in Honour of Trevor H. Levere*, edited by Jed Buchwald and Larry Stewart, 65–79. Cham: Springer, 2017.

Miller, David Philip. "Of Patents, Principles, and the Construction of Heroic Invention: The Case of Neilson's Hot Blast in Iron Production." *Proceedings of the American Philosophical Society* 160 (2016): 361–422.

Miller, David Philip. "Was Matthew Boulton a Scientist? Operating between the Abstract and the Entrepreneurial." In *Matthew Boulton, Enterprising Industrialist of the Enlightenment*, edited by Kenneth Quickenden, Sally Baggott, and Malcolm Dick, 51–65. Farnham: Ashgate, 2013.

Miller, David Philip. "Testing Power and Trust: The Steam Indicator, The 'Reynolds Controversy,' and the Relations of Engineering Science and Practice in Late Nineteenth-Century Britain." *History of Science* 50 (2012): 212–50.

Miller, David Philip. "The Mysterious Case of James Watt's '"1785" Steam Indicator': Forgery or Folklore in the History of an Instrument?" *International Journal for the History of Engineering and Technology* 81 (2011): 129–50.

Miller, David Philip. *James Watt, Chemist: Understanding the Origins of the Steam Age*. London: Pickering & Chatto, 2009.

Miller, David Philip. "Scales of Justice: Assaying Matthew Boulton's Reputation and the Partnership of Boulton and Watt." *Midland History* 34 (2009): 58–76.

Miller, David Philip. "Principle, Practice and Persona in Isambard Kingdom Brunel's Patent Abolitionism." *The British Journal for the History of Science* 41 (2008): 43‒72.

Miller, David Philip. "Watt in Court: Specifying Steam Engines and Classifying Engineers in the Patent Trials of the 1790s." *History of Technology* 27 (2006): 43‒76.

Miller, David Philip. *Discovering Water: James Watt, Henry Cavendish and the Nineteenth-Century "Water Controversy."* Aldershot: Ashgate, 2004.

Miller, David Philip. "True Myths: James Watt's Kettle, His Condenser and His Chemistry." *History of Science* 42 (2004): 333‒60.

Miller, David Philip. "Puffing Jamie: The Commercial and Ideological Importance of Being a 'Philosopher' in the Case of the Reputation of James Watt (1736‒1819)." *History of Science* 38 (2000): 1‒21.

Miller, David Philip. "The Usefulness of Natural Philosophy: The Royal Society of London and the Culture of Practical Utility in the Later Eighteenth Century." *The British Journal for the History of Science* 32 (1997): 185‒201.

Miller, David Philip, and Trevor H. Levere. "'Inhale it and See?' The Collaboration between Thomas Beddoes and James Watt in Pneumatic Medicine." *Ambix* 55 (2008): 5‒28.

Misa, Thomas. "The Compelling Tangle of Modernity and Technology." In *Modernity and Technology*, edited by Thomas J. Misa, Philip Brey, and Andrew Feenberg, 4‒30. Cambridge, MA: The MIT Press, 2003.

Mokyr, Joel. *An Enlightened Economy: An Economic History of Britain 1700‒1850.* New Haven: Yale University Press, 2009.

Mokyr, Joel. *The Gifts of Athena: Historical Origins of the Knowledge Economy.* Princeton & Oxford: Princeton University Press, 2002.

Morrell, Jack. "The University of Edinburgh in the Late Eighteenth Century: Its Scientific Eminence and Academic Structure." *Isis* 62 (1970): 158‒71.

Morrell, Jack, and Arnold Thackray, eds., *Gentlemen of Science: Early Correspondence of the British Association for the Advancement of Science.* London: The Royal Historical Society, 1984.

Morrell, Jack, and Arnold Thackray. *Gentlemen of Science: Early Years of the British Association for the Advancement of Science.* Oxford: Clarendon Press, 1981.

Morris, Robert J. "Lavoisier and the Caloric Theory." *The British Journal for the History of Science* 6 (1972): 1‒38.

Morrison-Low, Alison D. *Making Scientific Instruments in the Industrial Revolution*. Aldershot: Ashgate, 2007.

Muir, James. *John Anderson, Pioneer of Technical Education and the College He Founded*. Glasgow: J. Smith, 1950.

Muirhead, James Patrick. *The Life of James Watt with Selections from His Correspondence*. London: John Murray, 1858.

Muirhead, James Patrick. *Origin and Progress of the Mechanical Inventions of James Watt*, 3 vols. London: John Murray, 1854.

Muirhead, James Patrick, ed. *Correspondence of the Late James Watt on His Discovery of the Theory of the Composition of* Water. London: John Murray, 1846.

Murray, David. *Robert & Andrew Foulis and the Glasgow Press, with Some Account of the Glasgow Academy of Fine Arts*. Glasgow: James Maclehose and Sons, 1913.

Musson, A. E., and Eric Robinson. *Science and Technology in the Industrial Revolution*. Manchester: Manchester University Press, 1969.

Mutch, Alistair. "Religion and Accounting Texts in Eighteenth-Century Scotland." *Accounting, Auditing & Accountability Journal* 29 (2016): 926–46.

Naegel, Paul, and Pierre Teissier. "Obtaining a Royal Privilege in France for the Watt Engine, 1776–1786." *International Journal for the History of Engineering and Technology* 83 (2013): 96–118.

Nef, John U. *The Rise of the British Coal Industry*. Abingdon: Frank Cass, 1966.

Noszlopy, George. *Public Sculpture of Birmingham*. Liverpool: Liverpool University Press, 1998.

Nuvolari, Alessandro. "Collective Invention during the British Industrial Revolution: The Case of the Cornish Pumping Engine." *Cambridge Journal of Economics* 28 (2004): 347–63.

Nuvolari, Alessandro, and Bart Verspagen. "*Lean's Engine Reporter* and the Development of the Cornish Engine: A Reappraisal." *Transactions of the Newcomen Society* 77 (2007): 167–89.

Outram, Dorinda. "The Language of Natural Power: The Eloges of Georges Cuvier." *History of Science* 16 (1978): 153–78.

Oxford Dictionary of National Biography. 60 vols. Oxford: Oxford University Press, 2004. Continually updated at http://www.oxforddnb.com/.

Pacey, A. J. "Some Early Heat Engine Concepts and the Conservation of Heat." *The British Journal for the History of Science* 7 (1974): 135–45.

Parker, J. F. "Some Notes on the Tangye Family." *Transactions of the Newcomen Society* 45 (1972–73): 191–204.

Partington, Charles F. *An Historical and Descriptive Account of the Steam Engine.* London: J. Taylor, 1826.

Partington, J. R. *History of Chemistry*, 4 vols. London: Macmillan, 1962–1964.

Paskins, Matthew. "Sentimental Industry: The Society of Arts and the Encouragement of Public Useful Knowledge." PhD diss., University College, London, 2014.

Paterson, Len. *From Sea to Sea: A History of the Scottish Lowland and Highland Canals.* Glasgow: Neil Wilson Publishing, 2006.

Pemberton, T. Edgar. *James Watt of Soho and Heathfield: Annals of Industry and Genius.* Birmingham: Cornish Brothers, 1905.

Playfair, John. "Account of Steam Engines—from A Treatise on Mechanics, Theoretical, Practical and Descriptive. By Olinthus Gregory, A.M. Second Edition. London 1807." *Edinburgh Review* 13 (Oct 1808–Jan 1809): 311–33.

Playfair, William. *The Statistical Breviary.* London: T. Bensley, 1801.

Pole, William. *A Treatise on the Cornish Pumping Engine; in Two Parts.* London: John Weale, 1844.

Porter, Roy S. "Science, Provincial Culture and Public Opinion in Enlightenment England." *Journal for Eighteenth-Century Studies* 3 (1980): 20–46.

Priestley, Joseph. *The Doctrine of Phlogiston Established, and That of the Composition of Water Refuted.* Northumberland, PA: The Author, 1800.

Priestley, Joseph. *Experiments on the Generation of Air from Water.* London: J. Johnson, 1793.

Priestley, Joseph. *Experiments and Observations on Different Kinds of Air . . . In Three Volumes.* Birmingham: Thomas Pearson, 1790.

Priestley, Joseph. "Experiments Relating to Phlogiston, and the Seeming Conversion of Water into Air." *Philosophical Transactions of the Royal Society of London* 73 (1783): 398–434.

Priestley, Joseph. *Experiments and Observations Relating to Various Branches of Natural Philosophy with a Continuation of the Observations on Air.* Birmingham: J. Johnson, 1781.

Pumphrey, Stephen. "Ideas above His Station: A Social Study of Hooke's Curatorship of Experiments." *History of Science* 29 (1991): 1–37.

Pye, Charles. *A Description of Modern Birmingham; Whereunto Are Annexed, Observations Made during an Excursion around the Town, in the Summer of 1818.* Birmingham: John Lowe, 1820.

Quickenden, Kenneth, Sally Baggott, and Malcolm Dick, eds. *Matthew Boulton: Enterprising Industrialist of the Enlightenment*. Farnham: Ashgate, 2013.

Quinault, Roland. "The Cult of the Centenary, c. 1784–1914." *Historical Research* 71 (1998): 303–23.

Report and Minutes of Evidence taken before the Select Committee of the House of Lords appointed to consider the Bill intituled 'An Act further to Amend the Law Touching Letters Patent for Inventions' . . . London: House of Commons, 1851.

Rhodes, Barbara, and William Streeter. *Before Photocopying: The Art and History of Mechanical Copying, 1780–1938*. Delaware: Oak Knell Press, 1999.

Roberts, Lissa. "Full Steam Ahead: Entrepreneurial Engineers as Go-Betweens during the Late Eighteenth Century." In *The Brokered World: Go-Betweens and Global Intelligence, 1770–1820*, edited by Simon Schaffer, Lissa Roberts, Kapil Raj, and James Delbourgo, 193–238. Sagamore Beach: Science History Publications, 2009.

Roberts, Lissa, Simon Schaffer, and Peter Dear, eds. *The Mindful Hand: Inquiry and Invention from the Late Renaissance to Early Industrialization*. Amsterdam: Koninklijke Nederlandse Akademie van Wetenschappen, 2007.

Robertson, Frances. "Ruling the Line: Learning to Draw in the First Age of Mechanical Reproduction." PhD diss., Glasgow School of Art, 2011.

Robinson, Eric. "James Watt and the Law of Patents." *Technology and Culture* 13 (1972): 115–39.

Robinson, Eric. "James Watt and Early Experiments in Alkali Manufacture." In *Science and Technology in the Industrial Revolution*, edited by A. E. Musson and Eric Robinson, 352–71. Manchester: Manchester University Press, 1969.

Robinson, Eric. "The Origins and Life-Span of the Lunar Society." *University of Birmingham Historical Journal* 11 (1967): 5–16.

Robinson, Eric. "II. Matthew Boulton and the Art of Parliamentary Lobbying." *The Historical Journal* 7 (1964): 209–29.

Robinson, Eric. "The Lunar Society: Its Membership and Organization." *Transactions of the Newcomen Society* 35 (1962–63): 153–77.

Robinson, Eric. "James Watt and the Tea Kettle: A Myth Justified." *History Today* 6 (1956): 261–65.

Robinson, Eric, and A. E. Musson. *James Watt and the Steam Revolution*. London: Adams and Dart, 1969.

Robinson, Eric, and Douglas McKie, eds. *Partners in Science: Letters of James Watt and Joseph Black*. Cambridge, MA: Harvard University Press, 1970.

Robison, John. *A System of Mechanical Philosophy with Notes by David Brewster*, 4 vols. Edinburgh: John Murray, 1822.

Robison, John. "Steam Engine." In *Supplement to the Third Edition of the Encyclopaedia Britannica*, II: 522–24. Edinburgh: Bonar, 1801.

Robison, John. "Steam-Engine." In *Encyclopaedia Britannica*, 3rd ed., 18 vols., 36, XVII, 743–72. Edinburgh: A. Bell and C. Macfarquhar, 1797.

Robison, John. "Attempts Towards the Improvement of a Machine as Useful as the Contrivance of It Is Ingenious." *The Universal Magazine of Knowledge and Pleasure* 21 (1757): 229–31.

Robison, John Jr. "Account of the Flexible Water Main, Contrived by the Late Mr Watt for the Glasgow Water-Work Company." *The Edinburgh Philosophical Journal* 3 (1820): 60–62.

Rogers, Nicholas. *The Press Gang: Naval Impressment and Its Opponents in Georgian Britain*. London: Continuum, 2007.

Roll, Eric. *An Early Experiment in Industrial Organisation: Being a History of the Firm of Boulton & Watt, 1775–1805*. Abingdon: Frank Cass, 1930.

Rolt, L. T. C., and J. S. Allen. *The Steam Engine of Thomas Newcomen*. New York: Science History Publications, 1977.

Ross, Sidney. "*Scientist:* The Story of a Word." *Annals of Science* 18 (1962): 65–85.

Rudwick, M. J. S. "Jean-André De Luc and Nature's Chronology." In *The Age of the Earth from 4004 BC to AD 2002*, edited by C. L. E. Lewis and S. J. Knell, 51–60. London: The Geological Society, 2001.

Russell, Ben. *James Watt: Making the World Anew*. London: Reaktion Books, 2014.

Russell, Ben. "Preserving the Dust: The Role of Machines in Commemorating the Industrial Revolution." *History & Memory* 26 (2014): 106–32.

Schofield, Robert E. *The Enlightened Joseph Priestley: A Study of his Life and Work from 1773–1804*. University Park, PA: Pennsylvania State University Press, 2004.

Schofield, Robert E. *The Lunar Society of Birmingham: A Social History of Provincial Science and Industry in Eighteenth-Century England*. Oxford: Clarendon Press, 1963.

Schrantz, Kristen M. "The Tipton Chemical Works of Mr James Keir: Networks of Conversants, Chemicals, Canals and Coal Mines." *International Journal for the History of Engineering and Technology* 84 (2014): 248–73.

Schuster, John. "The Scientific Revolution." In *Companion to the History of Modern*

Science, edited by R. C. Olby, G. N. Cantor, J. R. R. Christie, and M. J. S. Hodge, 217–42. London: Routledge, 1990.

Scott, Walter. *The Monastery: A Romance*. Edinburgh: Longman, Hurst, Rees, Orme and Brown, 1820.

Selgin, George, and John L. Turner. "Strong Steam, Weak Patents, or, the Myth of Watt's Innovation-Blocking Monopoly Exploded." *Journal of Law and Economics* 54 (2011): 841–61.

Selgin, George, and John L. Turner. "James Watt as Intellectual Monopolist: Comment on Boldrin and Levine." *International Economic Review* 47 (2006): 1341–48.

Seymour, Edward. *History of the Wars Resulting from the French Revolution: With a Relation of the Circumstances Which Led to That Important Event*, 2 vols. London: Thomas Crabb, 1815.

Shapin, Steven. *The Scientific Revolution*. Chicago: The University of Chicago Press, 1996.

Shapin, Steven. "Who Was Robert Hooke?" In *Robert Hooke: New Studies*, edited by Michael Hunter and Simon Schaffer, 253–85. Woodbridge: The Boydell Press, 1990.

Shapin, Steven. "The Invisible Technician." *American Scientist* 77 (1989): 554–63.

Shawe-Taylor, Desmond. *The Georgians: Eighteenth-Century Portraiture and Society*. London: Barrue & Jenkins, 1990.

Sher, Richard B. *Church and University in the Scottish Enlightenment: The Moderate Literati of Edinburgh*, 2nd ed. Edinburgh: Edinburgh University Press, 2015.

Sher, Richard B. "Commerce, Religion and the Enlightenment in Eighteenth-Century Glasgow." In *Glasgow: Beginnings to 1830*, edited by Thomas M. Devine and Gordon Jackson, 312–59. Manchester: Manchester University Press, 1995.

Sher, Richard B., and Andrew Hook, eds. *The Glasgow Enlightenment*. East Linton: Tuckwell Press, 1995.

Sher, Richard B., and Andrew Hook. "Introduction." In *The Glasgow Enlightenment*, edited by Richard B. Sher and Andrew Hook. East Linton: Tuckwell Press, 1995.

Siemens, C. William. "Presidential Address, the Mechanical Section." *Report of the Thirty-Ninth Meeting of the British Association for the Advancement of Science* (London: John Murray, 1870), 200–06.

Simpson, A. D. C., ed. *Joseph Black 1728–1799: A Commemorative Symposium.* Edinburgh: The Royal Scottish Museum, 1982.

Sinclair, J. B., and R. W. D. Fenn, "Mr Watt buys some farms, Part II." *Transactions of the Radnorshire Society* 65 (1995): 85–108.

Sinclair, J. B. and R. W. D. Fenn. "Mr Watt buys some farms, Part I." *Transactions of the Radnorshire Society* 64 (1994): 79–98.

Sinclair, J. B., and R. W. D. Fenn. "James Watt, father & son." *Transactions of the Radnorshire Society* 62 (1992): 51–65.

Sinclair, John. *The Statistical Account of Scotland,* 21 vols. Edinburgh: William Creech, 1791–1799.

Skempton, A. W., et al., eds. *A Biographical Dictionary of Civil Engineers in Great Britain and Ireland, I: 1500–1830.* London: Thomas Telford Publishing Ltd, 2002.

Skempton, A. W. *John Smeaton F.R.S.* London: Thomas Telford, 1981.

Smeaton, W. A. "Some Comments on James Watt's Published Account of His Work on Steam and Steam Engines." *Notes and Records of the Royal Society of London* 26 (1971): 35–42.

Smiles, Samuel. *Lives of the Engineers: The Steam Engine. Boulton and Watt.* London: J. Murray, 1878.

Smiles, Samuel. *Lives of the Engineers: Harbours–Lighthouses–Bridges. Smeaton and Rennie.* Rev. ed. London: John Murray, 1874.

Smiles, Samuel. *The Lives of Boulton and Watt: Principally from the Original Soho MSS.* London: John Murray, 1865.

Smith, Crosbie. "'Nowhere but in a Great Town': William Thomson's Spiral of Classroom Credibility." In *Making Space for Science: Territorial Themes in the Shaping of Knowledge,* edited by Crosbie Smith and Jon Agar, 118–46. Basingstoke: Macmillan, 1998).

Smith, Crosbie. *The Science of Energy: A Cultural History of Energy Physics in Victorian Britain.* London: The Athlone Press, 1998.

Smith, Robert Murray. *The History of Greenock.* Greenock: Orr, Pollock & Co., 1921.

Sorrenson, Richard. *Perfect Mechanics: Instrument Makers at the Royal Society of London in the Eighteenth Century.* Boston, MA: Docent Press, 2013.

Sotheby's. *The James Watt Sale: Art and Science.* London: Sotheby's, 2003.

Southey, C. C., ed. *The Life and Correspondence of Robert Southey,* 6 vols. London: Longman, Brown, Green and Longmans, 1850.

Spence, Ian, and Howard Wainer. "Who Was Playfair?" *Chance* 10 (1997): 35–7.

Spence, Ian, and Howard Wainer. "William Playfair: A Daring Worthless Fellow." *Chance* 10 (1997): 31–34.

Stanley, Arthur Penrhyn. *Historical Memorials of Westminster Abbey*. 3rd rev. ed. London: John Murray, 1886.

Stewart, Larry. "James Watt's Paine: Mob Rules, Democrats and Demons." In *James Watt (1736–1819): Culture, Innovation and Enlightenment*, edited by Malcolm Dick and Caroline Archer. Liverpool: Liverpool University Press, forthcoming 2019.

Stewart, Larry. "At the Medical Edge or, The Beddoes Effect." In *The Romance of Science: Essays in Honour of Trevor H. Levere*, edited by Jed Buchwald and Larry Stewart, 47–64. Cham: Springer, 2017.

Stewart, Larry. "A Jacobin Cloven Foot." In Trevor Levere, Larry Stewart, and Hugh Torrens with Joseph Wachelder, *The Enlightenment of Thomas Beddoes: Science, Medicine and Reform*, 116–70. Abingdon: Routledge, 2017.

Stewart, Larry. "'Ordinary' People and Philosophers in the Laboratories and Workshops of the Early Industrial Revolution." In *In Praise of Ordinary People: Early Modern Britain and the Dutch Republic*, edited by Margaret C. Jacob and Catherine Secretan, 95–122. Basingstoke: Palgrave Macmillan, 2013.

Stewart, Larry. "Assistants to Enlightenment: William Lewis, Alexander Chisholm and Invisible Technicians in the Industrial Revolution." *Notes and Records: The Royal Society Journal of the History of Science* 62 (2008): 17–30.

Strang, John. *Glasgow and Its Clubs*. London: Richard Griffin, 1857.

Stuart, Robert. *A Descriptive History of the Steam Engine*. London: John Knight and Henry Lacey, 1824.

Symons, David. "Matthew Boulton and the Royal Mint." In *Matthew Boulton: A Revolutionary Player*, edited by Malcolm Dick, 170–84. Studley: Brewin Books, 2009.

Tann, Jennifer. "Matthew Boulton—Innovator." In *Matthew Boulton, Enterprising Industrialist of the Enlightenment*, edited by Kenneth Quickenden, Sally Baggott, and Malcolm Dick, 33–50. Farnham: Ashgate, 2013.

Tann, Jennifer. "Steam and Sugar: The Diffusion of the Stationary Steam Engine to the Caribbean Sugar Industry 1779–1840." *History of Technology* 19 (1998): 63–84.

Tann, Jennifer. "Two Knights at Pandemonium: A Worm's-Eye View of Boulton, Watt & Co, c. 1800–1820." *History of Technology* 20 (1998): 47–72.

Tann, Jennifer. "Riches from Copper: The Adoption of the Boulton & Watt Engine by Cornish Mine Adventurers." *Transactions of the Newcomen Society* 67 (1995): 27–51.

Tann, Jennifer. "Fixed Capital Formation in Steam Power, 1775–1825: A Case Study of the Boulton and Watt Engine." In *Studies in Capital Formation in the United Kingdom 1750–1920*, edited by Charles H. Feinstein and Sidney Pollard, 164–81. Oxford: Clarendon Press, 1988.

Tann, Jennifer, ed. *The Selected Papers of Boulton & Watt, Volume 1: The Engine Partnership 1775–1825*. London: Diploma Press, 1981.

Tann, Jennifer. "Mr Hornblower and His Crew: Watt Engine Pirates at the End of the 18th Century." *Transactions of the Newcomen Society* 51 (1979–80): 95–107.

Tann, Jennifer. "Marketing Methods in the International Steam Engine Market: The Case of Boulton and Watt." *The Journal of Economic History* 38 (1978): 363–91.

Tann, Jennifer. "Boulton and Watt's Organization of Steam Engine Production before the Opening of Soho Foundry." *Transactions of the Newcomen Society* 49 (1977–78): 41–53.

Thackray, Arnold. "Natural Knowledge in Cultural Context: The Manchester Model." *American Historical Review* 79 (1974): 672–709.

Thomson, George. "James Watt and the Monkland Canal." *The Scottish Historical Review* 29 (1950): 121–33.

Timbs, John. *Stories of Inventors and Discoverers in Science and the Useful Arts*. London: Kent and Co., 1860.

Timmins, Samuel, ed. *The Resources, Products and Industrial History of Birmingham and the Midland Hardware District*. London: Robert Hardwicke, 1866.

Todd, Arthur C. *Beyond the Blaze: A Biography of Davies Gilbert*. Truro: Bradford Barton, 1967.

Tomory, Leslie. "Fostering a New Industry in the Industrial Revolution: Boulton & Watt and Gaslight 1800–1812." *The British Journal for the History of Science* 46 (2013): 199–229.

Tomory, Leslie. *Progressive Enlightenment: The Origins of the Gaslight Industry, 1780–1820*. Cambridge, MA: The MIT Press, 2012.

Torrens, Hugh S. "The Geological Work of Gregory Watt, His Travels with William Maclure in Italy (1801–1802), and Watt's 'Proto-geological' Map of Italy." In *The Origins of Geology in Italy*, edited by Gian Battista Vai and W. Glen E. Caldwell, 179–97. Boulder, CO: Geological Society of America, 2006.

Torrens, Hugh. "Jonathan Hornblower (1753–1815) and the Steam Engine: A Historiographic Analysis." In *Perceptions of Great Engineers: Fact and Fantasy*, edited by Denis Smith, 23–34. London/Liverpool: Science Museum/University of Liverpool, 1994.

Torrens, Hugh, and Joseph Wachelder. "Models, Toys, and Beddoes' Struggle for Educational Reform, 1790–1800." In Trevor Levere, Larry Stewart, and Hugh Torrens with Joseph Wachelder, *The Enlightenment of Thomas Beddoes: Science, Medicine and Reform*, 206–37. Abingdon: Routledge, 2017.

Toynbee, Arnold. *Lectures on the Industrial Revolution in England*. London: Rivingtons, 1884.

Tredgold, Thomas. *The Steam Engine*. London: J. Taylor, 1827.

Trevithick, Francis. *Life of Richard Trevithick, with an Account of His Inventions*, 2 vols. London: E. & F. N. Spon, 1872.

Tunbridge, P. A. "Jean André De Luc, F.R.S." *Notes and Records of the Royal Society of London* 26 (1971): 15–33.

Tungate, Sue. "Matthew Boulton's Mints: Copper to Customer." In *Matthew Boulton: Selling What All the World Desires*, edited by Shena Mason, 80–88. New Haven: Yale University Press, 2009.

Turner, Charles. "On the Invention, Progress and Advantages of the Art of Engraving in Mezzotint upon Steel." *Transactions of the Society Instituted at London for the Encouragement of Arts, Manufactures and Commerce* 42 (1823): 53–57.

Tyas, G. F. "Matthew Murray: A Centenary Appreciation." *Transactions of the Newcomen Society* 6 (1925): 111–43.

Uglow, Jenny. *The Lunar Men: The Friends Who Made the Future 1730–1810*. London: Faber & Faber, 2002.

Verbruggen, Jan Adrianus. "The Correspondence of Jan Daniel Huichelbos Van Liender." PhD diss., University of Twente, 2005.

Wade, H. T. "The James Watt Centenary." *Scientific American* 121 (30 August 1919): 206, 218–19.

Walcot, Patrick. *A Sketch of the Life of Dr William Small and His Relationship with Matthew Boulton & James Watt*. Sutton Coldfield: Privately printed, November 2015.

Walker, Fred M. *Ships and Shipbuilders: Pioneers of Design and Construction*. Barnsley: Seaforth Publishing, 2010.

Wardhaugh, Benjamin Sutherland. "Charles Hutton and the 'Dissensions' of 1783–

84: Scientific Networking and Its Failures." *Notes and Records: The Royal Society Journal of the History of Science* 71 (2017): 41–60.

Waters, Colin N., et al. "The Anthropocene Is Functionally and Stratigraphically Distinct from the Holocene." *Science* 351, no. 6269 (8 January 2016): 137. DOI: 10.1126/science.aad2622.

Watt, Gregory. "Observations on Basalt, and on the Transition from the Vitreous to the Stony Texture, Which Occurs in the Gradual Refrigeration of Melted Basalt; With some Geological Remarks." *Philosophical Transactions of the Royal Society of London* 94 (1804): 279–314.

Watt, James. "Memorandum concerning Mr Boulton, commencing with my first acquaintance with him." Glasgow, September 17, 1809, printed as *Memoir of Matthew Boulton by James Watt*. Birmingham: College of Arts and Crafts, 1943.

Watt, James. "Sequel to the Thoughts on the Constituent Parts of Water and Dephlogisticated Air." *Philosophical Transactions of the Royal Society of London* 74 (1784): 354–57.

Watt, James. "Thoughts on the Constituent Parts of Water and of Dephlogisticated Air; With an Account of Some Experiments on That Subject. In a Letter from Mr James Watt, Engineer, to Mr De Luc, F.R.S." *Philosophical Transactions of the Royal Society of London*, 74 (1784): 329–53.

Watters, Brian. "Charles Gascoigne of Carron Company," Falkirk Local Historical Society, 2005. http://www.falkirklocalhistorysociety.co.uk/home/index.php?id=108.

Watters, Brian. *Where Iron Runs Like Water! A New History of Carron Iron Works 1759–1982*. Edinburgh: John Donald, 1998.

Weir, Daniel. *History of the Town of Greenock*. Greenock: Privately printed, 1829.

Werrett, Simon. "Household Oeconomy and Chemical Inquiry." In *Compound Histories: Materials, Governance and Production, 1760–1840*, edited by Lissa Roberts and Simon Werrett, 35–56. Leiden/Boston: Brill, 2018.

West, Shearer. *Portraiture*. Oxford: Oxford University Press, 2004.

Whatley, Christopher A. *Scottish Society, 1707–1830: Beyond Jacobitism, Towards Industrialisation*. Manchester: Manchester University Press, 2000.

Williams, Robert. "Management Accounting Practice and Price Calculation at Boulton and Watt's Soho Foundry: A Late 18th Century Example." *The Accounting Historians Journal* 26 (1999): 65–88.

Williamson, George. *Memorials of the Lineage, Early Life, Education, and Development of the Genius of James Watt*. Greenock: Printed for the Watt Club, 1856.

Williamson, George. *Letters Respecting the Watt Family*. Greenock: W. Johnston & Son, 1840.

Wilson, David B. *Seeking Nature's Logic: Natural Philosophy in the Scottish Enlightenment*. University Park, PA: Pennsylvania State University Press, 2009.

Wilson, George. *The Life of the Honourable Henry Cavendish*. London: The Cavendish Society, 1851.

Wilson, Thomas. *A Comparative Statement of the Effects of Messrs. Boulton and Watt's Steam Engines with Newcommen's and Mr. Hornblower's*. Truro: W. Harry, 1792.

Wise, M. Norton, and Crosbie Smith. "Work and Waste: Political Economy and Natural Philosophy in Nineteenth-Century Britain, Parts I, II and III." *History of Science* 27 (1989): 263–301, 391–449; 28 (1990): 221–61.

Wood, Paul. "Introduction." In John Robison, *A System of Mechanical Philosophy*, I: v–xxiii. Reprint, London: Thoemmes Continuum, 2005.

Wood, Paul. "Jolly Jack Phosphorus in the Venice of the North, or, Who Was John Anderson." In *The Glasgow Enlightenment*, edited by Richard B. Sher and Andrew Hook, 111–32. East Linton: Tuckwell Press, 1995.

Wood, Paul. "Science, the Universities and the Public Sphere in Eighteenth-Century Scotland." *History of Universities* 13 (1994): 99–135.

Woolrich, A. P. "John Farey and His *Treatise on the Steam Engine* of 1827." *History of Technology* 22 (2000): 63–106.

Woolrich, A. P. "Hornblower and Maberley Engines in London, 1805." *Transactions of the Newcomen Society* 56 (1984–1985): 159–68.

Wright, Michael. "James Watt: Musical Instrument Maker." *The Galpin Society Journal* 55 (2002): 104–29.

Wyke, Terry. *Public Sculpture of Greater Manchester*. Liverpool: Liverpool University Press, 2005.

Yeo, Richard. *Encyclopaedic Visions: Scientific Dictionaries and Enlightenment Culture*. Cambridge: Cambridge University Press, 2001.

Index

Evans, Oliver, 113
Exeter, 221–22
Eyre, James, 198, 201

Fairbairn, William, 298
Faraday, Michael, 303
Fehr, Henry Charles, 296, 297
Fordyce, George, 27
Fothergill, John, 63, 66, 272
Foulis Press, 12, 20–21
Foulis, William, 304–5, 377n129
Fox, Charles James, 141
France: 17, 23, 106, 109, 218, 237, 271, 274, 307, 320, 360n47; Boulton visits, 180; chemistry in, 118, 144, 146–49; National Institute of, 122, 179, 229, 245–46; Restoration and Napoleon's return to, 247; Revolution in, 118, 189–90, 219, 268, 271, 354n61, 361n65; Wars with, 118, 214, 218–19, 247, 288; Watt visits, 108–9, 180, 219, 228, 245. *See also* steam engine; Watt Jr., James: in France
Franklin, Benjamin, 68
Freer, Thomas Lane, 270–71, 370n7
Freiberg School of Mines, 188
Fry, Joseph, 132, 142, 346n69

Garbett, Francis, 71, 215, 359n38
Garbett, Samuel, 51–52, 63, 67
Gardner, John, 125, 167, 329n82, 343n21, 343n22, 351n2
garret workshop. *See* Heathfield House; Watt, James
gaslight, 145, 163–65, 175, 314
Geneva, 121, 187
George II, King, 17
George III, King, 107, 217
George IV, King, 289
Gibson, Agnes (nee Miller, granddaughter), 277–78, 357n10
Gibson, James M.D., 278, 357n10
Gibson-Watt, David, 290–91, 293; death of 280;
Gibson-Watt, James (great-grandson), 277, 279–80, 281, 357n10, 372n33, 373n52
Gibson-Watt, James Miller (great-great-grandson), 282, 373n56
Giddy (Gilbert), Davies, 112, 177, 355n76
Glasgow, xi, xiii, xvii, 3–4, 6, 12–14, 18–33, 53, 57, 61, 68, 75, 61, 68, 75, 81–82, 85, 87, 116, 119, 124–25, 146, 157–58, 163, 167, 182, 184, 186–88, 204–5, 216–17, 224, 227, 232, 237, 248, 291–92, 294, 296; and anniversaries, 304–5, 307; Central Station, 292; Exhibition of Industrial Power in, 303; George Square statue in, 292, 294; Green and

separate condenser, 48–49, 313; Literary Club, 13; Mechanics' Institute, 294; Philosophical Society, 304; Watt assists with waterworks for, 242–43; Watt's shop in, 19, 21, 25, 29–31, 41, 205
Glasgow College, xiv, 12–14, 17, 219–20, 232, 277; chemistry at 23–24, 125, 128–29; demolition of, 20, 294; inner court of, 20; map of, 19; Observatory, 18; teaching methods at 22; Watt instrument maker at, 20–22, 30–31; Watt Prize at, 247–48; Watt receives Doctor of Laws from, 247, 353n40
Godfrey, Peter, 197
Greenock, xiii, 3–8, 11, 14, 16, 18, 20, 26, 29, 31, 73, 81, 205, 216, 258, 272, 291; and 1936 Anniversary celebrations in, 308; Grammar School, 8; James Watt Tavern in, 303; memorials to Watt in, xvii, 202, 295–96, 299, 309; Philosophical Society, 304, 377n125; water pipes in, 82, 242; Watt Club of, 11, 303–4; Watt Monument Library in, 304
Gregory, Olinthus, 250–54; *Treatise on Mechanics* by, 250, 252
Guernsey, Lord, (Heneage Finch), 77–79

Hall, A. Rupert, 154
Hamilton, Gilbert, 72, 82, 94, 102, 139, 146, 167, 195, 214, 219, 221–22, 242, 248; death of, 224; intermediary with Annie McGrigor, 85–87; Watt's agent in Scotland, 28, 81, 83, 85, 97, 180, 208, 210–11
Handley, Thomas, 79, 168–69, 351n5
Harrison, Joseph, 79, 168–69, 351n5
Hart, Robert, 48–49, 247, 331n29, 343n21
Hassenfratz, Jean Henri, 143–44, 219
Heathfield House, xvi, 88, 187, 191, 206, 211–12, 213, 217, 219, 229–30, 232, 238, 265–66, 269, 277, 306, 357n10, 359n27, 365n56, 370n126; books at, 363n21; building of, 211–12; demolition of, 278, 283, 373n60; description of, 213; gardens at, 347n87; garret workshop at, 31, 223, 227, 230, 278–83, 279; heirlooms at, 278–79, 373n60; inventory of (1791), 258; laboratory at, 148, 158, 213, 347n87, 359n31; leased, 279–80, 281–82; purchase by Birmingham Corporation planned, 282; relics at, 278–79, 373n51; and sale of estate, 283; Watt's will and, 206; Watt Jr.'s will and, 277–78
Hendricks, Mr, 239
Herschel, William, 197, 200, 259
Hills, Richard L., xvii–xviii, 12, 16, 22, 28, 35, 41, 48, 51–52, 57, 124, 146, 160, 176, 208, 336n59, 349n116
Hoogendijk, Steven, 109
Hornblower: family, 94, 195–96, 205, 290, 355n73, 356n88

Soho Manufactory (*cont.*): Watt impressed by, 67–68
Sotheby's "The James Watt Sale," 280, 325n23
Southern, John, 170, 175–77, 181, 195, 197, 244,
    343n24; collaboration with Watt, 125, 158, 179;
    death of, 179; and steam indicator, 123, 177, *178*,
    179
Southampton, 218, 220
Speer, William, 241, 364n38
St. Mary's Church, Handsworth, 224, 249, 316; Boulton
    buried in, 224; Memorial Chapel in, 278, 298, 316,
    370n6; Watt buried in, 249, 269–70; Watt memori-
    al service in, 306, 308
St. Paul's Cathedral, 293
St. Petersburg, 130, 200
Stadtfeld (Upper Saxony), 187
Stahl, Georg Ernst, 117
Stanley, Arthur Penrhyn, 287–88
steam engine, 41–62, *91*, *103*; at Albion Mill, 105–6; at
    Bloomfield Colliery, 90–92, 170; boiler design for,
    127; at Chacewater mine, 94; the "Cornish," 112;
    crank used in, 100, 169; double-acting, 101; draw-
    ings, 59, 79, 88–90, *91*, 95–96, 171, 197, 256,
    260, 261, 352n13; earnings from, 204–11, 358n14,
    358n19; in France, 108–9, 180; governor and, 104,
    313; high-pressure, xii, xv, 111–15, 257, 290–91,
    313, 341n153, 341n155; indicator, 123, *178*, 297;
    indicator diagram, 123, 276, 343n14; at Kinneil,
    61–62, 76, 131; marine, 113, 284; and natural
    philosophy, 153–57, 160; Newcomen, 42–46, 47,
    54, 68, 93, 95, 100–101, 110, 119, 125, 127–28,
    153–56, 168, 194; at New Willey ironworks, 90,
    92; rotative, 99–102, *103*, 104–6; Savery, 68, 79; in
    Scotland, 45, 51, 67, 90, *91*, 92; separate condens-
    er and, 44, 48–49, 50, 156–57, 313; sources of econo-
    my in Watt's, 50; Smethwick, 306; Stratford-le Bow,
    92; and sun-and-planet gear system, 100–101, 104,
    105–6, 169, 174–76; Ting Tang, 93–94; Torryburn,
    92; Wanlockhead, *91*; Wheal Bussy, 93–94; wheel,
    70, 75; writers on, 254. *See also* patent system;
    patent trials; Watt, James
steam indicator. *See* Southern, John; steam engine; Watt,
    James
Stewart, Iain M., 291–92, 374n84, 375n88
Stewart, Larry, 154, 160, 214
Stone, Marcus, *Watt Discovers the Condensation of
    Steam* (1863), 7, 8, 267–68, 309

Tangye, George, 281–82, 285, 306, 373n48
Tann, Jennifer, 217–18, 360n43
technology, xiv, xviii, 44, 53, 308, 318, 321, 323n2;
    studies, 332n63, 355n77. *See also* industrial arts

Tenby, 172, 290n621
Thomson, William, 276, 327n55, 349n116, 377n125,
    377n127
Timmins, Samuel, 281, 372n46
Tomory, Leslie, 163–64
Torrens, Hugh, 236, 355n73, 366n73
Trevithick, R.E., 290, 374n82
Trevithick, Richard, 111–14, 198, 290, 340n148,
    341n153
Tuffen, John, 265, 369n116
Turner, Charles, 262–63, *264*, 369n121
Turner, Charles Hampden, 288

University of Glasgow, 247, 276, 294, 295, 305, 308,
    365n58. *See also* Glasgow College

Versailles, Palace of, 109
Von Guericke, Otto, 42

Wales, 80, 93, 211, 215–16, 218, 219, 230. *See also*
    Watt, James: and Welsh estates
Walker, T & R, 189–91
Walker, Thomas, 189
Warltire, John, 139
Wasborough, Matthew, 100, 176
water controversy, xviii, 136, 138–43, 166, 275,
    345n58
waterpower, xii, 52, 104, 106; at Soho Manufactory, 68;
    versus steampower, 99–100, 339n122
Watt, Agnes (nee Muirhead, mother), 5, 9, 12, 205, 258;
    death of, 8, 11; portrait of, 258
Watt, Anne ("Annie," nee McGrigor, second wife), xi,
    28, 94, 107, 109, 165, 176, 182, 189–90, 201, 210,
    211–12, 265, 316–17, 339n115; and bleaching
    experiments, 143–44; on Boulton, 224, 353n33;
    and business, 179–80, 195, 208; courtship and
    marriage of, 83–87, 182, 201; death of, 278; death
    of children of, 219–20; portrait of, *84*; suffers
    rheumatism, 218, 230; visits Scotland, 94, 143,
    184–85, 201, 216–17, 230, 232, 258, 353n33; and
    stepchildren, 184–87 189, 191, 210–11, 225–26;
    and travel, 217–18, 221; and characterization of
    Watt, 230, 232, 316, 339n115, 360n41; in Watt's
    will, 206–7; will of, 357n10; urges Watt to enjoy
    life, 205–6, 360n41
Watt, Gregory (son): 74, 185–86, 201, 210, 223, 258;
    birth of, 94, 184; buried in St. Peter's Cathedral,
    Exeter, 221; character of, 219; in Cornwall, 220,
    223; meets Davy, 220; death of, 146, 219–22; at
    Glasgow College, 219–20, 223, 232; homoerotic
    correspondence of, 223; illness of, 146, 220–21;

Watt, James (*cont.*): offered Russian employment, 80, 336n58; as sage, xvi, 230–31, 232; and sculpture machines, 227–29, 230, 263, 279, 280, 315, 319; and separate condenser, xii, xiv, 44, 48–49, 53–58, 60, 73, 100, 120, 156, 168, 193, 238, 247, 250–52, 257, 304–5, 313, 355n74; avoids service as Sheriff, 216, 248–49, 316; and specific heat, 50, 128, 155; statues of, xvii, 243, 276, 286, 287–88, 290–92, 293, 294–96, 297, 298–99, 304, 308, 370n6, 374n81; and steam engine, xii, xiv, 18–19, 23–25, 32, 33–34, 41–62, 67–68, 75–76, 79–80, 91, 95, 100–102, 103, 104–5, 119, 121–25, 130–31, 156–57, 159, 165, 192–93, 234–35, 243, 250–51, 276, 301, 313; and steam indicator, 105, 123, 177, 178, 179, 276, 285, 296, 297, 299; and steam kettle, xvii, 8–10, 128–29, 268, 309, 318; and "Team Watt," xv, xix, 49, 157, 167–203, 204, 252, 277, 313, 314; and thermodynamics, xviii, 50, 123, 155–56, 177–79, 276, 296, 349n116, 353n30; on water, xv, 77, 108, 117, 120–21, 124–25, 131, 132–38, 152–53, 161; and "Watt's Law," 130, 132, 137; and Welsh estates, xv, 206, 211, 215–16, 218, 246, 268; Westminster Abbey statue of, xvii, 256, 286, 287–90, 298; his will, xvi, 187, 206–7, 211, 216, 247, 265, 269, 337n75; workshops in Birmingham, 88 (*see also* Heathfield House); workshops in Glasgow, 21–22, 31–32, 53, 125, 294, 329nn82–83, 343nn21–22. *See also* Boulton, Matthew; Boulton & Watt; steam engine
Watt, James Jr. (son), xv–xvii, 11, 31, 49, 74, 81, 85, 114, 170–71, 182–92, 201–2, 209, 262, 316, 350n144; and Arago's Éloge, 274–76; and Aston Hall, 216, 266, 277, 281, 370n127; business tactics of, 256; business taken over by, 167, 190–91, 201, 205, 211, 214, 216, 284–85; caught stealing as child, 182; and copying machine, 97; death of, 277; education of, 121, 185–88; and engines, 111, 163, 168, 216, 284–85; father disciplines, 182, 184–85, 314; father's biography for *Encylopaedia Britannica*, 273; and father's death, 265–66, 269, 270–71; father's reputation defended by, xvii, 192, 247, 253, 271–73, 326n29, 374n68; and "filial project," xvii, 168, 192, 267–78, 287–88, 295, 298, 304, 325n23, 370n6; in France, 189–90, 195; heirlooms and, xvii, 267, 278–81, 28284, 373n60; as High Sheriff of Warwickshire, 216; inheritance of, 206; and legal actions, 192–93, 199, 202, 209; "Olynthiad" written by, 250, 252–53, 269; in Manchester, 188–89; marriage plans with Anne Boulton, 225–27, 361n76; as militia volunteer, 248; portrait

of, 183; radicalism of, 148, 189–90; returns to England, 191; and Royal Society, 163–65; on Soho engineers, 250; and stepmother, 88, 184–86, 210; and "water controversy," 133, 274; will of, 277–78, 357n10. *See also* Muirhead, James Patrick
Watt, James Sr. (father): 5–7, 8–9, 11, 14, 35, 81; cares for grandchildren, 73, 81, 182; health of, 82, 85; his house cleared, 258; as merchant, 6–7, 26, 205; portrait of, 258
Watt, Janet ("Jessy," daughter), 145, 184–86, 210, 219, 258, 354n47; death of, 148, 191, 201–2, 213
Watt, John ("Jockey," brother), 7–8, 18, 31; death at sea of, 8, 317, 325n16
Watt, John (uncle), xiii, 35, 118, 258
Watt, Margaret ("Peggy," daughter), 31, 73, 81, 83, 87, 94, 180, 182–86, 212, 222, 258; children born to, 187; death of, 187, 201–2; lives with Aunts in Glasgow, 186–87, 336n59; marries, 187; Watt's treatment of, 187, 223, 316
Watt, Margaret ("Peggy," first wife), xi, 18, 30–31, 33, 38, 81, 225; helps with business, 179–80; death of, xi, xiv, 72–73, 81, 130, 182, 204
Watt, Thomas (grandfather), xiii, 4–5, 118; portrait of, 258
Wedderburn, Alexander, 76, 79
Wedgwood, Josiah, 66, 74, 98, 103, 154, 227, 366n70
West Indies, 108
Westminster Abbey, xvii, 256, 286, 287–90, 294, 298; 1936 Anniversary memorial service in, 307; Watt memorial removed from, 290–91
Westminster Hall, 196–97
Weston, Ambrose, 164, 193, 197, 202, 214, 228, 230, 248–49; uses Watt as patent advisor, 240–41
Weston, James, 269
Whitbread, Samuel, 103, 107–9
Wilkinson, John, 76, 90, 92, 98, 102–3, 187, 251, 335n40
Williamson, George, 7, 11, 304
Wilson, Alexander, 20–21, 25, 35
Wilson, Thomas, 94–96, 191, 193–95, 196–98, 202, 209, 220, 338n104, 356n88
Winsor, Frederick, 164
Withering, William, 74, 211, 259
Woodcroft, Bennet, 280–81, 300, 372n44
Woodmason, James, 97
Woolf, Arthur, 111, 113
Wray, Mrs Mary, 258, 368n102
Wyatt, Samuel, 105, 212

Yorke, Joseph, 244